Solid-Phase Microextraction

Solid-Phase Microextraction

Special Issue Editors

Constantinos K. Zacharis
Paraskevas D. Tzanavaras

MDPI • Basel • Beijing • Wuhan • Barcelona • Belgrade

Special Issue Editors
Constantinos K. Zacharis
Aristotle University of Thessaloniki
Greece

Paraskevas D. Tzanavaras
Aristotle University of Thessaloniki
Greece

Editorial Office
MDPI
St. Alban-Anlage 66
4052 Basel, Switzerland

This is a reprint of articles from the Special Issue published online in the open access journal *Molecules* (ISSN 1420-3049) from 2018 to 2020 (available at: https://www.mdpi.com/journal/molecules/special_issues/Solid_phase_microextraction).

For citation purposes, cite each article independently as indicated on the article page online and as indicated below:

LastName, A.A.; LastName, B.B.; LastName, C.C. Article Title. *Journal Name* **Year**, *Article Number*, Page Range.

ISBN 978-3-03928-262-3 (Pbk)
ISBN 978-3-03928-263-0 (PDF)

© 2020 by the authors. Articles in this book are Open Access and distributed under the Creative Commons Attribution (CC BY) license, which allows users to download, copy and build upon published articles, as long as the author and publisher are properly credited, which ensures maximum dissemination and a wider impact of our publications.

The book as a whole is distributed by MDPI under the terms and conditions of the Creative Commons license CC BY-NC-ND.

Contents

About the Special Issue Editors . . . . **vii**

Preface to "Solid-Phase Microextraction" . . . . **ix**

Constantinos K. Zacharis and Paraskevas D. Tzanavaras
Solid-Phase Microextraction
Reprinted from: *Molecules* **2020**, *25*, 379, doi:10.3390/molecules25020379 . . . . **1**

Madina Tursumbayeva, Jacek A. Koziel, Devin L. Maurer, Bulat Kenessov and Somchai Rice
Development of Time-Weighted Average Sampling of Odorous Volatile Organic Compounds in Air with Solid-Phase Microextraction Fiber Housed inside a GC Glass Liner: Proof of Concept
Reprinted from: *Molecules* **2019**, *24*, 406, doi:10.3390/molecules24030406 . . . . **5**

Bulat Kenessov, Jacek A. Koziel, Nassiba Baimatova, Olga P. Demyanenko and Miras Derbissalin
Optimization of Time-Weighted Average Air Sampling by Solid-Phase Microextraction Fibers Using Finite Element Analysis Software
Reprinted from: *Molecules* **2018**, *23*, 2736, doi:10.3390/molecules23112736 . . . . **23**

Luis F. C. Miranda, Rogéria R. Gonçalves and Maria E. C. Queiroz
A Dual Ligand Sol–Gel Organic-Silica Hybrid Monolithic Capillary for In-Tube SPME-MS/MS to Determine Amino Acids in Plasma Samples
Reprinted from: *Molecules* **2019**, *24*, 1658, doi:10.3390/molecules24091658 . . . . **37**

Stefano Dugheri, Alessandro Bonari, Matteo Gentili, Giovanni Cappelli, Ilenia Pompilio, Costanza Bossi, Giulio Arcangeli, Marcello Campagna and Nicola Mucci
High-Throughput Analysis of Selected Urinary Hydroxy Polycyclic Aromatic Hydrocarbons by an Innovative Automated Solid-Phase Microextraction
Reprinted from: *Molecules* **2018**, *23*, 1869, doi:10.3390/molecules23081869 . . . . **54**

Neus Jornet-Martínez, Adrián Ortega-Sierra, Jorge Verdú-Andrés, Rosa Herráez-Hernández and Pilar Campíns-Falcó
Analysis of Contact Traces of Cannabis by In-Tube Solid-Phase Microextraction Coupled to Nanoliquid Chromatography
Reprinted from: *Molecules* **2018**, *23*, 2359, doi:10.3390/molecules23092359 . . . . **69**

Jacqueline M. Hughes-Oliver, Guangning Xu and Ronald E. Baynes
Skin Permeation of Solutes from Metalworking Fluids to Build Prediction Models and Test A Partition Theory
Reprinted from: *Molecules* **2018**, *23*, 3076, doi:10.3390/molecules23123076 . . . . **81**

Rui Zhao, Lihua Lu, Qingxing Shi, Jian Chen and Yurong He
Volatile Terpenes and Terpenoids from Workers and Queens of *Monomorium chinense* (Hymenoptera: Formicidae)
Reprinted from: *Molecules* **2018**, *23*, 2838, doi:10.3390/molecules23112838 . . . . **100**

Andrzej Białowiec, Monika Micuda, Antoni Szumny, Jacek Łyczko and Jacek A. Koziel
Quantification of VOC Emissions from Carbonized Refuse-Derived Fuel Using Solid-Phase Microextraction and Gas Chromatography-Mass Spectrometry
Reprinted from: *Molecules* **2018**, *23*, 3208, doi:10.3390/molecules23123208 . . . . **114**

Hasan Al-Khshemawee, Xin Du, Manjree Agarwal, Jeong Oh Yang and Yong Lin Ren
Application of Direct Immersion Solid-Phase Microextraction (DI-SPME) for Understanding Biological Changes of Mediterranean Fruit Fly (*Ceratitis capitata*) During Mating Procedures
Reprinted from: *Molecules* **2018**, *23*, 2951, doi:10.3390/molecules23112951 128

Xiao-Wei Ma, Mu-Qing Su, Hong-Xia Wu, Yi-Gang Zhou and Song-Biao Wang
Analysis of the Volatile Profile of Core Chinese Mango Germplasm by Headspace Solid-Phase Microextraction Coupled with Gas Chromatography-Mass Spectrometry
Reprinted from: *Molecules* **2018**, *23*, 1480, doi:10.3390/molecules23061480 141

Huan Cheng, Jianle Chen, Peter J. Watkins, Shiguo Chen, Dan Wu, Donghong Liu and Xingqian Ye
Discrimination of Aroma Characteristics for Cubeb Berries by Sensomics Approach with Chemometrics
Reprinted from: *Molecules* **2018**, *23*, 1627, doi:10.3390/molecules23071627 162

Somchai Rice, Devin L. Maurer, Anne Fennell, Murlidhar Dharmadhikari and Jacek A. Koziel
Evaluation of Volatile Metabolites Emitted In-Vivo from Cold-Hardy Grapes during Ripening Using SPME and GC-MS: A Proof-of-Concept
Reprinted from: *Molecules* **2019**, *24*, 536, doi:10.3390/molecules24030536 176

Chenchen Wang, Wenjun Zhang, Huidong Li, Jiangsheng Mao, Changying Guo, Ruiyan Ding, Ying Wang, Liping Fang, Zilei Chen and Guosheng Yang
Analysis of Volatile Compounds in Pears by HS-SPME-GC×GC-TOFMS
Reprinted from: *Molecules* **2019**, *24*, 1795, doi:10.3390/molecules24091795 201

Jacek Łyczko, Klaudiusz Jałoszyński, Mariusz Surma, Klaudia Masztalerz and Antoni Szumny
HS-SPME Analysis of True Lavender (*Lavandula angustifolia* Mill.) Leaves Treated by Various Drying Methods
Reprinted from: *Molecules* **2019**, *24*, 764, doi:10.3390/molecules24040764 211

Francesca Ieri, Lorenzo Cecchi, Elena Giannini, Clarissa Clemente and Annalisa Romani
GC-MS and HS-SPME-GC×GC-TOFMS Determination of the Volatile Composition of Essential Oils and Hydrosols (By-Products) from Four *Eucalyptus* Species Cultivated in Tuscany
Reprinted from: *Molecules* **2019**, *24*, 226, doi:10.3390/molecules24020226 224

Lucas J. Leinen, Vaille A. Swenson, Hope L. Juntunen, Scott E. McKay, Samantha M. O'Hanlon, Patrick Videau and Michael O. Gaylor
Profiling Volatile Constituents of Homemade Preserved Foods Prepared in Early 1950s South Dakota (USA) Using Solid-Phase Microextraction (SPME) with Gas Chromatography–Mass Spectrometry (GC-MS) Determination
Reprinted from: *Molecules* **2019**, *24*, 660, doi:10.3390/molecules24040660 239

About the Special Issue Editors

Constantinos K. Zacharis (Assistant Professor of Pharmaceutical Analysis) is currently an Assistant Professor in School of Pharmacy (Aristotle University of Thessaloniki, Greece). Dr. Zacharis received his BSc in Chemistry in 2001 and his PhD in Analytical Chemistry in 2006 from the Department of Chemistry (AUTh). Since 2010, he has been the Editor of the Analytical Chemistry section of the international scientific journal Open Chemistry (formerly Central European Journal of Chemistry) and serves on the Editorial Boards of Molecules, SEJ Pharmaceutical Analysis, Instrumentation Science & Technology, Advances in Analytical Chemistry and Current Analysis on Chemistry. He has authored and co-authored more than 45 scientific articles and reviewed more than 650 papers in 50 scientific journals. He also acted as Guest Editor of 10 Special Issues. His current research is mainly focused on the development and application of novel analytical methodologies using separation techniques (LC, GC, CE) for pharmaceutical analysis.

Paraskevas D. Tzanavaras (Assistant Professor of Analytical Chemistry) is currently an Assistant Professor of Analytical Chemistry in the Department of Chemistry of Aristotle University of Thessaloniki in Greece. He received his PhD in 2003 from this same department, in automated flow injection analysis. He has published more than 90 research and review articles in international journals (H-index = 26 / Google Scholar), contributed 7 book chapters in international edited volumes and has organized, as an editor, 9 Special Issues in international journals. He is the assistant editor of flow injection analysis in Analytical Letters and a member of the Editorial Board of numerous journals. His current research is focused on the development of novel analytical methods based on zone fluidics and the coupling of liquid chromatography to on-line post-column derivatization for the determination of bioactive thiols and amines.

Preface to "Solid-Phase Microextraction"

We are pleased to announce the launch of the present Special Issue on "Solid Phase Microextraction", published in the international journal Molecules.

Among other sample preparation techniques, solid phase microextraction (SPME) is a state-of-the-art solvent-free technology. It was developed by the research group of Pawliszyn in 1989 [1]. Due to its versatility, reliability, low cost, and sampling convenience (on-site sampling), SPME has been widely used in combination with separation techniques (LC, GC, CE) in academic research and routine analyses. SPME has been successfully utilized for numerous applications in various scientific fields. As a result of its impact, a search in Scopus revealed almost 2000 publications (research articles, reviews, and book chapters) reporting research/applications using SPME.

The participation of colleagues from all over the world was immensely satisfactory and, following peer-review and evaluation of the articles, the Special Issue will consist of 16 research articles. The articles originate from eight countries: Australia (1), Brazil (1), China (3), Italy (2), Kazakhstan (1), Poland (2), Spain (1) and the USA (5). The readers of the Special Issue will find interesting and informative up-to-date articles on various topics.

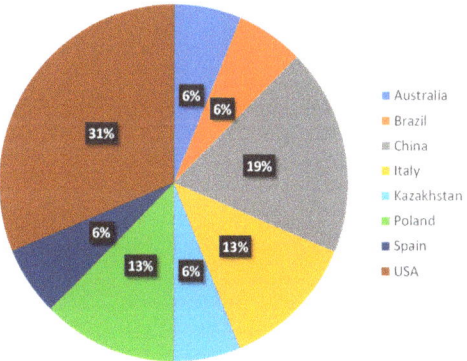

As the Guest Editors of this Special Issue, we would like to thank all the contributing authors for the interest and enthusiasm that enabled them to complete their articles in a timely manner. We are also grateful to the reviewers for their constructive and detailed criticism that improved the quality of the accepted articles. We would especially like to express our gratitude and appreciation for the editorial staff of the journal Molecules and Mrs. Katie Zhang for her continuous interest, help, and guidance throughout all the stages of our editorial work.

Constantinos K. Zacharis and Paraskevas D. Tzanavaras
Guest Editors

References

1. Arthur, C.L.; Pawliszyn, J. solid phase microextraction with thermal desorption using fused silica optical fibers. *Anal. Chem.* **1990**, *62*, 2145–2148.

Editorial

Solid-Phase Microextraction

Constantinos K. Zacharis [1,*] and Paraskevas D. Tzanavaras [2,*]

1 Laboratory of Pharmaceutical Analysis, Department of Pharmaceutical Technology, School of Pharmacy, Aristotle University of Thessaloniki, GR-54124 Thessaloniki, Greece
2 Laboratory of Analytical Chemistry, School of Chemistry, Faculty of Sciences, Aristotle University of Thessaloniki, GR-54124 Thessaloniki, Greece
* Correspondence: czacharis@pharm.auth.gr (C.K.Z.); ptzanava@chem.auth.gr (P.D.T.)

Received: 15 January 2020; Accepted: 16 January 2020; Published: 17 January 2020

Undoubtedly, sample preparation is one of the most important steps in the analytical process. It is estimated that approximately 60% of workload, time, and cost are devoted to sample preparation. Among other modern sample preparation techniques, solid-phase microextraction (SPME) is a state-of-the-art, solvent-free technology that was developed by the research group of Pawliszyn [1] in 1989. Due to the fact of its versatility, reliability, low cost, and sampling convenience (i.e., on-site sampling), SPME has been widely used in combination with separation techniques (e.g., LC, GC, and CE) in academic research and routine analyses. Up to now, SPME has been successfully utilized in numerous applications in various scientific fields. As a result of its impact, a search in Scopus revealed almost 2000 publications (e.g., research articles, reviews, book chapters) reporting research/applications using SPME.

The present Special Issue includes sixteen outstanding contributions covering the latest research trends and applications of SPME. The advantages of SPME were exploited by the research group of Koziel [2] in their research work "Development of Time-Weighted Average Sampling of Odorous Volatile Organic Compounds in Air with Solid-Phase Microextraction Fiber Housed Inside a GC Glass Liner: Proof of Concept". One of the primary goals of the work was to fabricate a rugged SPME-based sampler that can be deployed for longer periods in remote locations. Acetic acid was utilized as a model compound, since it is the most abundant VOC in any animal facility including swine farms. The researchers concluded that Car/PDMS material provided superior performance compared to other materials.

An analogous topic is published by Kenessov et al. [3] in their research regarding "Optimization of Time-Weighted Average Air Sampling by Solid-Phase Microextraction Fibers Using Finite Element Analysis Software". An SPME model was developed with both absorptive and adsorptive fibers located inside a protective needle using a finite element analysis-based software. This model was utilized to determine the potential sources of quantification inaccuracies of the time-weighted average sampling of VOCs in ambient air. Various SPME parameters were investigated and optimized. Using the modeling results, alternative sampling geometries were proposed.

The research group of Queiroz [4] present "A Dual Ligand Sol-Gel Organic-Silica Hybrid Monolithic Capillary for In-Tube SPME-MS/MS to Determine Amino Acids in Plasma Samples". A hybrid organic-silica monolithic capillary column with amino and cyano groups was fabricated and evaluated as an extraction device for in-tube SPME. The manufactured material was characterized using various techniques including scanning electron microscopy, Fourier transform infrared spectroscopy, nitrogen sorption experiments, among others. The in-tube SPME was utilized for the extraction of amino acids and neurotransmitters from plasma samples obtained from schizophrenic patients. The detection of the analytes was carried out using tandem mass spectrometry.

An automated SPME method is proposed by Dugheri et al. [5] in their article entitled "High-Throughput Analysis of Selected Urinary Hydroxy Polycyclic Aromatic Hydrocarbons by an Innovative Automated Solid-Phase Microextraction". A commercially available xyz robotic apparatus was employed to facilitate direct-immersion SPME in combination with GC-QqQ-MS for the determination of hydroxy-based metabolites of polycyclic aromatic hydrocarbons in urine samples. These analytes were used as biomarkers of internal doses to access recent exposure to polycyclic aromatic hydrocarbons. An on-fiber derivatization protocol using N-tetr-butyldimethylsilyl-N-methyltrifluoroacetamide was followed in order to enhance the gas chromatographic properties (i.e., volatility) of the analytes.

Herraez-Hernandez and co-workers [6] in their article "Analysis of Contact Traces of Cannabis by In-Tube Solid-Phase Microextraction Coupled to Nanoliquid Chromatography" reported a powerful tool based on in-tube SPME in combination with nano-LC for the quantitation of contact traces of drugs (e.g., cannabis). A set of cannabinoids was tested on various surfaces involving aluminum foil, office paper, hand skin, etc. The main difficulty in the analysis of contact traces of drugs is the low amount of available sample that is often only visible through microscopy. A relatively simple extraction protocol was employed after sampling using cotton swabs.

Research data regarding the permeation of chemical compounds through skin is provided by the group of Baynes [7] in the publication "Skin Permeation of Solutes from Metalworking Fluids to Build Prediction Models and Test a Partition Theory". A membrane-coated SPME fiber was utilized simulating skin permeation. The work aimed at the investigation of the permeation of 37 analytes through the membrane under certain conditions associated with skin exposure to several fluids (mineral oil, polyethylene glycol 200, synthetic oil, etc.) widely used in the metalworking industry.

Headspace SPME is an interesting alternative offering several advantages. These features were exploited by Chen et al. [8] in their research work entitled "Volatile Terpenes and Terpenoids from Workers and Queens of *Monomorium chinense* (Hymenoptera: Formicidae)". The objective of this study was to identify certain terpenes and terpenoids, determine their glandular origins, and study the effect of diet on terpene composition. The obtained data helped the authors to find out whether de novo terpene and terpenoid synthesis occurs in this species of ant.

Bialowiec et al. [9], in their study "Quantification of VOC Emissions from Carbonized Refuse-Derived Fuel Using Solid-Phase Microextraction and Gas Chromatography-Mass Spectrometry", present quantitative data from the analysis of VOCs from carbonized refuse-derived fuel using headspace SPME combined with GC-MS. The analyzed samples were generated from the torrefaction of municipal waste. A commercially available three-component SPME fiber was used for the analytes' extraction. The authors concluded that the VOC emitted from the torrefied samples were different from that emitted by other types of pyrolyzed samples, produced from either different types of feedstock or under different pyrolysis conditions.

The biological changes of the Mediterranean fruit fly during mating procedures were the main target of the work of Ren et al. [10] entitled "Application of Direct Immersion Solid-Phase Microextraction (DI-SPME) for Understanding Biological Changes of Mediterranean Fruit Fly (*Ceratitis capitata*) During Mating Procedures". This report investigates the feasibility of using DI-SPME high-resolution metabolism for the profiling of fruit fly tissues at various stages of adulthood. The obtained results were statistically treated using principal component analysis.

Characterization of the aromatic profile of mango germ sperm using headspace SPME is reported by the research group of Wang [11] in the paper "Analysis of the Volatile Profile of Core Chinese Mango Germplasm by Headspace Solid-Phase Microextraction Coupled with Gas Chromatography-Mass Spectrometry". A standard SPME protocol was followed for the extraction of aroma volatiles from the samples. The authors found that there were quantitative and qualitative differences in the volatile compounds among Chinese mango cultivars.

A sensomics approach combined with principal component analysis was exploited by Ye and co-workers [12] in their work entitled "Discrimination of Aroma Characteristics for Cubeb Berries by Sensomics Approach with Chemometrics". The aroma profiles of cubeb berries were evaluated by different extraction approaches involving hydro-distillation, simultaneous distillation/extraction, and SPME followed by GC-MS-olfactometry. The experimental parameters affecting the performance of SPME were studied and optimized. Almost 90 volatile compounds were identified in the studied samples.

Koziel and colleagues [13], in their communication "Evaluation of Volatile Metabolites Emitted In Vivo from Cold-Hardy Grapes during Ripening Using SPME and GC-MS: A Proof-of-Concept", reported the exploitation of SPME coupled with GC-MS in order to evaluate the volatile metabolites produced from cold-hardy grapes. Special glassware in conjunction with SPME was employed for the non-destructive sampling of biogenic volatiles emitted by the grape cluster.

A two-dimensional GC × GC-TOF-MS method was developed by Chen et al. [14] for the analysis of volatile compounds in pears. Their results are published in the article entitled "Analysis of Volatile Compounds in Pears by HS-SPME-GC×GC-TOFMS". After the optimization of the SPME conditions, the authors finally identified 241 compounds in the tested samples consisting of esters, alcohols, and aldehydes. Cluster analysis was used for the treatment of the results.

An application of headspace SPME is reported by Lyczko et al. [15] in the article "HS-SPME Analysis of True Lavender *(Lavandula angustifolia* Mill.) Leaves Treated by Various Drying Methods". The main objectives of this work was to determine the volatile profile composition of true lavender leaves and also the effect of three drying protocols applied. The analyses were carried out using GC-MS. An interesting finding was that the drying process may decrease the share of camphor while increasing the share of linalool and linalyl acetate which are the most desirable components in true lavender aroma.

An interesting application of headspace SPME is contributed by Cecchi and colleagues [16] in "GC-MS and HS-SPME-GC×GC-TOFMS Determination of the Volatile Composition of Essential Oils and Hydrosols (By-Products) from Four *Eucalyptus* Species Cultivated in Tuscany". In this report, a preliminary characterization of the volatile profile of samples obtained from various *Eucalyptus* species was carried out. After SPME sampling, GC×GC-TOF/MS was employed for fingerprint analysis.

Last but not least, Videau et al. [17], in the article "Profiling Volatile Constituents of Homemade Preserved Foods Prepared in Early 1950s South Dakota (USA) Using Solid-Phase Microextraction (SPME) with Gas Chromatography-Mass Spectrometry (GC-MS) Determination", presented the development of a novel analytical method based on SPME sampling in argon-filled gas sampling bags with direct GC-MS determination. The main scope of this work focused on the volatile profiling of 31 homemade preserves prepared in South Dakota (USA) during the period 1950–1953.

Generally, publishing a Special Issue always puts considerable pressure on the authors and reviewers. The Guest Editors of this Special Issue would like to thank them for their timely responses and constructive comments and hope that this issue will prove to be full of substance for the readers.

Funding: This research received no external funding.

Conflicts of Interest: The authors declare no conflict of interest.

References

1. Arthur, C.L.; Pawliszyn, J. Solid phase microextraction with thermal desorption using fused silica optical fibers. *Anal. Chem.* **1990**, *62*, 2145. [CrossRef]
2. Tursumbayeva, M.; Koziel, J.; Maurer, D.; Kenessov, B.; Rice, S. Development of Time-Weighted Average Sampling of Odorous Volatile Organic Compounds in Air with Solid-Phase Microextraction Fiber Housed inside a GC Glass Liner: Proof of Concept. *Molecules* **2019**, *24*, 406. [CrossRef] [PubMed]
3. Kenessov, B.; Koziel, J.; Baimatova, N.; Demyanenko, O.; Derbissalin, M. Optimization of Time-Weighted Average Air Sampling by Solid-Phase Microextraction Fibers Using Finite Element Analysis Software. *Molecules* **2018**, *23*, 2736. [CrossRef] [PubMed]
4. Miranda, L.F.C.; Gonçalves, R.R.; Queiroz, C.; Maria, E. A Dual Ligand Sol–Gel Organic-Silica Hybrid Monolithic Capillary for In-Tube SPME-MS/MS to Determine Amino Acids in Plasma Samples. *Molecules* **2019**, *24*, 1658. [CrossRef] [PubMed]

5. Dugheri, S.; Bonari, A.; Gentili, M.; Cappelli, G.; Pompilio, I.; Bossi, C.; Arcangeli, G.; Campagna, M.; Mucci, N. High-Throughput Analysis of Selected Urinary Hydroxy Polycyclic Aromatic Hydrocarbons by an Innovative Automated Solid-Phase Microextraction. *Molecules* **2018**, *23*, 1869. [CrossRef] [PubMed]
6. Jornet-Martínez, N.; Ortega-Sierra, A.; Verdú-Andrés, J.; Herráez-Hernández, R.; Campíns-Falcó, P. Analysis of Contact Traces of Cannabis by In-Tube Solid-Phase Microextraction Coupled to Nanoliquid Chromatography. *Molecules* **2018**, *23*, 2359. [CrossRef] [PubMed]
7. Hughes-Oliver, J.M.; Xu, G.; Baynes, R.E. Skin Permeation of Solutes from Metalworking Fluids to Build Prediction Models and Test A Partition Theory. *Molecules* **2018**, *23*, 3076. [CrossRef] [PubMed]
8. Zhao, R.; Lu, L.; Shi, Q.; Chen, J.; He, Y. Volatile Terpenes and Terpenoids from Workers and Queens of Monomorium chinense (Hymenoptera: Formicidae). *Molecules* **2018**, *23*, 2838. [CrossRef] [PubMed]
9. Białowiec, A.; Micuda, M.; Szumny, A.; Łyczko, J.; Koziel, J. Quantification of VOC Emissions from Carbonized Refuse-Derived Fuel Using Solid-Phase Microextraction and Gas Chromatography-Mass Spectrometry. *Molecules* **2018**, *23*, 3208. [CrossRef] [PubMed]
10. Al-Khshemawee, H.; Du, X.; Agarwal, M.; Yang, J.; Ren, Y. Application of Direct Immersion Solid-Phase Microextraction (DI-SPME) for Understanding Biological Changes of Mediterranean Fruit Fly (Ceratitis capitata) During Mating Procedures. *Molecules* **2018**, *23*, 2951. [CrossRef] [PubMed]
11. Ma, X.-W.; Su, M.-Q.; Wu, H.-X.; Zhou, Y.-G.; Wang, S.-B. Analysis of the Volatile Profile of Core Chinese Mango Germplasm by Headspace Solid-Phase Microextraction Coupled with Gas Chromatography-Mass Spectrometry. *Molecules* **2018**, *23*, 1480. [CrossRef] [PubMed]
12. Cheng, H.; Chen, J.; Watkins, P.; Chen, S.; Wu, D.; Liu, D.; Ye, X. Discrimination of Aroma Characteristics for Cubeb Berries by Sensomics Approach with Chemometrics. *Molecules* **2018**, *23*, 1627. [CrossRef] [PubMed]
13. Rice, S.; Maurer, D.; Fennell, A.; Dharmadhikari, M.; Koziel, J. Evaluation of Volatile Metabolites Emitted In-Vivo from Cold-Hardy Grapes during Ripening Using SPME and GC-MS: A Proof-of-Concept. *Molecules* **2019**, *24*, 536. [CrossRef] [PubMed]
14. Wang, C.; Zhang, W.; Li, H.; Mao, J.; Guo, C.; Ding, R.; Wang, Y.; Fang, L.; Chen, Z.; Yang, G. Analysis of Volatile Compounds in Pears by HS-SPME-GC×GC-TOFMS. *Molecules* **2019**, *24*, 1795. [CrossRef] [PubMed]
15. Łyczko, J.; Jałoszyński, K.; Surma, M.; Masztalerz, K.; Szumny, A. HS-SPME Analysis of True Lavender (Lavandula angustifolia Mill.) Leaves Treated by Various Drying Methods. *Molecules* **2019**, *24*, 764. [CrossRef] [PubMed]
16. Ieri, F.; Cecchi, L.; Giannini, E.; Clemente, C.; Romani, A. GC-MS and HS-SPME-GC×GC-TOFMS determination of the volatile composition of essential oils and hydrosols (By-products) from four Eucalyptus species cultivated in Tuscany. *Molecules* **2019**, *24*, 226. [CrossRef] [PubMed]
17. Leinen, L.; Swenson, V.; Juntunen, H.; McKay, S.; O'Hanlon, S.; Videau, P.; Gaylor, M. Profiling Volatile Constituents of Homemade Preserved Foods Prepared in Early 1950s South Dakota (USA) Using Solid-Phase Microextraction (SPME) with Gas Chromatography–Mass Spectrometry (GC-MS) Determination. *Molecules* **2019**, *24*, 660. [CrossRef] [PubMed]

© 2020 by the authors. Licensee MDPI, Basel, Switzerland. This article is an open access article distributed under the terms and conditions of the Creative Commons Attribution (CC BY) license (http://creativecommons.org/licenses/by/4.0/).

Article

Development of Time-Weighted Average Sampling of Odorous Volatile Organic Compounds in Air with Solid-Phase Microextraction Fiber Housed inside a GC Glass Liner: Proof of Concept

Madina Tursumbayeva [1,2], Jacek A. Koziel [1,*], Devin L. Maurer [1], Bulat Kenessov [3] and Somchai Rice [1]

1. Department of Agricultural and Biosystems Engineering, Iowa State University, Ames, IA 50011, USA; madina@iastate.edu (M.T.); dmaurer@iastate.edu (D.L.M.); somchai@iastate.edu (S.R.)
2. Department of Meteorology and Hydrology, Al-Farabi Kazakh National University, Almaty 050040, Kazakhstan
3. Center of Physical Chemical Methods of Research and Analysis, Al-Farabi Kazakh National University, Almaty 050012, Kazakhstan; bkenesov@cfhma.kz
* Correspondence: koziel@iastate.edu; Tel.: +1-515-294-4206

Academic Editors: Constantinos K. Zacharis and Paraskevas D. Tzanavaras
Received: 31 December 2018; Accepted: 21 January 2019; Published: 23 January 2019

Abstract: Finding farm-proven, robust sampling technologies for measurement of odorous volatile organic compounds (VOCs) and evaluating the mitigation of nuisance emissions continues to be a challenge. The objective of this research was to develop a new method for quantification of odorous VOCs in air using time-weighted average (TWA) sampling. The main goal was to transform a fragile lab-based technology (i.e., solid-phase microextraction, SPME) into a rugged sampler that can be deployed for longer periods in remote locations. The developed method addresses the need to improve conventional TWA SPME that suffers from the influence of the metallic SPME needle on the sampling process. We eliminated exposure to metallic parts and replaced them with a glass tube to facilitate diffusion from odorous air onto an exposed SPME fiber. A standard gas chromatography (GC) liner recommended for SPME injections was adopted for this purpose. Acetic acid, a common odorous VOC, was selected as a model compound to prove the concept. GC with mass spectrometry (GC–MS) was used for air analysis. An SPME fiber exposed inside a glass liner followed the Fick's law of diffusion model. There was a linear relationship between extraction time and mass extracted up to 12 h ($R^2 > 0.99$) and the inverse of retraction depth ($1/Z$) ($R^2 > 0.99$). The amount of VOC adsorbed via the TWA SPME using a GC glass liner to protect the SPME was reproducible. The limit of detection (LOD, signal-to-noise ratio (S/N) = 3) and limit of quantification (LOQ, S/N = 5) were 10 and 18 $\mu g \cdot m^{-3}$ (4.3 and 7.2 ppbV), respectively. There was no apparent difference relative to glass liner conditioning, offering a practical simplification for use in the field. The new method related well to field conditions when comparing it to the conventional method based on sorbent tubes. This research shows that an SPME fiber exposed inside a glass liner can be a promising, practical, simple approach for field applications to quantify odorous VOCs.

Keywords: SPME; retracted SPME; TWA SPME; GC–MS; on-site sampling; air quality; air monitoring; VOCs; odor; environmental analysis

---

1. Introduction

Offensive odors dispersed from animal feeding operations are a common concern for neighboring communities [1]. These odors originate mainly from manure and other organic matters in livestock operations and are a complex mixture of many gases, of which the largest portion (by number) are volatile organic compounds (VOCs). VOCs are complex chemicals distinguished by their ability to evaporate easily at room temperatures. VOCs originating from industry and transportation are studied extensively. Less attention is focused on VOCs found in animal production systems. However, the research in this area, especially research on the mitigation of odor emissions, is still limited [1–5] especially in regards to farm-scale proven technologies. Public concerns and research interest are focused mainly on solving odor nuisance.

Addressing public concerns about odorous emissions from livestock operations is challenging since many of these VOCs usually have a low odor detection threshold. Even at low concentrations (ppbV, pptV), they can be potent and objectionable odorants [6]. Thus, sampling and analysis of VOCs associated with animal operations are still challenging. Methods to detect and quantify VOCs from animal facilities are important for measuring air quality, developing and testing technologies that can mitigate odorous emissions. Many approaches used for sampling and analysis of VOCs are effective for qualitative analysis, but many standard methods developed for urban air are typically either not suitable for typical odorous VOCs or not sensitive enough to quantify trace concentrations.

Numerous VOCs can be found at animal facilities. Starting from 1965 when stearic acid was first identified [7], the list of known VOCs at animal facilities is constantly expanding. The results of the most recent studies show that more than 512 VOCs in total are found at swine facilities [7]. VOCs found in animal facilities can be classified into several groups. They are acids, alcohols, aldehydes, amines, hydrocarbons, indoles, nitrogen-containing compounds, phenols, sulfur-containing compounds, volatile fatty acids, and others [8]. However, sulfur-containing VOCs (S-VOCs) and volatile fatty acids (VFAs) were identified as the most dominant classes of VOCs at animal facilities which are responsible for those offensive odors [6]. A derivative of phenolics, *p*-cresol, was reported to be one of the main compounds responsible for characteristic odor at swine barns [6,9]. In order to test sampling methods, most studies focused on 10–15 odorous VOCs, which were used to sample emissions from livestock farms or to simulate them in a laboratory [6,10,11]. Some of the odorous VOCs include acetic, propionic, butyric, and isovaleric acids; methyl, ethyl, and butyl mercaptans; dimethyl sulfide, *p*-cresol, and others.

Acetic acid is considered the most abundant VOC in any animal facility, including swine farms. It is a colorless liquid that can be easily evaporated, and it has a strong and distinct pungent and vinegar-like smell. It was reported that the concentration of acetic acid in gaseous emissions from swine and dairy farms in the United States (US) could range from ~1 to 617 mg·m^{-3} [11]. Due to its abundance, it is reasonable to consider acetic acid as a model compound to validate concepts involving new VOC sampling methods and for testing the effectiveness of odorous VOC mitigation technologies in the context of livestock agriculture.

1.1. Air Sampling of Odorous VOCs

Most odorous VOCs are found at low concentrations [11]. VOC quantification requires reliable air sampling techniques and analytical methods that are representative of the air at the monitored site. The time-weighted average (TWA) sampling approach can be useful in such cases. This approach is used to determine the average concentration of an air pollutant over periods that can extend from a few minutes to several weeks [12]. TWA concentrations are needed to estimate average exposure to a contaminant. A number of different sampling techniques were introduced to obtain TWA concentrations of VOCs in the field. To date, the most common techniques are whole-air sampling techniques and sorbent tubes [13,14]. A short summary of those methods is given in Appendix A. Those methods require specialized equipment [14–21] (cleaning and evacuation of canisters, flushing air sampling bags with ultra-pure air or nitrogen, thermal desorption, air sampling pump) which makes the methods laborious and expensive to work with. Thus, simpler and more reliable methods to quantify VOCs at animal feeding operations are needed.

1.2. The TWA SPME Approach

Solid-phase microextraction (SPME) combines passive air sampling and sampling preparation. SPME uses a compact-size sampler that consists of a polymeric fiber that is kept inside a hollow metallic needle. During air sampling, VOCs are collected on an SPME fiber. SPME was shown to provide low detection limits reaching parts-per-trillion levels. After VOCs are transferred to an analytical instrument (e.g., via hot gas chromatography (GC) injector), extracted VOCs are thermally desorbed from the fiber, which can be reused. Thus, SPME eliminates the need for solvents and works with existing analytical technologies.

SPME is applicable for assessment of TWA concentrations in continuous sampling mode where the SPME fiber is retracted into the needle at a known distance during the desired sampling time. In contrast to the exposed fiber where an analyte reaches an equilibrium with the SPME, extraction of VOCs via the retracted fiber is controlled by diffusion. Since the fiber is kept inside the needle and extraction of VOCs is controlled by diffusion, the extraction rates are lower. Thus, the fiber in the protecting needle can be used for longer periods before reaching an equilibrium with the environment [22]. Analytes accumulated on the SPME fiber enable the measurement of the average gas (e.g., a VOC or total VOCs) concentration to which the fiber was exposed [23].

Quantification of the TWA concentrations with a retracted SPME fiber follows Fick's first law of diffusion (Equation (1)): the mass extracted on the fiber is proportional to (1) the diffusion coefficient of the analyte (D_g), (2) the concentration of the analyte in the gas phase (C_{gas}), (3) sampling time (t), and (4) cross-sectional area of the SPME needle opening (A); it is inversely proportional to the diffusion path length (Z, i.e., the distance from the needle opening to the tip of retracted fiber).

$$n = D_g \frac{A}{Z} \int C_{gas}(t) dt \qquad (1)$$

1.3. Application of the TWA SPME for VOCs

Despite the advantages of the TWA SPME approach, comparatively few studies were conducted to bring the approach to the field. The studies [12,24–30] showed that SPME devices could be used as TWA samplers to access exposure to different volatile (hydrocarbons, formaldehyde, and others) and chlorinated semi-volatile organic compounds [12] at the source. VOCs were also quantified from biomass gasification process streams in fast-moving environments at elevated temperatures such as syngas stream [28,29] and idling vehicle exhaust [30,31]. A major challenge with the TWA SPME approach is the influence of the metallic SPME needle assembly on the VOC extraction process, as documented earlier [28–31]. The metallic surface of the SPME needle (studied using "broken fiber", i.e., fiber without coating) had adsorptive properties that were significant compared with the adsorption by the fiber itself. Similarly, Koziel et al. [32] evaluated the contribution of the metallic parts first before quantifying five biomarkers (VOCs) such as dimethyl disulfide, dimethyl trisulfide, pyrimidine, phenol, and p-cresol emitted during aerobic digestion of animal tissue. The current suggestion to overcome this issue for the TWA SPME for quantification of VOCs is the mandatory evaluation of the contribution of mass extracted by a "broken fiber" so this effect can be accounted for. Thus, while reproducible, the contribution of metallic SPME parts on the TWA SPME process adds more steps to method development.

Recent modeling of the TWA SPME process by Kenessov et al. [22] provided an insightful identification of limitations for the use of retracted SPME fibers and possible means to address them. In their study, they found that a Carboxen/polydimethylsiloxane (Car/PDMS) SPME fiber with a greater size of a protecting needle (23 ga; as opposed to 24 ga) extracted greater amounts of analytes (about 19% more) than the fiber with a smaller protecting needle gauge size. This study suggests that the space between the SPME coating and the inner wall of the protecting needle plays a crucial role in extracting mass, since it allows faster diffusion of analytes not only to the tip of the fiber but also to its sides. The paper also recommends using a 23-ga SPME fiber for quantification of analytes with lower detection limits. However, no research reported the quantification of major VOCs that are responsible for the characteristic offensive odor downwind from animal feeding operations using a TWA SPME approach.

In this research, we aimed at addressing two major needs and gaps in knowledge: (1) to minimize or eliminate the need to consider the effect of the metallic SPME needle on air sampling of VOCs with TWA SPME, and (2) to enable SPME technology to be used for odorous VOC quantification in farm environments.

1.4. Objectives

The goal of this work was to develop a method for the quantification of target odorous VOC (using acetic acid as a model compound) with a TWA SPME approach that is more accurate and less laborious. Unlike the previous TWA SPME approaches where an SPME fiber is retracted into a metallic needle, this research proposes to use an SPME fiber that is exposed inside the GC glass liner to achieve the effect of a traditional retracted fiber *without the need to estimate and account for the inherent adsorption of VOCs onto metallic parts of SPME needle during sampling*. Since a GC glass liner has a greater cross-sectional area than a traditional retracted SPME fiber (Figure 1), the new approach should allow for greater amounts of the analyte extracted on the fiber, and increased exposure of the side surfaces of the coating to the sample, resulting in lower detection limits and greater accuracy.

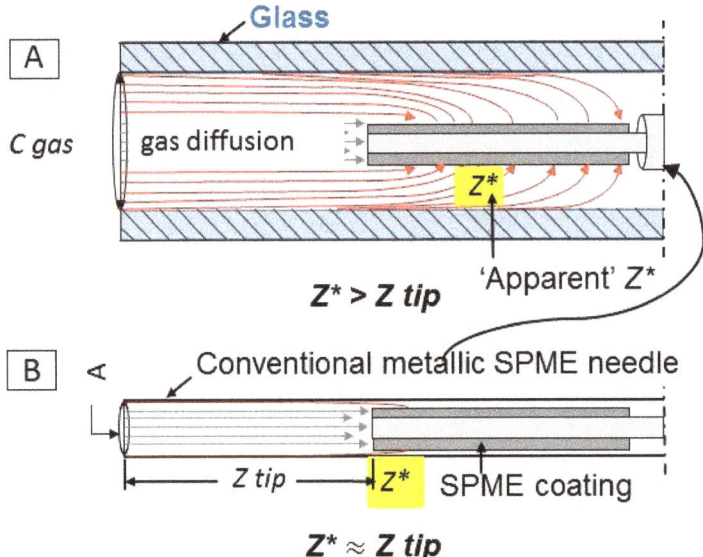

Figure 1. Time-weighted average (TWA) gas sampling with solid-phase microextraction (SPME). Comparison of proposed (**A**) and conventional (**B**) TWA SPME. (**A**) Sampling with SPME fiber exposed and retracted inside of a glass liner; (**B**) a typical case of TWA SPME where the SPME fiber is retracted inside of a conventional SPME needle. Gray arrows represent the diffusion path between bulk gas (left side) and the retracted fiber tip (Z tip). Red arrows represent the "apparent" diffusion path extending beyond the tip to the SPME fiber coating side. The "apparent" diffusion path represents the extracting process enhanced by the sides of the SPME coating. Z^* may continue to increase after the tip is saturated.

Our working hypothesis is that the glass liner enclosure might be less affected by the apparent departure from the ideal quantification model (Fick's Law, Equation (1)) that is associated with the use of metallic needle enclosures to facilitate TWA SPME. The inside of the glass liner serves as a diffusion path. Thus, extraction of VOCs is controlled by diffusion, and potentially can be used for sampling of VOCs in remote locations. The method utilizes GC glass liners that are readily available in many analytical laboratories. As the most abundant VOC in livestock operations, acetic acid was chosen as a model compound to prove the concept.

The specific objectives of this research were to (1) build and verify a standard gas generation system for odorous VOC that simulates typical dynamic animal facility air in the lab; (2) test the performance of an SPME fiber retracted into a glass liner and the adherence of this air sampling concept to the Fick's Law; (3) test the new method for quantification of acetic acid on a typical Iowa swine facility and evaluate its feasibility; and (4) compare the developed method side-by-side to a standard method under field conditions.

2. Results and Discussions

2.1. Standard Gas Stability Check

The stability of standard gas generated by the standard gas generation system is shown in Figure 2. For the purpose of checking stability, the standard gas was simultaneously measured with exposed and "retracted" SPME fibers and sorbent tubes several times per day for three consecutive days.

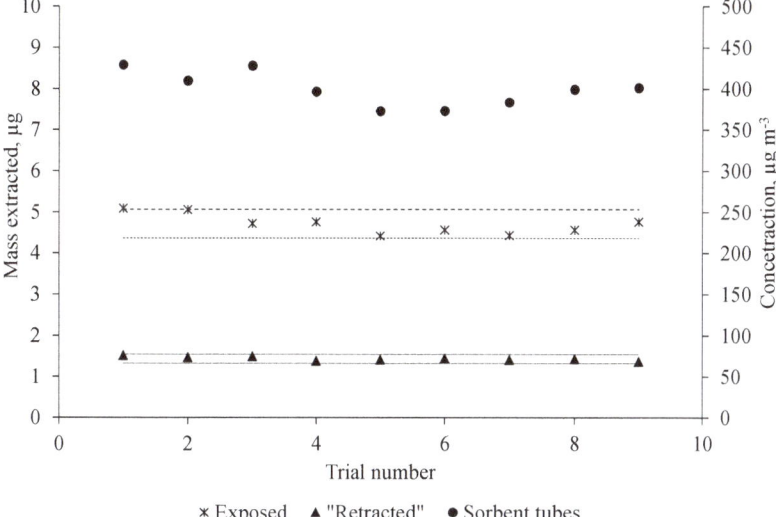

Figure 2. Standard gas stability needed for simulating steady-state conditions for TWA SPME sampling. Extraction conditions: two 85-μm Car/PDMS SPME fibers (one was a standard exposed fiber, and the other was an exposed fiber that was kept inside of a glass liner). Both were exposed to the standard gas (acetic acid, C_{gas} = 617 μg·m^{-3}). Retraction depth was 17.5 mm. Gas sampling was performed every hour for three consecutive days. Sampling times were 20 s for the exposed SPME fiber and 1 h for the retracted SPME fiber. The dashed lines on the graph indicate a ±7.5% band from the average. The concentration of acetic acid was verified with sorbent tubes. The concentration of acetic acid in the system obtained by sorbent tubes is shown on the right y-axis. Selected ion monitoring (SIM) mode at m/z 60.0 was used for acetic acid detection and quantification.

The result of daily extractions with exposed and "retracted" SPME fibers and sorbent tubes shows that the standard gas generation system was successful in generating a continuous supply of acetic acid. As can be seen in Figure 2, the exposed SPME fiber responses were more variable (relative standard deviation, RSD 5.6%) than the "retracted" SPME fiber (RSD 3.2%) in terms of extracted mass. Because the exposed SPME fiber was fully in contact with the moving gas, it resulted in more than two orders of magnitude higher extraction rates than the "retracted" SPME fiber. These results are consistent with the findings from Baimatova et al. [30]. The limits of detection (LOD, signal-to-noise ratio (S/N) = 3) and limits of quantification (LOQ, S/N = 5) were 10 and 18 µg·m^{-3} (4.3 and 7.2 ppbV), respectively.

2.2. Effects of SPME Fiber Type on TWA Sampling with Glass Liner

A glass liner facilitating TWA SPME was used. Two adsorptive SPME coatings were tested, i.e., 85-µm Car/PDMS and 50/30-µm (divinylbenzene, DVB) DVB/Car/PDMS, and both types of coatings effectively extracted acetic acid (Figure 3) for up to 12 h. Mass extracted by the fibers showed a linear response with sampling time ($R^2 > 0.99$). However, the results show that the average masses extracted by both SPME fibers were higher than the theoretical value (Equation (1)) by 11.1% and 3.7% on average for Car/PDMS and DVB/Car/PDMS fibers, respectively. This (relatively small and reproducible) discrepancy from theory (Equation (1)) could be considered excellent, considering that no effects of metallic SPME fiber assembly were taken into account.

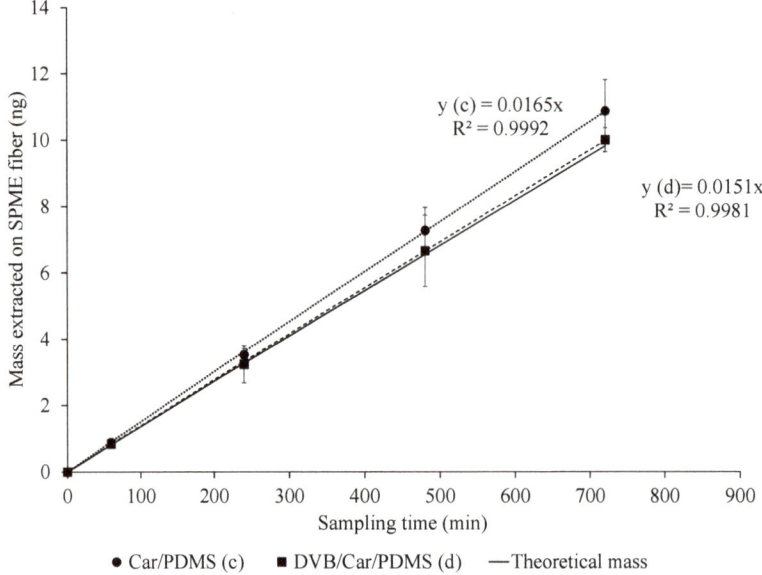

Figure 3. TWA SPME where fiber is retracted into a glass liner. Comparison of the extraction efficiency of acetic acid by 85-µm Car/PDMS and 50/30-µm DVB/Car/PDMS SPME fibers. The theoretical mass on the SPME fiber (shown as a solid line) was calculated using Equation (1) (Fick's law of diffusion). The experimental masses are shown as dotted and dash lines for Car/PDMS and DVB/Car/PDMS fibers, respectively. Extraction conditions: 85-µm Car/PDMS fiber exposed inside a glass liner, standard gas (acetic acid, $C_{gas} = 617$ µg·m^{-3}). Retraction depth was 1.75 cm. SIM mode at m/z 60.0 was used for acetic acid detection and quantification. Experiments were completed in triplicate.

The 85-µm Car/PDMS fiber provided a slightly higher response than the DVB/Car/PDMS fiber for acetic acid, which is consistent with the studies of Kenessov et al. [22] and Abalos et al. [33]. The total mass extracted by the SPME fibers was reproducible. The RSDs of MS responses with Car/PDMS (ranging from 2.3% to 12.2%) were lower in comparison with the DVB/Car/PDMS fiber (ranging from 3.2% to 14.7%). A linear regression model with a log-transformed response showed that masses extracted were not significantly different between the two SPME fibers (p-value = 0.44), as well as between both fibers and theoretical values (p-value = 0.43). The differences in mass extracted with 50/30-µm DVB/Car/PDMS at every sampling time were 9% less than the mass extracted with 85-µm

Car/PDMS, respectively. Log-transformation of mass extracted on SPME fiber was performed because there was non-constant variance in the residuals.

2.3. Effect of the Glass Liner Conditioning

There was no apparent effect of the glass liner conditioning on sampling of acetic acid. TWA sampling of acetic acid using exposed SPME fibers inside cleaned (new, unused) and "saturated" (exposed to standard gas for an extended period) glass liners was carried out. The rationale for testing the "saturated" glass liner was to test if that kind of conditioning is needed for practical air sampling. Resulting total masses extracted on SPME fibers were reproducible. RSDs ranged from 4.3% to 8.2% with cleaned and from 1.6% to 7.9% with "saturated" liners. A two-sample t-test did not show a statistically significant difference in mass extracted on the SPME fiber exposed inside cleaned and "saturated" liners. To determine if the rates of increase were different, a linear regression model with a log-transformed response was used. Log-transformation of masses extracted on SPME fiber was performed because there was non-constant variance in residuals. Fitting of the model showed no significant difference in interaction between the condition of a glass liner and time (p-value = 0.74). Upon analyzing the means of mass extracted on an SPME fiber with different glass liners at each time point, the p-values were not significant (from 0.68 to 0.93 for each time point, respectively). However, one of the interesting findings was that the percentage difference between both glass conditions was the highest at a sampling time of 1 h (15.0%). Then, the percentage difference decreased to 2.6% at a sampling time of 4 h and continued to decrease at longer sampling times. Figure 4 summarizes the results of previous experiments with both SPME fibers, and both glass liner conditions (clean 85-μm Car/PDMS vs. "saturated").

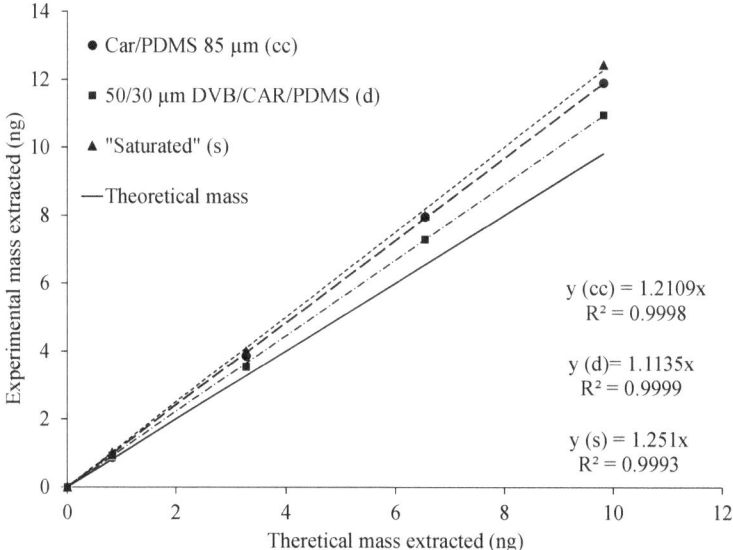

Figure 4. Comparison of the theoretical mass with the experimental masses extracted using 85-μm Car/PDMS (with clean and saturated glass liners) and 50/30-μm DVB/Car/PDMS fibers. The theoretical mass extracted was calculated using Fick's first law of diffusion (Equation (1)).

The result of the previous analysis shows that the SPME fibers extracted reproducible amounts of the target compound. Thus, the theoretical mass extracted on the fiber was proportional to the diffusion coefficient of the acetic acid, the concentration of the acetic acid in the gas phase, sampling time, and cross-sectional area of the glass liner opening, and it was inversely proportional to the diffusion path (i.e., distance between glass liner opening and the SPME fiber tip) length.

We also showed that the mass of extracted VOC on the SPME fiber remains inversely proportional to the retraction depth as a prerequisite for using Equation (1) for quantification. Thus, it was decided to investigate the possible influence of SPME fiber retraction depths inside a glass liner on extracted mass. Several diffusion path lengths (5, 10, 30, and 35 mm) were tested and compared to the fixed retraction depth of 17 mm that was used in the previous experiments. The aim of these new tests was to identify if different retraction depths would affect the mass extraction process inside of a glass liner. The results of the effect of retraction depth are shown in Figure 5.

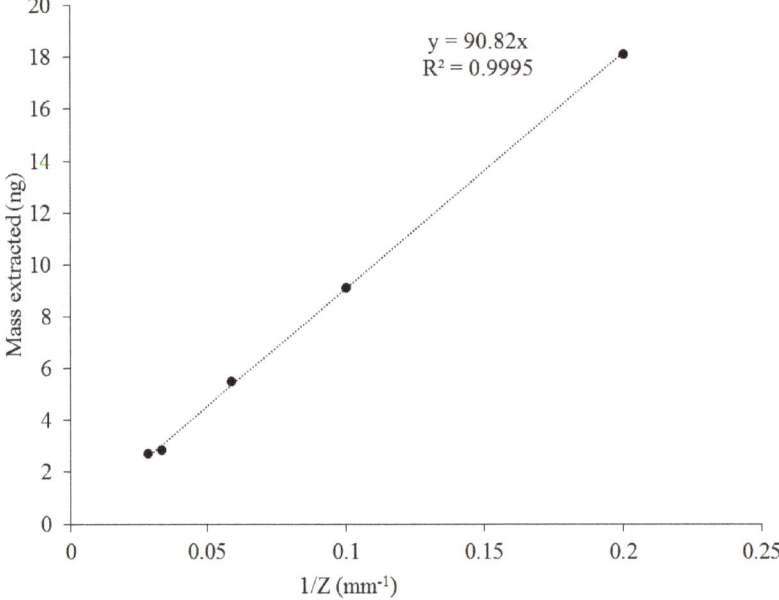

Figure 5. Effect of diffusion path length (Z) on the extracted mass of acetic acid. Extraction conditions: "retracted" 85-μm Car/PDMS, standard gas (acetic acid). SIM mode at m/z 60.0 was used for detection and quantification of the target compound. A sampling time of 4 h was used.

Extracted masses at the diffusion path lengths followed a power-law distribution. RSDs for extracted masses did not exceed 10% (7.6, 2.0, 5.5, 1.7, and 3.3% for 5, 10, 17, 30, and 35 mm, respectively). Thus, a diffusion path length can be adjusted, e.g., for achieving lower detection limits and/or higher accuracies at higher sampling times [22].

2.4. Verification of Glass-Liner-Facilitated TWA SPME via a Side-by-Side Comparison with the Sorbent-Tube-Based Method

The new method was compared with sorbent-tube-based sampling (a conventional method). Table 1 shows the comparison of measured concentrations of acetic acid in indoor air (laboratory, office space) and at a commercial swine farm in Iowa. Triplicates were taken at each sampling site.

Table 1. Comparison of measured acetic acid concentrations in different locations using time-weighted average (TWA) solid-phase microextraction (SPME) (85 μm Car/PDMS) facilitated with a glass liner and sorbent-tube-based measurement.

Location	Measured Concentration ($\mu g \cdot m^{-3}$)		% Difference (TWA SPME vs. Sorbent Tubes)	p-Value
	TWA SPME (Glass Liner)	Sorbent Tubes		
Office	17.7 (±2.7)	9.7 (±1.0)	58	0.002
Laboratory	15.2 (±0.8)	6.6 (±0.7)	78	0.0001
Farm 1, Day 1	3620 (±430)	755 (±20)	131	0.0004
Farm 1, Day 2	2400 (±310)	750 (±180)	104	0.0008
Farm 2, Day 1	685 (±70)	340	67	0.002
Farm 2, Day 2	750 (±90)	375 (±20)	67	0.0001

The concentration of acetic acid in the air was calculated using Fick's first law of diffusion (Equation (2)):

$$C_{gas} = \frac{m \cdot Z}{D_g \cdot t \cdot A} \quad (2)$$

Generally, the masses extracted by the SPME fibers were reproducible. In comparison with sorbent tubes, SPME fibers were much simpler to operate and did not require a thermal desorption system and additional instruments (a flowmeter and a pump) for VOC sampling in the field. It was also convenient to use in quiet places such as an office; the noise of the running pump caused a little discomfort to graduate students.

A comparison of the two methods showed that the concentrations obtained using the SPME fibers were much higher than the result based on the sorbent tubes. The difference between those methods varied depending on the sampling site. For example, in an indoor air setting, the differences between the two methods in resulting concentrations of acetic acid were 58% and 78% in the office and the laboratory, respectively. The differences between the two methods in the indoor setting were statistically significant ($p < 0.002$). Both indoor sampling sites had nearly similar concentrations of acetic acid. The small difference in concentrations between those two sampling sites could be explained by the more efficient ventilation system in the laboratory, which helped keep the concentration of the compound low, whereas the doors of the office were kept closed during the sampling, so there was less air mixing between the office and the hallway.

The TWA concentration of acetic acid in swine barns was approximately 50–200 times higher compared to indoor air environments. A sampling of acetic acid at Farm 1 for two days revealed larger differences in results produced by "retracted" SPME and sorbent tubes. The differences were statistically significant ($p < 0.001$). Sampling with tubes was much shorter over the entire period and, thus, not capable of measuring variations. During the first day of sampling, the glass tubes housing SPME fibers were placed in the direction facing the barn air flow. On that day, the differences between both methods were the highest (130%). On the following day, when SPME fibers were placed pointing in the direction of exhaust fans (i.e., glass liner opening faced the other direction), the discrepancies decreased (by nearly 26%), but remained high. The effect of TWA SPME sampler positioning requires additional research. An interesting fact is that the concentrations measured by the two methods were higher than previously reported in the literature. At Farm 2, both methods showed less differences than at the first farm. The differences between them did not exceed 70%. The RSDs of masses extracted for both methods were under 11%. In Table 1, at Farm 2, only one sample with sorbent tubes was taken on Day 1, so SD could not be calculated.

3. Materials and Methods

All materials and methods are described in greater detail by Tursumbayeva's (2017) [34] graduate thesis. Below is a summary of key details.

3.1. Chemicals and Materials

Chemicals used in this study included acetic acid and helium. Acetic acid, glacial (certified by ACS (American Chemical Society) ≥ 99.7%) was purchased from Fisher Chemical (Fair Lawn, NJ, USA), and helium (≥99.99%) was purchased from Air Gas (Des Moines, IA, USA). The 85-µm Car/PDMS and 50/30-µm DVB/Car/PDMS SPME fibers and manual SPME holders were obtained from Supelco (Bellefonte, PA, USA).

3.2. Standard Gas Generation and Sampling System

The standard gas generation and sampling system were built to simulate typical air flow rates through swine facilities (Figure 6).

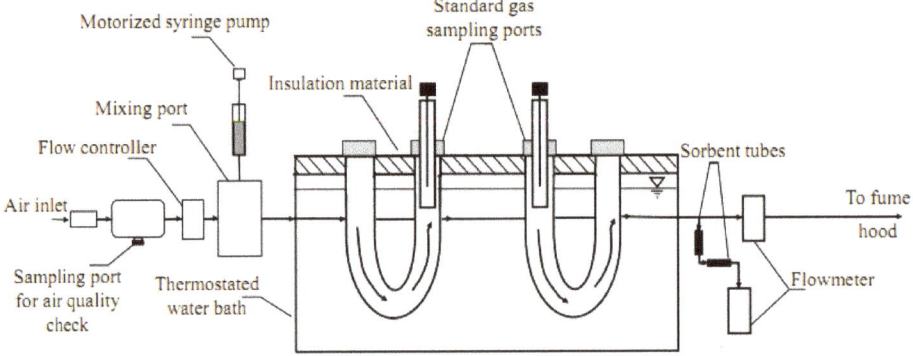

Figure 6. Schematic of standard mixture flow in the system. Passive gas sampling was completed with SPME retracted inside a gas chromatography (GC) injector glass liner.

The standard gas generation system included sampling ports for air quality check, a mass flow controller (Aalborg, Orangeburg, NY, USA), a motorized syringe pump (KD Scientific, Holliston, MA, USA), a 50-µL gastight syringe (Hamilton, Reno, NV, USA), a mixing port, polytetrafluoroethylene (PTFE) tubing (Thermo Scientific, Rochester, NY, US), and compression fittings. After the clean compressed air was introduced into the standard gas generation system, it flowed through the air quality check to be purified. Air flow (150 mL·min^{-1}) was managed by a mass flow controller. The rate of the target compound injection was controlled by a motorized syringe pump. Known volumes of the target compound were introduced to clean air in a heated mixing port to produce the desired concentrations. After standard gas (C_{gas}) was generated, it passed through the gas sampling system.

The gas sampling system consisted of two U-shaped gas bulbs submerged inside of a thermostated water bath. Gas bulbs were filled with solid glass balls to help evenly distribute acetic acid in clean air. Both sides of the bulbs were sealed with lids. A sampling port was installed on one of the lids of a bulb. Sampling ports included an SPME fiber enclosed in a glass liner (Figure 6). The distance between the opening of the liner and the tip of the fiber was fixed at 1.75 cm. As can be seen in the inset of Figure 7A,B (close-up), a glass liner was inserted into the gas bulb. The PTFE tubing was slid around the top of the glass liner. A septum was inserted into the PTFE tubing to close the top of the glass liner and for SPME needle insertion. The water bath was covered with insulation material to avoid excessive water evaporation. The temperature of the water in the bath was held at 25 °C. After passing through the gas sampling system, air flow was checked with a volumetric flowmeter (Bios Defender 520, MesaLabs, Butler, NJ, USA) to detect possible leaks in the system, and then exhausted to the fume hood.

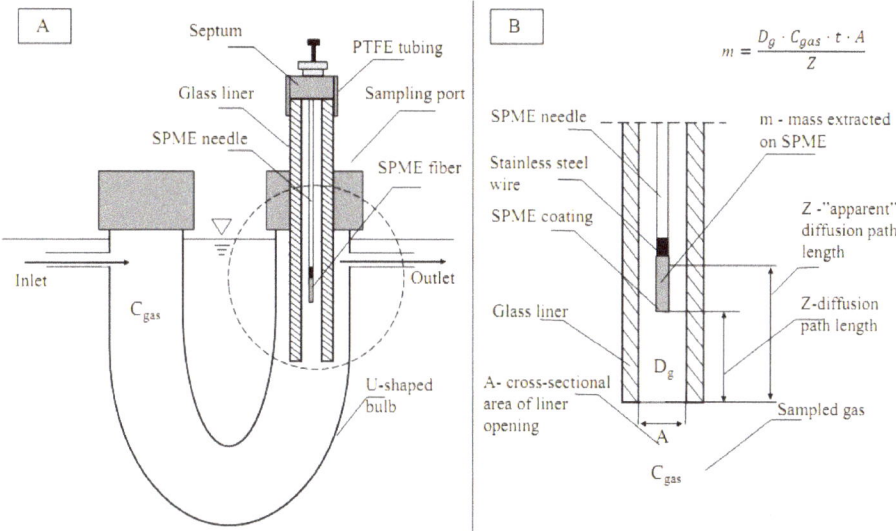

Figure 7. Passive gas sampling with SPME fiber retracted inside a GC injector glass liner. (**A**) Design of sampling port in the standard gas generation system. (**B**) Terms in Fick's first law of diffusion used for quantifications. The SPME fiber is exposed inside of a GC glass liner; thus, the walls of the liner serve as a protective needle in the conventional retracted mode.

The mass flow controller and the motorized syringe pump were used to produce the desired concentration of acetic acid in the gas generation system. The maximum concentration of acetic acid (617 µg·m^{-3}) which was reported by Cai et al. [11] was chosen in our research to assess the method. To achieve the desired concentration, the rate of acetic acid injection into a heated mixing port was calculated using Equations (4)–(6) described in the study by Baimatova et al. [30]. Since the calculated injection rate to generate 617 µg·m^{-3} acetic acid in the system was small (5.553 µg·h^{-1}), it was decided to dilute acetic acid with distilled water at the ratio of 5 to 1000. The syringe with the acetic acid standard solution was refilled every day. The dilution with water also helped avoid big fluctuations in the concentration of acetic acid since the dilution increased the number of solution injections into the system (Figure 2). A description of quality assurance and quality control measured pertaining to the liquid injection and flow rate verification are provided in Appendix B.

3.3. MS Detector Calibration with an Acetic Acid Standard Solution

To convert the peak area count of acetic acid extracted from the SPME fiber, we needed to know the response factor. The response factor was obtained by injecting different volumes (0.1–0.3 µL) of the analyte solution in hexane into the GC inlet working in splitless mode and determining the corresponding peak areas. Direct injections were conducted in triplicate. The calibration curve was constructed using four data points of average masses of acetic acid (from 500 to 5000 ng) that were injected into the GC. Response factor was calculated from the average mass injections, and corresponding peak area counts (Equation (3)).

$$RF = \frac{PA}{m} \quad (3)$$

where RF is the response factor, PA is the peak area count, and m is the known mass introduced into a column. Taking into account that the instrumental responses were linear over the tested period and the intercept was statistically zero, the response factor was equal to 14,400 peak area units·ng^{-1}. Knowing the response factor, the quantification of acetic acid mass extracted on the SPME fiber was done using the same equation.

3.4. SPME Fiber Conditioning

A new SPME fiber was thermally cleaned in a heated GC injection port according to the manufacturer's instructions. Before each sampling, the SPME fiber was cleaned in the GC injector port. This was done by holding the SPME fiber in the heated GC injection port at 240 °C for 3 min. Then, the fiber was introduced into the glass liner at the sampling port. After adsorption of the target compound, the SPME fiber was quickly transported to the GC injection port, where it was kept for 3 min for desorption. Between injections, the SPME fiber was kept in aluminum foil to avoid the absorption of VOCs present in the laboratory air.

3.5. Conditions of GC–MS

A gas chromatograph coupled with a mass spectrometer (6890N/5975C, Agilent, Santa Clara, CA, USA) was used in this study. Helium was selected as a carrier. The constant flow of helium in the column was 7.5 mL·min^{-1}. The flow was relatively high for an MS because the instrument was fitted with an olfactometry port/open split interface (human panelists were not used in this research). Temperatures of the ion source, quadrupole, and MS interface were 230, 150, and 240 °C, respectively. Splitless mode on the GC injection port at 240 °C was used. The oven temperature was initially set at 40 °C for 3 min, followed by heating rate increments of 7 °C·min^{-1} up to 125 °C, and 30 °C·min^{-1} up to a final 240 °C (held for 2 min). Total GC run time was 29.41 min. The retention time of acetic acid was 12.7 min. The MS detector was autotuned daily.

3.6. Standard Gas Stability Check

The standard gas that was generated by the gas generation system was checked for stability. For this purpose, the standard gas was checked several times for three consecutive days. The standard gas was sampled with an SPME fiber every hour after injection with an exposed 85-µm Car/PDMS fiber. A sampling time of 20 s was sufficient. Simultaneously, the concentration of acetic acid was monitored with the same type of fiber, but in a "retracted" position. The sampling time for the "retracted" fiber was 1 h. This stability check provided the information that the system was capable of producing stable responses over time and that the data which were going to be collected in the future would be reproducible. Furthermore, before starting a new set of experiments, the concentration of acetic acid was verified with an exposed fiber. At the same time, the standard method (sorbent tubes) was used to verify the concentration of acetic acid in the system. After 24 h, the syringe was refilled with an acetic acid solution (50 µL).

3.7. Experimental Design

Calibration of the SPME fiber was conducted by exposing the fiber inside a glass liner to the air with an acetic acid concentration of 617 µg·m^{-3} at 25 °C generated by the standard gas generation system. Retraction depth was fixed at 1.7 cm. The inner diameter of the glass liner (a standard GC liner recommended for SPME injections) was measured using a digital microscope (CC-HDMI-CD1, New Haven, CT, USA) and was equal to 0.844 mm. As an adsorptive fiber [35], the SPME fiber required testing of different sampling times to make sure that the fiber did not reach its sorptive capacity. Thus, sampling times of 1, 4, 8, and 12 h were examined to determine the longest sampling time before the sorptive capacity limit of the fiber was reached. All experiments were completed in triplicate. To improve the S/N ratio, quantification of acetic acid was performed using SIM mode at m/z 60.0. Limits of detection (LOD) and quantification (LOQ) were calculated by estimating concentrations corresponding to signal-to-noise (S/N) ratios 3:1 and 5:1, respectively.

3.8. SPME Fiber Selection

Two commercially available SPME fibers, 85-μm Car/PDMS and 50/30-μm DVB/Car/PDMS, were tested to select the most suitable fiber for extracting the target compound. Both SPME fibers were inserted in each sampling port (Figure 6) and exposed inside a glass liner. Before every SPME fiber injection, glass liners were washed and baked overnight. Extractions of acetic acid with the two different fibers were conducted simultaneously. Three replicate samples were taken with each fiber. Sampling times between 1 and 12 h were examined. Constant dry air flow at 150 mL·min^{-1} with a diluted acetic acid injection rate of 5.55 μg·h^{-1} was used to generate the desired concentration.

3.9. Effect of Glass Liner

The possible effect of glass liner conditioning was examined because of the rationale based on previous studies of Baimatova et al. [30,31] and Koziel et al. [32], which accounted for adsorption to the SPME metallic assembly. In their work, SPME needle assembly was shown to extract a significant portion of VOCs. To minimize or possibly eliminate this effect, the exposed SPME fiber was inserted into a protective glass liner (Figure 1). Two different conditions of a glass liner were tested. In the "cleaned" condition, glass liners were washed and baked overnight to evaporate all remaining VOCs. Cleaned liners were inserted into the sampling port in the standard gas generation system immediately before the SPME fiber insertion. In the "saturated" condition, glass liners remained in the sampling port of the standard gas generation system for at least an hour before SPME fiber insertion. A t-distribution was used to test the null hypothesis that the two population means (mass extracted on the SPME fiber exposed to cleaned and saturated liners) had no statistical difference at the 95% confidence interval (CI) (two-tailed test).

3.10. Sorbent Tubes

Sorbent tubes packed with Tenax TA were used to compare the results of the exposed SPME fiber inside a glass liner. The sorbent-tube-based method was used as a "benchmark" for the new method. Table A1 summarizes the pros and cons of compared and available methods. The procedure of sampling with sorbent tubes was completed as described in the work of Zhang et al. [10]. Firstly, sorbent tubes were thermally cleaned at 260 °C under a 100-mL·min^{-1} N$_2$ flow for 5 h; then, before subsequent uses, they were pre-conditioned at 260 °C under a 100-mL·min^{-1} N$_2$ flow for 30 min. In the field, sorbent tubes with two sections, sampling and breakthrough (against saturation), were connected to an air sampling pump (SKC Inc., Eighty Four, PA, USA) at a 50 mL·min^{-1} set flow rate. The sampling flow rate was monitored with a flow meter.

3.11. Application in the Field

After validating the described method in the lab, sampling of acetic acid was performed in indoor and livestock settings. Indoor air sampling included two sites: a manure treatment laboratory and an office space at Iowa State University. In the livestock setting, air sampling of acetic acid was carried out inside of the barns. Livestock air samples were taken at two swine farms: a typical swine farm located in Central Iowa (Farm 1) and a new farm with air scrubber and filtration technology for odor reduction (Farm 2). Both the new method (i.e., an SPME fiber exposed inside of a glass liner) and the conventional method (i.e., the sorbent tubes) were used at the sampling sites. The samplers were placed upstream of exhaust fans. The opening of the "retracted" fibers and sorbent tubes were pointed in the direction of the exhaust fans.

Three 85-μm Car/PDMS fibers were used at each site. Every fiber was thermally cleaned in a GC injector port as described earlier. Then, the fiber was assessed for residuals. For SPME fiber protection in the field, a "retracted" SPME fiber was placed inside of a 40-mL thermally cleaned vial. This was done to make an additional barrier between the dusty and odorous environment and the TWA SPME sampler (Figure 8). Thus, only the opening of the glass liner was exposed to the environment. Vials with a "retracted" SPME fiber were kept in thermally clean aluminum foil to prevent any interaction with the environment before actual sampling. Depending on anticipated concentrations at each monitoring site, the sampling time for the "retracted" SPME fiber was adjusted. For the quantification of acetic acid in the indoor setting, a sampling time of 12 h was used. For testing the method in the livestock setting, a sampling time of 40 min was sufficient. The diffusion coefficient was equal to 1.1×10^{-5} $m^2 \cdot s^{-1}$ at 25 °C [36].

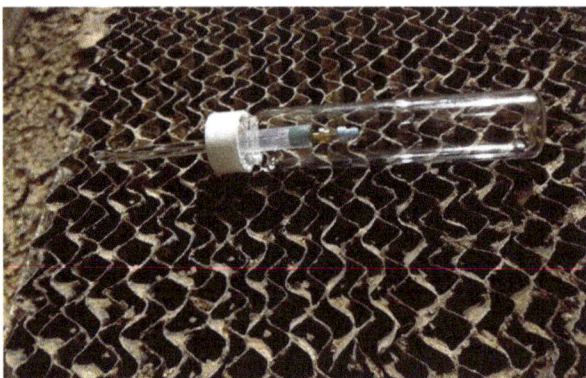

Figure 8. Field air sampling in TWA SPME mode on a commercial swine farm (Farm 2) in Iowa. The air sample diffuses through the opening in the GC glass liner (left side of the photo) onto an SPME fiber fully exposed inside the liner. A short section of Teflon tubing and a half-hole septum seal the liner and facilitate SPME insertion. A clear glass vial encloses the SPME assembly from dust and other gases in the sampled air.

Quantification of acetic acid was also performed with Tenax sorbent tubes. The sorbent tubes were thermally cleaned as described earlier. Multiple air samples were taken with two adjacent sorbent tubes, and the results were averaged for the indoor setting. The sampling time was 20 min. For the swine farm setting, a sampling time of 40 min was used.

After samples were taken, SPME fibers and sorbent tubes were covered with thermally cleaned aluminum foil and placed in clean glass vials and then transported for further analysis. All samples were analyzed within 5 h of sample collection. A *t*-distribution was used to test the null hypothesis that the sample means received with the two methods were equal at the 95% CI (two-tailed test).

4. Conclusions

A novel and simple TWA SPME-based method for the quantification of acetic acid in ambient air was developed. The following conclusions can be drawn:

- An SPME fiber exposed inside a glass liner followed Fick's law of diffusion. There were linear relationships between mass of the analyte extracted and extraction time up to 12 h ($R^2 > 0.99$), and mass extracted and the inverse of retraction depth ($1/Z$) ($R^2 > 0.99$). The amount of VOC adsorbed via the TWA SPME using a GC glass liner to protect the SPME was reproducible.
- There was no statistically significant difference between cleaned and "saturated" (equilibrated) glass liners. Thus, no special precautions are recommended for a practical application of this approach.
- The 85-μm Car/PDMS fiber revealed a higher response than the DVB/Car/PDMS fiber. The mass extracted by Car/PDMS was 8.9% higher than the mass extracted by the DVB/Car/PDMS fiber coating.
- The limit of detection (LOD, S/N = 3) and limit of quantification (LOQ, S/N = 5) were 10 and 18 $\mu g \cdot m^{-3}$ (4.3 and 7.2 ppbV), respectively.

- The new method was evaluated under field conditions by comparing it to the standard method (sorbent tubes) in four different locations. The TWA SPME sampling with a glass liner showed a reasonable match with the sorbent tubes.

The method shown is a relatively simple and practical, yet accurate sampling technique for the quantification of acetic acid in both an indoor workplace and a swine farm building. The method is reusable. Further research should be done to extend the number of odorous VOCs that can be used with this method, allowing further improvement of TWA SPME modeling (e.g., Reference [22]), and the incorporation of temporal changes in sampled air on TWA SPME [37].

Author Contributions: Investigation, M.T. and J.A.K.; methodology, M.T. and J.A.K.; software, D.L.M. and S.R.; resources, D.L.M. and J.A.K.; writing—original draft preparation, M.T.; writing—review and editing, M.T., J.A.K., B.K., and S.R.; visualization, M.T. and S.R.; supervision, J.A.K.; funding acquisition, M.T. and J.A.K.

Funding: This research was partially supported by the Iowa Agriculture and Home Economics Experiment Station, Ames, Iowa. Project No. IOW05556 (Future Challenges in Animal Production Systems: Seeking Solutions through Focused Facilitation) is sponsored by the Hatch Act and State of Iowa funds.

Acknowledgments: The authors would like to thank the Ministry of Education and Science of the Republic of Kazakhstan for supporting Madina Tursumbayeva with an M.S. study scholarship via the Bolashak Program.

Conflicts of Interest: The authors declare no conflicts of interest.

Appendix A. VOC Sampling Methods

There are numerous techniques for sampling VOCs, the most common of which are whole-air sampling techniques and sorbent tubes. The choice of air sampling technique depends on the chemical and physical properties of the VOCs of interest and on preferences motivated by regulatory reasons [9].

Whole-air sampling tools come in two common forms: evacuated stainless-steel canisters and sampling bags. In the US, evacuated canisters were introduced in the 1980s [15]. Today, canisters are applicable for sampling of up to 150 polar and nonpolar VOCs [16]. Canisters are equipped with flow controllers, particulate matter filters, and a vacuum gauge. For TWA sampling of VOCs in the field, the flow controller is pre-calibrated for the desired sampling time. Canister walls can modify the original content of sampled gas, as VOCs adsorb and undergo reactions, and samples can have poor recoveries [15,17].

Air sampling bags are used for the sampling of odorous gases [14,15]. Sampling bags are simple (consisting of a polymer film and a connector) and inexpensive to use. Despite their simplicity and cost-effectiveness, there are several limitations with regards to poor sample recoveries [14]. For example, Tedlar bags can desorb acetic acid and phenol, and absorb indole, p-cresol, nonanoic and octanoic acids, and some other VOCs resulting in an increased or decreased total mass of those VOCs in every sample [14]. Metalized bags can improve sample recovery for selected odorous VOCs [18]. Nalophane is the least expensive material; however, the material is not recommended for benzene and other petrochemicals, and it is not recommended to store samples for more than 6 h [19]. Teflon FEP bags are considered the most chemically inert among other bags, but they have a higher cost [20].

Sorbent tubes can be a good alternative to canisters and bags. Sorbents can be selected for application to a wider range of analytes including odorous VOCs [10]. Unlike canisters, sorbent tubes are compact and are easier to transport and store. Moreover, sorbent tubes have greater stability when exposed to polar compounds (most odorous VOCs are polar). In this approach, contaminated air passes through a tube containing sorbent material, which adsorbs VOCs. Usually, to facilitate this process, the contaminated air passes through the tube at a constant rate with the help of an air sampling pump. Sampling with sorbent tubes is one of the conventional sampling procedures for VOC quantification in ambient air [10,13,14,21]. All methods are summarized in Table A1.

Table A1. Comparison of sampling methods available for volatile organic compound (VOC) sampling. TWA—time-weighted average; SPME—solid-phase microextraction.

Sampling Technique	Whole-Air Sampling (Sampling Bags and Canisters)	Active Sorbent Tubes Sampling	SPME in Grab Sampling Mode	SPME in Continuous (TWA) Sampling Mode
Measurements in TWA mode	Possible	Possible	Possible	Possible
Advantages	Simple, accurate	Simple, accurate	Simple, accurate, fast, no pre-concentration and pump needed, low detection limits	Simple in operation, reusable, low cost, no pre-concentration and pump needed
Disadvantages	Relatively high cost; difficulties in transportation and storage; pump and pre-concentration required; the need for evacuation and cleaning in lab prior sampling; could be problematic to reuse bags	Pump and thermal desorption system required	Several grab samples needed for TWA concentration; mass extracted greatly affected by environmental variables	Complicated standard gas generation system and calibration required

Appendix B. Liquid Injection and Flow Rate Verification

Since the motorized syringe pump and the mass flow controller are key instruments to generate the concentration of acetic acid in the standard gas generation system, the reliability of these instruments was verified. To verify that the motorized syringe pump provided a correct rate of injection, a known volume of water was injected into the empty vial. The mass of the vial was weighed before and after injection. The results of the mass of injected liquid and the set point were compared, and the difference between them did not exceed 3%. The rate of injection was constantly verified visually during the experiments.

A similar verification for flow rate was completed to assure that the system did not leak. Measurements for three different flow rates were compared with the mass flow controller and the flowmeter. The difference between readings on the flow controller and the flowmeter depended on the flow rate. Smaller flow rates yielded a higher difference between readings on the two instruments.

References

1. Maurer, D.; Koziel, J.A.; Harmon, J.D.; Hoff, S.J.; Rieck-Hinz, A.M.; Andersen, D.S. Summary of performance data for technologies to control gaseous, odor, and particulate emissions from livestock operations: Air Management Practices Assessment Tool (AMPAT). *Data Brief* **2016**, *7*, 1413–1429. [CrossRef] [PubMed]
2. Maurer, D.; Koziel, J.A.; Bruning, K.; Parker, D.B. Farm-scale testing of soybean peroxidase and calcium peroxide for surficial swine manure treatment and mitigation of odorous VOCs, ammonia, hydrogen sulfide emissions. *Atmos. Environ.* **2017**, *166*, 467–478. [CrossRef]
3. Zhu, W.; Koziel, J.A.; Maurer, D.L. Mitigation of livestock odors using a black light and a new titanium dioxide-based catalyst. Proof-of-concept. *Atmosphere* **2017**, *6*, 103. [CrossRef]
4. Maurer, D.; Koziel, J.A.; Kalus, K.; Andersen, D.; Opalinski, S. Pilot-scale testing of non-activated biochar for swine manure treatment and mitigation of ammonia, hydrogen sulphide, odorous VOCs, and greenhouse gas emissions. *Sustainability* **2017**, *6*, 929. [CrossRef]
5. Kalus, K.; Opalinski, S.; Maurer, D.; Rice, S.; Koziel, J.A.; Korczynski, M.; Dobrzanski, Z.; Kolacz, R.; Gutarowska, B. Odour reducing microbial-mineral additive for poultry manure treatment. *Front. Environ. Sci. Eng.* **2017**, *3*, 7. [CrossRef]
6. Yang, X.; Zhu, W.; Koziel, J.A.; Cai, L.; Jenks, W.S.; Laor, Y.; van Leeuwen, H.J.; Hoff, S.J. Improved quantification of livestock associated odorous volatile organic compounds in a standard flow-through system using solid-phase microextraction and gas chromatography–mass spectrometry. *J. Chromatogr. A* **2015**, *1414*, 31–40. [CrossRef] [PubMed]
7. Ni, J.; Robarge, W.P.; Xiao, C.; Heber, A.J. Volatile organic compounds at swine facilities: A critical review. *Chemosphere* **2012**, *89*, 769–788. [CrossRef] [PubMed]
8. Lo, Y.C.; Koziel, J.A.; Cai, L.; Hoff, S.J.; Jenks, W.S.; Xin, H. Simultaneous chemical and sensory characterization of VOCs and semi-VOCs emitted from swine manure using SPME and multidimensional gas chromatography-mass spectrometry-olfactometry system. *J. Environ. Qual.* **2008**, *37*, 521–534. [CrossRef]

9. Bulliner, E.A.; Koziel, J.A.; Cai, L.; Wright, D. Characterization of livestock odors using steel plates, solid phase microextraction, and multidimensional-gas chromatography–mass spectrometry–olfactometry. *J. Air Waste Manag. Assoc.* **2006**, *56*, 1391–1403. [CrossRef]
10. Zhang, S.; Cai, L.; Koziel, J.A.; Hoff, S.J.; Schmidt, D.R.; Clanton, C.J.; Heber, A.J. Field air sampling and simultaneous chemical and sensory analysis of livestock odorants with sorbent tubes and GC–MS/olfactometry. *Sens. Actuators B Chem.* **2010**, *146*, 427–432. [CrossRef]
11. Cai, L.; Koziel, J.A.; Zhang, S.; Heber, A.J.; Cortus, E.L.; Parker, D.B.; Hoff, S.J.; Sun, G.; Heathcote, K.Y.; Jacobson, L.D.; et al. Odor and Odorous Chemical Emissions from Animal Buildings: Part 3. Chemical Emissions. *Trans. ASABE* **2015**, *5*, 1333–1347. [CrossRef]
12. Paschke, A.; Vrana, B.; Popp, P.; Schüürmann, G. Comparative application of solid-phase microextraction fibre assemblies and semi-permeable membrane devices as passive air samplers for semi-volatile chlorinated organic compounds. A case study on the landfill "Grube Antonie" in Bitterfeld, Germany. *Environ. Pollut.* **2006**, *144*, 414–422. [CrossRef] [PubMed]
13. Woolfenden, E. Monitoring VOCs in Air Using Sorbent Tubes Followed by Thermal Desorption-Capillary GC Analysis: Summary of Data and Practical Guidelines Monitoring VOCs in Air Using Sorbent Tubes Followed by Thermal Desorption-Capillary GC Analysis: Summary of Data and Practical Guidelines. *J. Air Waste Manag. Assoc.* **1997**, *47*, 20–36. [CrossRef]
14. Koziel, J.A.; Spinhirne, J.P.; Lloyd, J.D.; Parker, D.B.; Wright, D.W.; Kuhrt, F.W. Evaluation of sample recovery of malodorous livestock gases from air sampling bags, solid-phase microextraction fibers, Tenax TA sorbent tubes, and sampling canisters. *J. Air Waste Manag. Assoc.* **2005**, *55*, 1147–1157. [CrossRef] [PubMed]
15. Krol, S.; Zabiegala, B.; Namiesnik, J. Monitoring VOCs in atmospheric air II. Sample collection and preparation. *Trends Anal. Chem.* **2010**, *9*, 1101–1112. [CrossRef]
16. Wang, D.K.W.; Austin, C.C. Determination of complex mixtures of volatile organic compounds in ambient air: Canister methodology. *Anal. Bioanal. Chem.* **2006**, *386*, 1099–1120. [CrossRef] [PubMed]
17. Watson, N.; Davies, S.; Wevill, D. Air monitoring: New advances in sampling and detection. *Sci. World J.* **2011**, *11*, 2582–2598. [CrossRef]
18. Zhu, W.; Koziel, J.A.; Cai, L.; Wright, D.; Kuhrt, F. Testing odorants recovery from a novel metallized fluorinated ethylene propylene gas sampling bag. *J. Air Waste Manag. Assoc.* **2015**, *65*, 1434–1445. [CrossRef]
19. Scentroid Sampling Bags. Available online: http://scentroid.com/wp-content/uploads/2015/07/Scentroid-Sampling-Bags.pdf (accessed on 29 December 2018).
20. Harreveld, A.P. Odor concentration decay and stability in gas sampling bags. *J. Air Waste Manag. Assoc.* **2003**, *1*, 51–60. [CrossRef]
21. Koziel, J.A.; Pawliszyn, J. Air sampling and analysis of volatile organic compounds with solid phase microextraction. *J. Air Waste Manag. Assoc.* **2001**, *51*, 173–184. [CrossRef]
22. Kenessov, B.; Koziel, J.A.; Baimatova, N.; Demyanenko, O.P.; Derbissalin, M. Optimization of time-weighted average air sampling by solid-phase microextraction fibers using finite element analysis software. *Molecules* **2018**, *23*, 2736. [CrossRef]
23. Martos, P.A.; Pawliszyn, J. Time-weighted average sampling with solid-phase microextraction device: Implications for enhanced personal exposure monitoring to airborne pollutants. *Anal. Chem.* **1999**, *8*, 1513–1520. [CrossRef]
24. Koziel, J.A.; Jia, M.; Khaled, A.; Noah, J.; Pawliszyn, J. Field air analysis with SPME device. *Anal. Chim. Acta* **1999**, *400*, 153–162. [CrossRef]
25. Koziel, J.A.; Noah, J.; Pawliszyn, J. Field sampling and determination of formaldehyde in indoor air with solid-phase microextraction and on-fiber derivatization. *Environ. Sci. Technol.* **2001**, *35*, 1481–1486. [CrossRef] [PubMed]
26. Chen, Y.; Pawliszyn, J. Time-weighted average passive sampling with a Solid-Phase Microextraction device. *Anal. Chem.* **2003**, *75*, 2004–2010. [CrossRef]
27. Khaled, A.; Pawliszyn, J. Time-weighted average sampling of volatile and semi-volatile airborne organic compounds by the solid-phase microextraction device. *J. Chromatogr. A* **2000**, *892*, 455–467. [CrossRef]
28. Woolcock, P.J.; Koziel, J.A.; Cai, L.; Johnston, P.A.; Brown, R.C. Analysis of trace contaminants in hot gas streams using time-weighted average solid-phase microextraction: Proof of concept. *J. Chromatogr. A* **2013**, *1281*, 1–8. [CrossRef]

29. Woolcock, P.J.; Koziel, J.A.; Johnston, P.A.; Brown, R.C.; Broer, K.M. Analysis of trace contaminants in hot gas streams using time-weighted average solid-phase microextraction: Pilot-scale validation. *Fuel* **2015**, *153*, 552–558. [CrossRef]
30. Baimatova, N.; Koziel, J.A.; Kenessov, B. Quantification of benzene, toluene, ethylbenzene and o-xylene in internal combustion engine exhaust with time-weighted average solid phase microextraction and gas chromatography-mass spectrometry. *Anal. Chim. Acta* **2015**, *873*, 38–50. [CrossRef]
31. Baimatova, N.; Koziel, J.A.; Kenessov, B. Passive sampling and analysis of naphthalene in internal combustion engine exhaust with retracted SPME device and GC-MS. *Atmosphere* **2017**, *8*, 130. [CrossRef]
32. Koziel, J.A.; Nguyen, L.T.; Glanville, T.D.; Ahn, H.K.; Frana, T.S.; van Leeuwen, J.H. Method for sampling and analysis of volatile biomarkers in process gas from aerobic digestion of poultry carcass using time-weighted average SPME and GC-MS. *Food Chem.* **2017**, *232*, 799–807. [CrossRef] [PubMed]
33. Abalos, M.; Bayona, J.M.; Pawliszyn, J. Development of a headspace solid-phase microextraction procedure for the determination of free volatile fatty acids in waste waters. *J. Chromatogr. A* **2000**, *873*, 107–115. [CrossRef]
34. Tursumbayeva, M. Simple and Accurate Quantification of Odorous Volatile Organic Compounds in Air with Solid Phase Microextraction and Gas Chromatography—Mass Spectrometry. Master's Thesis, Iowa State University, Ames, IA, USA, 2017. [CrossRef]
35. Pawliszyn, J. SPME method development. In *Solid Phase Microextraction: Theory and Practice*; Wiley-VCH: New York, NY, USA, 1997; ISBN 0-471-19034-9.
36. Hafner, S.D.; Montes, F.; Rotz, C.A. Modeling emissions of volatile organic compounds from silage. In Proceedings of the American Society of Agricultural and Biological Engineers Annual International Meeting, Reno, NV, USA, 21–24 June 2009; Volume 3, pp. 1895–1911. [CrossRef]
37. Semenov, S.N.; Koziel, J.A.; Pawliszyn, J. Kinetics of solid-phase extraction and solid-phase microextraction in thin adsorbent layer with saturation sorption isotherm. *J. Chromatogr. A* **2000**, *873*, 39–51. [CrossRef]

Sample Availability: Samples of the compounds are not available from the authors.

© 2019 by the authors. Licensee MDPI, Basel, Switzerland. This article is an open access article distributed under the terms and conditions of the Creative Commons Attribution (CC BY) license (http://creativecommons.org/licenses/by/4.0/).

Article

Optimization of Time-Weighted Average Air Sampling by Solid-Phase Microextraction Fibers Using Finite Element Analysis Software

Bulat Kenessov [1,*], Jacek A. Koziel [2], Nassiba Baimatova [1], Olga P. Demyanenko [1] and Miras Derbissalin [1]

1 Center of Physical Chemical Methods of Research and Analysis, Al-Farabi Kazakh National University, Almaty 050012, Kazakhstan; baimatova@cfhma.kz (N.B.); demyanenko@cfhma.kz (O.P.D.); derbissalin@cfhma.kz (M.D.)
2 Department of Agricultural and Biosystems Engineering, Iowa State University, Ames, IA 50011, USA; koziel@iastate.edu
* Correspondence: bkenesov@cfhma.kz; Tel.: +7-727-2390624

Received: 15 September 2018; Accepted: 15 October 2018; Published: 23 October 2018

Abstract: Determination of time-weighted average (TWA) concentrations of volatile organic compounds (VOCs) in air using solid-phase microextraction (SPME) is advantageous over other sampling techniques, but is often characterized by insufficient accuracies, particularly at longer sampling times. Experimental investigation of this issue and disclosing the origin of the problem is problematic and often not practically feasible due to high uncertainties. This research is aimed at developing the model of the TWA extraction process and optimization of TWA air sampling by SPME using finite element analysis software (COMSOL Multiphysics, Burlington, MA, USA). It was established that sampling by porous SPME coatings with high affinity to analytes is affected by slow diffusion of analytes inside the coating, an increase of their concentrations in the air near the fiber tip due to equilibration, and eventual lower sampling rate. The increase of a fiber retraction depth (Z) resulted in better recoveries. Sampling of studied VOCs using 23 ga Carboxen/polydimethylsiloxane (Car/PDMS) assembly at maximum possible Z (40 mm) was proven to provide more accurate results. Alternative sampling configuration based on 78.5 × 0.75 mm internal diameter SPME liner was proven to provide similar accuracy at improved detection limits. Its modification with the decreased internal diameter from the sampling side should provide even better recoveries. The results obtained can be used to develop a more accurate analytical method for determination of TWA concentrations of VOCs in air using SPME. The developed model can be used to simulate sampling of other environments (process gases, water) by retracted SPME fibers.

Keywords: solid-phase microextraction; air sampling; air analysis; volatile organic compounds; COMSOL; time-weighted average

1. Introduction

Analysis of time-weighted average (TWA) concentrations of volatile organic compounds (VOCs) in outdoor and indoor (occupational) air is an important part of environmental monitoring programs aiming at chronic exposure or background concentrations. Such analysis is commonly conducted using gas chromatography (GC) in combination with various sampling and sample preparation approaches. Passive sampling is a common approach for determination of TWA concentrations because of its simplicity and low cost. However, most techniques require additional sample preparation and thermal desorption in a separate unit connected to a GC [1].

Solid-phase microextraction (SPME) is the only TWA sampling technique, that does not require additional stages and/or equipment [2]. It is based on sampling via the passive VOCs extraction by a fiber coating retracted inside a protecting needle followed by thermal desorption inside a GC injection port [3,4]. Desorption of VOCs from the SPME coating is fast and does not require cryogenic or another type of focusing as is the case with whole air- or sorbent tube-based samples [5]. In the TWA mode, the SPME device with retracted fiber is deployed into a sampling location for the desired period (e.g., 24 h for daily average sampling), then isolated from possible interferences during storage and transport to a laboratory and analyzed (Figure 1). The method can be considered "green" because it fulfills all the requirements of green analytical chemistry [6,7].

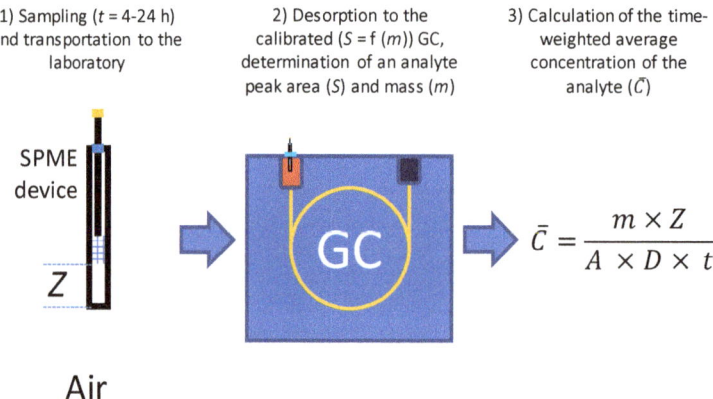

Figure 1. The typical procedure of time-weighted average sampling and analysis using retracted solid-phase microextraction fiber.

Calibration is relatively simple compared with a "classic" exposed SPME fiber that is subject to variable thickness of the boundary layer that affects the rate of extraction [8,9]. TWA sampling by retracted SPME fibers is described by the simplified version of the Fick's law of diffusion [3]:

$$\bar{C} = \frac{n \times Z}{A \times D \times t} \quad (1)$$

where $\bar{C}$—TWA concentration of an analyte, mol·m^{-3}; n—amount of an analyte extracted by a coating, mol; Z—diffusion path length (distance between the needle opening to the tip of the retracted fiber), m; A—internal cross-section area of a protecting needle, m^2; D—gas-phase molecular diffusion coefficient for a VOC, m$^2 \cdot$s^{-1}; t—sampling time, s.

Equation (1) can also be interpreted by an extraction process, i.e., the amount of analyte extracted is proportional to TWA concentration outside of the SPME needle opening, needle opening area, sampling time, and the gas-phase molecular diffusion coefficient, and inversely proportional to retraction depth.

Several important assumptions are made with the application of Equation (1) to TWA-SPME, i.e., (1) the fiber coating acts as a "zero sink" (without desorption of analytes) and does not affect the rate of sampling; (2) the SPME fiber coating is consistent and reliably responding to changing concentrations in the bulk gas-phase outside of the needle opening; and (3) the gas-phase concentration in the bulk are the same as at the face of the fiber needle opening.

To date, all published research on TWA-SPME has used Equation (1) as the basis of quantification [3,4,10–19] of VOCs in an outdoor air, laboratory air, pyrolysis reactor air, engine exhaust, and process air. Equation (1) predicted measured gas concentrations with reasonable accuracy and precision. However, more evidence suggests that the discrepancies between the model and experimental data exist. Woolcock et al. [17] reported a significant departure from the zero-sink assumption and from Equation (1) suggesting "apparent" diffusion coefficient (D) dependent on both sampling time (*t*) and retraction depth (*Z*). Baimatova et al. [11] reported significant differences in the extracted mass of naphthalene gas for different SPME coatings, i.e., that Equation (1) does not incorporate. Recent research by Tursumbayeva [20] shows that the discrepancy between Equation (1) and experimental data are amplified when a wide-bore glass liner is used for passive sampling with SPME fiber retracted inside it. Work by Tursumbayeva [20] suggests that not only the tip of the fiber coating (at the physical retraction depth *Z*) is involved in extraction, but the whole fiber coating surface with an "apparent" *Z* that is ~55% longer. Apparent saturation sorption kinetics might also be involved as predicted by Semenov et al. (2000) [21]. Thus, research is warranted to address apparent problems with the use of Equation (1).

Despite the simplicity, quantification of TWA concentrations of VOCs in ambient air using SPME can be associated with poor accuracy and precision [19]. Possible problems are variability of extraction efficiencies associated with inherent and acquired variability between individual SPME fibers, adsorption of analytes by metallic surfaces [16,19], effects of sampled air temperature and humidity.

Experimental optimization of the gas sampling process is very time-consuming, particularly at longer extraction times (>24 h). Such experimental setups are quite complex, and difficult to build and properly maintain in steady-state conditions (e.g., without leaks and with minimal impact of sorption onto the system itself). During experiments, the sensitivity of the analytical instrument can change leading to additional uncertainties. Uncertainties during experimental method optimization do not allow studying effects of parameters having potentially minor impacts on accuracy and precision.

Numerical simulation could provide useful data at various sampling parameters in a much faster and more accurate way. It could also allow modeling of the sensitivity of Equation (1) to ranges of practical (user controlled) parameters for air sampling with retracted SPME. COMSOL Multiphysics allowed efficient numerical modeling of the SPME process using a finite element analysis-based model [22–27] for liquid-phase extraction and absorption by SPME coating. Using this approach, it was possible to predict sampling profiles of analytes, which were consistent with experimental data.

The goal of this research was to develop a model for SPME with both absorptive and adsorptive fibers located (retracted) inside a protecting needle using a finite element analysis-based model (COMSOL Multiphysics) and use it to disclose potential sources of inaccuracies in the quantification of time-weighted average concentrations of VOCs in ambient air. Specifically, the effects of SPME sampling time, coating type, diffusion coefficient, fiber coating-gas distribution constant, the internal diameter of protecting needle, and SPME retraction depth on extraction were modeled for several common VOCs. Based on the results of the modeling, alternative sampling geometries were proposed.

2. Results and Discussion

2.1. Time-Weighted Average (TWA) Sampling Profiles of Benzene from Air Using Different Coatings

A sampling of VOCs from the air via retracted SPME has been described using a simplified form of the Fick's first law of diffusion (Equation (1)). However, this equation works only when a SPME fiber acts as a "zero sink" sorbent. Modeling using COMSOL Multiphysics software (methodology is provided in the Materials and Methods section) allowed obtaining sampling profiles for benzene (Figure 2). Closer inspection of Figure 2 illustrates that none of the studied coatings behave as "zero sink" sorbent adhering to Equation (1), an effect amplified by extended sampling time. After 100,000 s of sampling, Carboxen/polydimethylsiloxane (Car/PDMS), polydimethylsiloxane/divinylbenzene (PDMS/DVB), and polydimethylsiloxane (PDMS) extracted 77, 38 and 2.7%, respectively, of the theoretically required for a passive sampling technique. Even if sampling time is decreased to 10,000 s, recoveries for these three SPME fiber coatings were 91, 69 and 12.6%, respectively. At sampling time 1000 s, recoveries were 97, 88 and 32% for Car/PDMS, PDMS/DVB and PDMS, respectively.

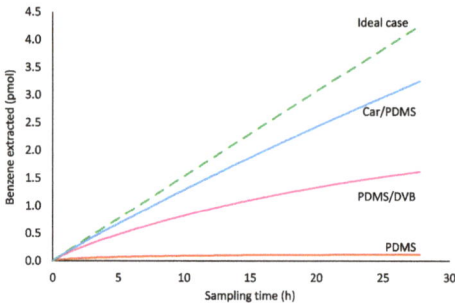

Figure 2. Benzene sampling profiles from ambient air (T = 298 K, Z = 10 mm, 24 ga needle, p = 1 atm, $C_{benzene}$ = 0.641 µmol·m^{-3}) obtained using different fiber coatings. The ideal case pertains to Equation (1).

One possible explanation for the departure from Equation (1) is that it can be caused by the increase of the analyte concentration in the air near the fiber tip (Figure 3a), which is directly proportional to the analyte concentration in the fiber tip continuously increasing during the sampling. The increase of analyte concentration in the air near the fiber tip results in the decrease of the analyte flux (i.e., the number of moles of analyte entering protecting needle per cross-sectional area and time) from the sampled air with time. This affects the sampling rate (i.e., number of moles of an analyte extracted by a coating per unit of time), which was previously assumed to be constant [3,4,10–19].

SPME fiber coating can affect the apparent rate of sampling. This was previously assumed to be negligible. According to Figure 3, Car/PDMS is the most efficient coating for TWA sampling of benzene because it provides the highest benzene extraction effectiveness indicated by the highest distribution constant. However, sampling by this coating is limited by the slow diffusion of an analyte via pores of the adsorbent (Figure 3b). At sampling time 100,000 s, the closest 1 mm of the Car/PDMS coating to the needle opening contains 41% of the total extracted analyte. Benzene concentration in the fiber tip is about 500 times higher than in its other end (furthest from the needle opening). For PDMS/DVB coating, the concentration in the tip is only about 24% higher. Slower diffusion of benzene via pores of Car/PDMS fiber is caused by the higher affinity of benzene to the surface of the solid phase (higher distribution constant), and lower porosity. Such non-uniform distribution of analytes in the Car/PDMS may be the reason of their slow desorption after TWA sampling and highly tailing peaks, particularly for most volatile analytes, which cannot be cold-trapped and refocused in a column front without cryogens. This problem also decreases the accuracy of the method.

The accuracy of the model was validated by increasing the pore diffusion coefficient of benzene inside Car/PDMS coating by three orders of magnitude. In this case, the benzene sampling profile was the same as predicted by Equation (1). This also confirms that an analyte diffusion coefficient inside a coating affects sampling profile and the accuracy of its quantification using TWA SPME. The model has also been validated in the 3D mode of COMSOL software, which is much slower compared to 2D (Video S1: COMSOL_TWA_SPME). The difference between the results of 2D and 3D modeling were below 2%, which confirms the accuracy of the 2D model.

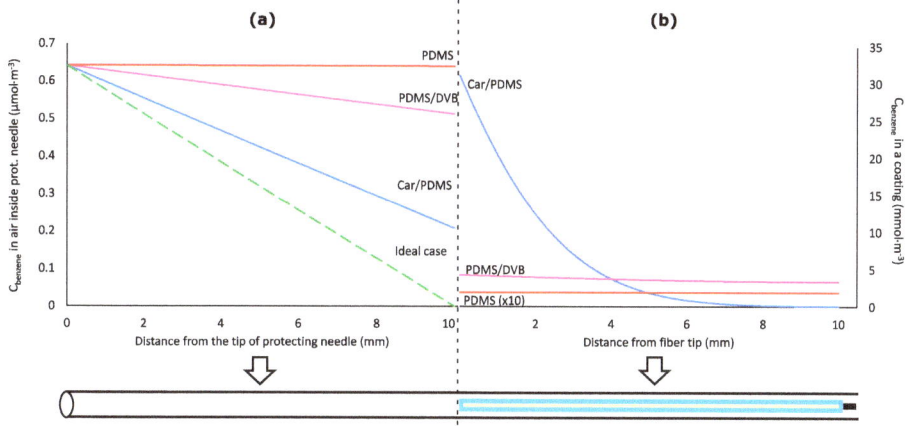

Figure 3. Concentrations of benzene in diffusion path air (**a**) and coating (**b**) of the retracted solid-phase microextraction (SPME) device after 100,000 s of time-weighted average (TWA) air sampling at $Z = 10$ mm.

2.2. Effect of the Diffusion Coefficient and Distribution Constant on Sampling of Analytes by 85-μm Carboxen/Polidimethylsiloxane Coating

The Car/PDMS coating was used for simulating extraction of other common VOCs associated with a wide range of diffusion coefficients and distribution constants. During 100,000 s, 3.3, 3.9, 3.5 and 3.3 pmol of dichloromethane, acetone, toluene, and benzene, respectively, were extracted, which corresponds to 68, 65, 82 and 77% of the theoretical values predicted by Equation (1) (Figure 4). The lowest value was observed for acetone having a distribution constant close to dichloromethane, and the highest diffusion coefficient among studied compounds. Highest recovery was observed for toluene having the lowest diffusion coefficient and the highest distribution constant. Thus, both diffusion coefficient and distribution constant affect the recovery of sampled analytes. Highest recovery can be achieved at the lowest diffusion coefficient and highest distribution constant. At sampling times 1000 and 10,000 s, recoveries are greater (95–98 and 85–93%, respectively) and less affected by the analyte's properties.

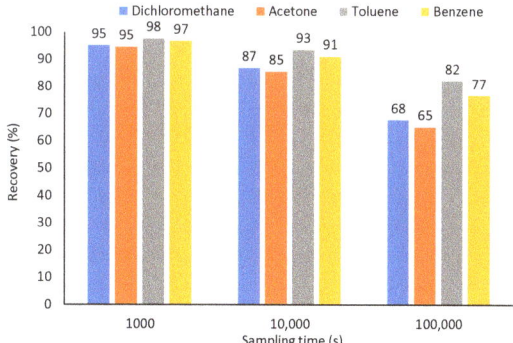

Figure 4. Effect of sampling time of TWA recoveries of analytes having different diffusion coefficients and distribution constants using 85-μm Car/PDMS fiber ($T = 298$ K, $Z = 10$ mm, 24 ga needle, $p = 1$ atm, $C = 0.641$ μmol·m^{-3}).

2.3. Effect of a Protecting Needle Gauge Size

Commercial SPME fiber assemblies are available with two different sizes of a protecting needle 24 ga and 23 ga having an internal diameter (I.D.) 310 and 340 µm, respectively. A cross-section area of the 23 ga needle is 20.3% greater than that of 24 ga needle, which (according to Equation (1)) should result in the proportionally greater amount of an analyte extracted by a 23 ga SPME assembly. However, as shown above, faster extraction rates result in a faster saturation of the coating and lower recovery at longer sampling times. According to the results of COMSOL simulations, despite ~19% greater amounts of extracted analytes compared to a 24 ga assembly, sampling with a 23 ga assembly provided similar recoveries of analytes. Such results can be explained by considering the effect of a greater space between the coating and the internal wall of the protecting needle allowing faster diffusion of analytes to the side and rear sides of the coating (Figure 5). This is consistent with recent experimental observations where straight glass GC liners were used (actual measured I.D. is ~0.84 mm compared with the nominal 0.75 mm I.D.) instead of SPME needle for sampling with retracted fiber [20]. Thus, TWA sampling using 23 ga SPME assembly is recommended over 24 ga for achieving lower detection limits without negative impact on the accuracy. All further modeling was conducted using a 23 ga SPME device.

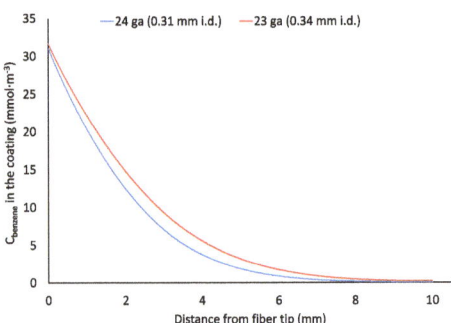

Figure 5. Effect of protecting needle gauge size concentration profile of benzene in the Car/PDMS coating after 100,000 s sampling.

2.4. Effect of Diffusion Path (Z) at Constant Analyte Concentration in Sampled Air

Diffusion path length is one of the two parameters that can easily be adjusted by users for achieving the optimal sampling conditions (the other one being sampling time). The increase of Z decreases the rate of sampling. It slows down the saturation of the fiber tip and increases the recoveries of analytes (Figure 6) at longer sampling times. For all studied analytes, at t = 100,000 s and Z = 40 mm, recovery was 86–93% compared to 66–82% at Z = 10 mm (Figure 6). The only major drawback of the increase of Z is the decrease of an analyte amount extracted by a coating and a lower analytical signal, which result in the increased detection limits. At Z = 40 mm, C = 50 µg·m^{-3} (0.641 µmol·m^{-3}) and t = 100,000 s, 23 ga Car/PDMS assembly extracts ~100 pg of benzene. For GC-mass spectrometry (MS), the detection limit of benzene is less than 2 pg [28] meaning that the detection limit will be ~1 µg·m^{-3}, which is five times lower than the maximum permissible annual average concentration of benzene in ambient air in the European Union (5 µg·m^{-3}). In other countries, permissible concentrations are even higher.

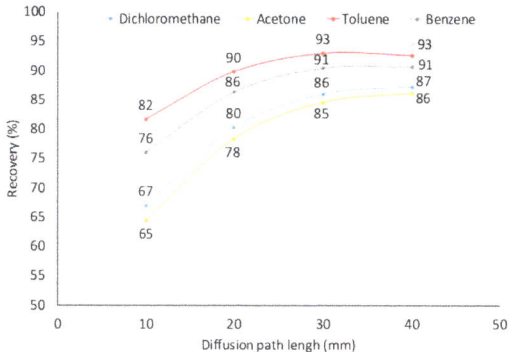

Figure 6. Effect of diffusion path length on recoveries of four analytes ($C = 0.641$ µmol·m^{-3}) after sampling for 100,000 s using 23 ga Car/PDMS fiber assembly.

2.5. Effect of Diffusion Path (Z) at Variable Analyte Concentration in Sampled Air (Worst-Case Scenario)

Time-weighted average sampling is conducted during long time periods (e.g., 24 h), during which concentrations of analytes in the sampled air can vary significantly. The apparent worst-case scenario can be when in the first half of sampling, concentration is much higher than during the second half. When the concentration of an analyte in the sampled air becomes close to or lower than the concentration near the fiber tip, the flux of analytes inside a protecting needle can go to a reverse direction resulting in desorption of analytes from a coating. However, this violates the main principle of TWA sampling: the rate of sampling should depend only on the concentration of an analyte in a sampled air. It means that if an analyte concentration in sampled air is zero, a rate of extraction should also be equal to zero. Thus, the aim of this part of the work was to model such a case and estimate the highest possible uncertainty of the TWA SPME sampling approach.

As was assumed, desorption of dichloromethane, acetone, and benzene from a fiber started after concentrations of analytes dropped from 1.176 to 0.1176 µmol·m^{-3} in the middle of the extraction process (Figure 7). Desorption of toluene was not observed because it has the highest distribution constant among all studied analytes. However, the toluene sampling rate after the drop of its concentration in sampled air was lower than theoretical. Recoveries of analytes at $Z = 10$ mm dropped from 65–82 to 52–70%, at $Z = 20$ mm from 78–90 to 67–79%, at $Z = 30$ mm from 85–93 to 73–82, at $Z = 40$ mm from 86–93 to 75–82% (Figure 7). Only at $Z = 40$ mm, it was possible to keep recovery of all analytes above 75%. Thus, if possible, for greater accuracy, sampling must be arranged so that no significant drop in concentration takes place. Such a drop can be observed, e.g., if the end of sampling is planned for the night when VOCs concentrations in ambient air are typically lower due to much lower road traffic and other human activities. Also, using shorter sampling times can minimize the risk of the reverse diffusion when ambient concentrations are predicted to drop significantly.

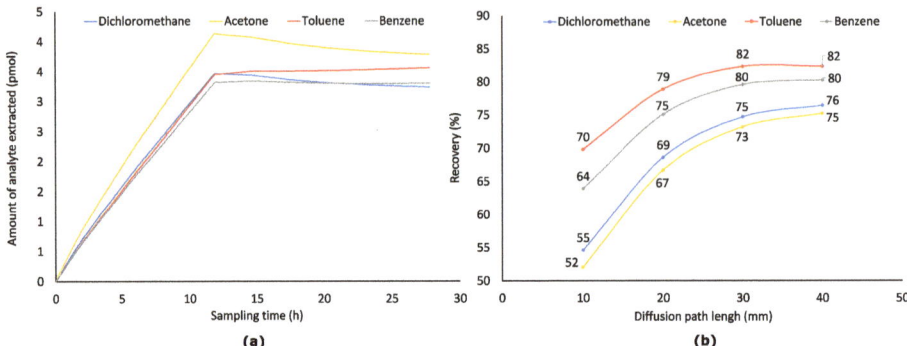

Figure 7. Sampling (Z = 10 mm) profiles (**a**) of four analytes from air having their varying concentrations ($C_{0-49,000\,s}$ = 1.176 µmol·m^{-3}, $C_{49,000-51,000\,s}$ = 1.176–0.1176 µmol·m^{-3}, $C_{51,000-100,000\,s}$ = 0.1176 µmol·m^{-3}) and recoveries of analytes (**b**) at t = 100,000 s and different Z.

2.6. Alternative Geometries for TWA SPME Sampling

As was shown above (Figure 5), an increase of the internal diameter of a protecting needle provides more space for analytes to diffuse around the coating and better reach the side of the coating. It decreases the controlling role of the fiber coating tip and should lead to more accurate and reproducible results.

Tursumbayeva [20] proposed using SPME liner for TWA SPME to avoid sorption of analytes onto metallic walls of a protecting needle. The same approach can be used to avoid equilibration of analytes between the fiber tip and the surrounding space after sampling over longer time periods. At variable concentrations of analytes (as simulated in the previous section), calculated recoveries for VOCs using Z = 67 mm (Figure 8a) are 73–84%, which are close to the values obtained using retracted fiber at Z = 40 mm. No improvement was observed because of 0.75-mm I.D. SPME liner has 4.9 times greater cross-sectional area than 23 ga protecting needle, which results in 2.9 times greater theoretical flux of analytes from sampled air to the coating under the set Z (67 and 40 mm, respectively). To decrease the flux of analytes, the liner can be manufactured with a lower I.D. (e.g., 0.34 mm as for 23 ga needle) from the sampling side almost to the expected location of the fiber as shown in Figure 8b. Under these conditions, recoveries increased to 88–91% (Figure 9).

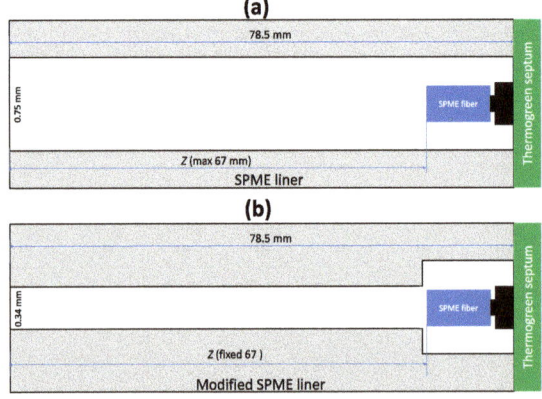

Figure 8. Alternative geometries for TWA SPME sampling: (**a**) used by Tursumbayeva [18], and (**b**) proposed in this research to minimize sources of deviation from Fick's law of diffusion calibration.

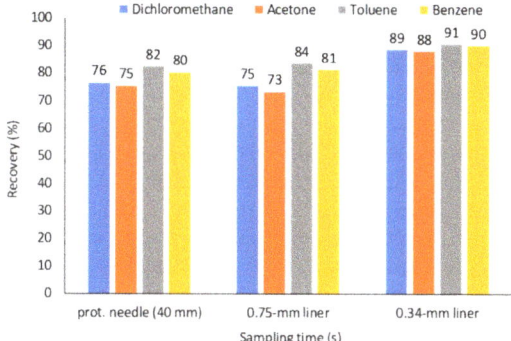

Figure 9. Effect of TWA SPME sampling geometry on recoveries (t = 100,000 s, $C_{0-49,000\text{ s}}$ = 1.176 µmol·m^{-3}, $C_{49,000-51,000\text{ s}}$ = 1.176–0.1176 µmol·m^{-3}, $C_{49,000-100,000\text{ s}}$ = 0.1176 µmol·m^{-3}).

The use of alternative geometries (Figure 10) resulted in a more uniform distribution of the analytes in the coating; for 0.75-mm I.D. SPME liner concentrations of analytes near the fiber tip were only 1.1–2.7 times greater than at another side of the coating. This should result in faster desorption of analytes, less pronounced peak tailing and greater accuracy of the method. A similar effect is achieved when using Radiello® passive air sampler [29], which provide a greater surface area of an adsorbent available for the diffusive air sampling.

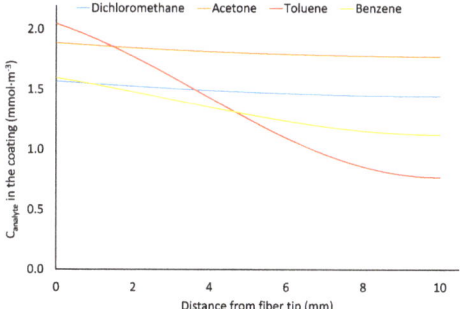

Figure 10. Profiles of analyte concentration in the Car/PDMS coating after sampling ambient air ($C_{0-49,000\text{ s}}$ = 1.176 µmol·m^{-3}, $C_{49,000-51,000\text{ s}}$ = 1.176–0.1176 µmol·m^{-3}, $C_{49,000-100,000\text{ s}}$ = 0.1176 µmol·m^{-3}) for 100,000 s using the geometry presented in Figure 8a.

3. Materials and Methods

3.1. General Parameters of Modeling

Simulations were completed using COMSOL Multiphysics 5.3a (Burlington, MA, USA) on a desktop computer equipped with quad-core Core i5 processor and 8 Gb of random-access memory. For modeling, "Chemical Species Transport" module ("Transport of diluted species" or "Transport of diluted species in porous media" physics) was used in "Time-Dependent" mode in two dimensions (axisymmetric). Fick's second law of diffusion was used by the module:

$$\frac{\partial c_i}{\partial t} = \nabla \times (D_i \times \nabla c_i) \quad (2)$$

Benzene, a ubiquitous air pollutant, was used as a model analyte for most initial calculations. Diffusion coefficients of benzene in the air and PDMS coating were set to 8.8×10^{-6} and 10^{-10} m^2·s^{-1}, respectively [30]. Distribution constant (K_d) for benzene and common SPME coatings was set to 150,000 (85 µm Car/PDMS) [31], 8300 (65 µm PDMS/DVB) [31], and 301 (PDMS) [5]. For dichloromethane, acetone and toluene, distribution constants between 85 µm Car/PDMS coating and air were set to 72,000, 71,000 and 288,000, respectively [31].

The geometry of a fiber assembly was built in as inputs based on the data provided by Pawliszyn [5]. Simulations were conducted for Stableflex® (Supelco, Bellefonte, PA, USA) fibers with a core diameter of 130 µm. For 85 µm Car/PDMS and 65 µm PDMS/DVB, total fiber diameters were set to 290 and 270 µm, respectively. Calculations were conducted for 24- and 23 ga coatings having internal diameter of 310 and 340 µm, respectively.

The extra fine free triangular mesh was used for the modeling. To provide better meshing at the coating−air interface, the resolution of narrow regions was increased to "2". The computation was completed in the range between 0 and 100,000 s at the step of 1000 s. The concentration of an analyte at the tip of the protecting needle was set to 0.641 µmol·m^{-3}, which corresponds to 50 µg·m^{-3} of benzene.

3.2. Sampling Using Absorptive Coatings

Inward (and outward) fluxes from (or backward to) air into an absorptive coating ($Flux_1$ and $Flux_2$, respectively) at the boundaries (marked by red lines in Figure 11) were simulated using the equation, previously proposed by Mackay and Leinonen [32] for the water−air interface:

$$Flux_1 = k \times \left(C_a - \frac{C_f}{K_d}\right); \quad Flux_2 = k \times \left(\frac{C_f}{K_d} - C_a\right) \quad (3)$$

where: k—flux coefficient, m·s^{-1}; C_a and C_f—concentrations of an analyte in air and coating at the boundary layer, respectively, mol·m^{-3}; K_d—distribution constant for a VOC between SPME coating and air.

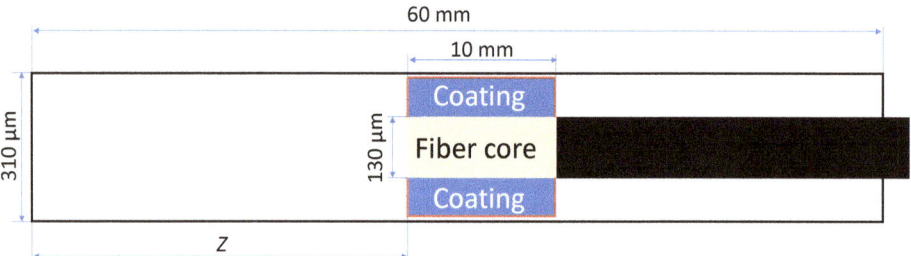

Figure 11. The geometry of SPME device (retracted inside a protective needle for TWA sampling) used for modeling. Note: red lines indicate the boundaries between air and coating.

The true value of the flux coefficient was unknown, but in this research, it was assumed to be sufficiently high for not affecting the flux, as was recently proposed by Alam et al. [23]. Thus, the flux coefficient was set to 1000 m·s^{-1}. A further increase of the flux coefficient did not affect the results of the modeling.

3.3. Sampling Using Adsorptive Coatings

For adsorptive coatings, the "Adsorption" mechanism was activated in the model. The isotropic diffusion coefficient (in the air inside pores) was the same as for air (set to 8.8, 8.7, 12.4 and 10.1 mm^2·s^{-1} for benzene, toluene, acetone, and dichloromethane, respectively). The approach proposed by Mocho and Desauziers [33] involving Knudsen diffusion in micro-pores was also tested. However, it was later rejected for model simplification because the diffusion of analytes inside coating is mainly driven by molecular diffusion inside macro-pores. The presence of PDMS binder was not considered in the model because: (1) it has much weaker affinity to analytes than Carboxen; and (2) the layer of PDMS in the coating is very thin and should not affect the diffusion of analytes [5]; (3) there is not enough published information about the exact structure of the coating.

Adsorption was set to "User defined" with a distribution constant (K_p, m^3·kg^{-1}) calculated as a dimensionless distribution constant divided by a coating density (K_d/ρ). Coating porosities (ε = 0.685 for Car/PDMS and 0.775 for PDMS/DVB) were calculated using intra-particle porosities (0.37 for Car, and 0.55 for DVB [34]) and inter-particle porosity. The exact value of the latter is proprietary and not available in the open literature. Taking into account, the spherical shape of particles and available scanning electron microscope (SEM) photos, the inter-particle porosity of both coatings was set to the maximum possible value (0.50). A particle porosity (ε) was calculated as the total volume of pores (0.78 mL for Car, and 1.54 mL for DVB) divided by the total volume of one gram of material (2.13 mL for Car, and 2.78 mL for DVB). Densities of the coatings were calculated using free fall densities of the particles (470 kg·m^{-3} for Car, and 360 kg·m^{-3} for DVB) [34] and inter-particle porosity. Effective diffusion coefficients were calculated during the calculations by the COMSOL software using the Tortuosity model [33]:

$$D_e = \frac{\varepsilon \times D_p}{\tau} \quad (4)$$

where: ε—porosity; τ—tortuosity calculated from the porosity [33]:

$$\tau = \varepsilon + 1.5 \times (1 - \varepsilon) \quad (5)$$

For Car/PDMS and PDMS/DVB coatings, tortuosity was set to 1.16 and 1.1125, respectively.

4. Conclusions

A finite element analysis-based model (based on COMSOL Multiphysics software) allowed efficient simulation of TWA air sampling of VOCs using retracted SPME fibers. It was possible to model the effects of sampling time, coating type (including adsorptive coatings for the first time) and composition, diffusion coefficient, the distribution constant, the internal diameter of a protecting needle and diffusion path on the recovery of analytes, their concentration profiles in the air inside protecting needle, and the coating. The advantages of such a simulation compared to an experiment are: (1) time and cost savings; (2) lower uncertainty and the possibility to discover minor impacts of sampling parameters on its performance; and (3) the possibility to understand and optimize a sampling process in greater detail. The results of this research allowed disclosing potential sources of the apparent departure from Fick's law of a diffusion-based model used for quantification of VOCs with retracted SPME.

It was established that sampling by porous coatings with high affinity to the analyte (Car/PDMS) is affected by the saturation of the fiber tip and slow diffusion of analytes in the coating. Highest recoveries are achieved for analytes having lowest diffusion coefficients and highest affinities to a coating. The increase of an internal diameter of a protecting needle from 24 to 23 ga allows proportionally greater responses to be obtained at similar recoveries.

The most important parameter of a sampling process that users can control is a retraction depth. The increase of Z allows slowing down the sampling and achieving higher recoveries of analytes. In this study, at Z = 40 mm and constant analyte concentration in a sampled air, recoveries of studied analytes reached 86–93% compared to 65–82% at Z = 10 mm. The developed model allowed simulation of the worst sampling case when analyte concentrations significantly drop in the middle of sampling. For the first time, it has been proven that at such sampling conditions and Z = 40 mm, recoveries of analytes can drop by ~10%, while at Z = 10 mm by ~15%.

According to the results of the simulation, it is optimal to conduct sampling of studied VOCs using a 23 ga Car/PDMS assembly at $Z = 40$ mm. Expected detection limits at these parameters are about 1 µg·m^{-3}.

Alternative geometries of a protective TWA SPME sampling devices could be used to increase recoveries of analytes. Sampling using 0.75-mm I.D. SPME GC liner at $Z = 67$ mm provides similar recoveries compared to sampling using a protecting needle at $Z = 40$ mm, but it provides greater amounts of analytes extracted and lower detection limits. To achieve greater recovery, part of the liner should have narrower I.D. (e.g., 0.34 mm). The increase of the diameter of the extraction zone where the coating is located results in a more uniform distribution of analytes, which should lead to faster desorption, less pronounced peak tailing and greater accuracy. Specific sampler parameters should be selected for particular sampling time and environmental conditions (temperature and atmospheric pressure) using the developed model.

The methodology used in this study could also be used for more accurate and simpler calibration of the method. It can be used to model the sampling of other environments (process gases, water) by retracted SPME fibers. Further modification of this model could allow simulation of soil and soil gas sampling.

Supplementary Materials: The following are available online at http://www.mdpi.com/1420-3049/23/11/2736/s1, Video S1: COMSOL_TWA_SPME.

Author Contributions: B.K. and J.A.K. developed the model and designed computational experiments, N.B. and O.P.D. undertook the literature review and validated the model, B.K. and M.D. conducted the modeling and processed the data, B.K. and J.A.K. described the results, B.K., N.B. and O.P.D. prepared the manuscript for submission. All authors have approved the final version of the manuscript.

Funding: This research was funded by the Ministry of Education and Science of the Republic of Kazakhstan, grant number AP05133158. This research was partially supported by the Iowa Agriculture and Home Economics Experiment Station, Ames, Iowa. Project No. IOW05400 (Animal Production Systems: Synthesis of Methods to Determine Triple Bottom Line Sustainability from Findings of Reductionist Research) is sponsored by Hatch Act and State of Iowa funds.

Acknowledgments: Authors would like to thank the Ministry of Education and Science of the Republic of Kazakhstan for supporting Miras Derbissalin and Olga Demyanenko with Ph.D. and M.S. scholarships, respectively.

Conflicts of Interest: The authors declare no conflict of interest.

References

1. Pienaar, J.J.; Beukes, J.P.; Van Zyl, P.G.; Lehmann, C.M.B.; Aherne, J. Passive diffusion sampling devices for monitoring ambient air concentrations. In *Comprehensive Analytical Chemistry*; Forbes, P.B.C., Ed.; Elsevier: Amsterdam, The Netherlands, 2015; Volume 70, pp. 13–52, ISBN 9780444635532.
2. Grandy, J.; Asl-Hariri, S.; Pawliszyn, J. Novel and emerging air-sampling devices. In *Comprehensive Analytical Chemistry*; Forbes, P.B.C., Ed.; Elsevier: Amsterdam, The Netherlands, 2015; Volume 70, ISBN 9780444635532.
3. Martos, P.A.; Pawliszyn, J. Time-weighted average sampling with solid-phase microextraction device: Implications for enhanced personal exposure monitoring to airborne pollutants. *Anal. Chem.* **1999**, *71*, 1513–1520. [CrossRef] [PubMed]
4. Khaled, A.; Pawliszyn, J. Time-weighted average sampling of volatile and semi-volatile airborne organic compounds by the solid-phase microextraction device. *J. Chromatogr. A* **2000**, *892*, 455–467. [CrossRef]
5. Pawliszyn, J. *Handbook of Solid Phase Microextraction*, 1st ed.; Elsevier: Amsterdam, The Netherlands, 2012; ISBN 9780124160170.
6. Gałuszka, A.; Migaszewski, Z.M.; Konieczka, P.; Namieśnik, J. Analytical Eco-Scale for assessing the greenness of analytical procedures. *Trends Anal. Chem.* **2012**, *37*, 61–72. [CrossRef]
7. Płotka-Wasylka, J. A new tool for the evaluation of the analytical procedure: Green Analytical Procedure Index. *Talanta* **2018**, *181*, 204–209. [CrossRef] [PubMed]
8. Koziel, J.; Jia, M.; Pawliszyn, J. Air sampling with porous solid-phase microextraction fibers. *Anal. Chem.* **2000**, *72*, 5178–5186. [CrossRef] [PubMed]

9. Augusto, F.; Koziel, J.; Pawliszyn, J. Design and validation of portable SPME-devices for rapid field air. *Anal. Chem.* **2001**, *73*, 481–486. [CrossRef] [PubMed]
10. Koziel, J.A.; Nguyen, L.T.; Glanville, T.D.; Ahn, H.; Frana, T.S.; van Leeuwen, J. Method for sampling and analysis of volatile biomarkers in process gas from aerobic digestion of poultry carcasses using time-weighted average SPME and GC–MS. *Food Chem.* **2017**, *232*, 799–807. [CrossRef] [PubMed]
11. Baimatova, N.; Koziel, J.; Kenessov, B. Passive sampling and analysis of naphthalene in internal combustion engine exhaust with retracted SPME device and GC-MS. *Atmosphere* **2017**, *8*, 130. [CrossRef]
12. Koziel, J.; Jia, M.; Khaled, A.; Noah, J.; Pawliszyn, J. Field air analysis with SPME device. *Anal. Chim. Acta* **1999**, *400*, 153–162. [CrossRef]
13. Koziel, J.A.; Pawliszyn, J. Air sampling and analysis of volatile organic compounds with solid phase microextraction. *J. Air Waste Manag. Assoc.* **2001**, *51*, 173–184. [CrossRef] [PubMed]
14. Koziel, J.A.; Noah, J.; Pawliszyn, J. Field sampling and determination of formaldehyde in indoor air with solid-phase microextraction and on-fiber derivatization. *Environ. Sci. Technol.* **2001**, *35*, 1481–1486. [CrossRef] [PubMed]
15. Chen, Y.; Pawliszyn, J. Time-weighted average passive sampling with a solid-phase microextraction device. *Anal. Chem.* **2003**, *75*, 2004–2010. [CrossRef] [PubMed]
16. Chen, Y.; Pawliszyn, J. Solid-phase microextraction field sampler. *Anal. Chem.* **2004**, *76*, 6823–6828. [CrossRef] [PubMed]
17. Woolcock, P.J.; Koziel, J.A.; Cai, L.; Johnston, P.A.; Brown, R.C. Analysis of trace contaminants in hot gas streams using time-weighted average solid-phase microextraction: Proof of concept. *J. Chromatogr. A* **2013**, *1281*, 1–8. [CrossRef] [PubMed]
18. Woolcock, P.J.; Koziel, J.A.; Johnston, P.A.; Brown, R.C.; Broer, K.M. Analysis of trace contaminants in hot gas streams using time-weighted average solid-phase microextraction: Pilot-scale validation. *Fuel* **2015**, *153*, 552–558. [CrossRef]
19. Baimatova, N.; Koziel, J.A.; Kenessov, B. Quantification of benzene, toluene, ethylbenzene and o-xylene in internal combustion engine exhaust with time-weighted average solid phase microextraction and gas chromatography mass spectrometry. *Anal. Chim. Acta* **2015**, *873*, 38–50. [CrossRef] [PubMed]
20. Tursumbayeva, M. Simple and Accurate Quantification of Odorous Volatile Organic Compounds in Air with Solid Phase Microextraction and Gas Chromatography—Mass Spectrometry. Master's Thesis, Iowa State University, Ames, IA, USA, 2017.
21. Semenov, S.N.; Koziel, J.A.; Pawliszyn, J. Kinetics of solid-phase extraction and solid-phase microextraction in thin adsorbent layer with saturation sorption isotherm. *J. Chromatogr. A* **2000**, *873*, 39–51. [CrossRef]
22. Zhao, W. Solid Phase Microextraction in Aqueous Sample Analysis. Ph.D. Thesis, University of Waterloo, Waterloo, ON, Canada, 2008.
23. Alam, M.N.; Ricardez-Sandoval, L.; Pawliszyn, J. Numerical modeling of solid-phase microextraction: binding matrix effect on equilibrium time. *Anal. Chem.* **2015**, *87*, 9846–9854. [CrossRef] [PubMed]
24. Alam, M.N.; Ricardez-Sandoval, L.; Pawliszyn, J. Calibrant free sampling and enrichment with solid phase microextraction: computational simulation and experimental verification. *Ind. Eng. Chem. Res.* **2017**, *56*, 3679–3686. [CrossRef]
25. Souza-Silva, E.A.; Gionfriddo, E.; Alam, M.N.; Pawliszyn, J. Insights into the effect of the PDMS-layer on the kinetics and thermodynamics of analyte sorption onto the matrix-compatible SPME coating. *Anal. Chem.* **2017**, *89*, 2978–2985. [CrossRef] [PubMed]
26. Alam, M.N.; Pawliszyn, J. Numerical simulation and experimental validation of calibrant-loaded extraction phase standardization approach. *Anal. Chem.* **2016**, *88*, 8632–8639. [CrossRef] [PubMed]
27. Alam, M.N.; Nazdrajić, E.; Singh, V.; Tascon, M.; Pawliszyn, J. Effect of transport parameters and device geometry on extraction kinetics and efficiency in direct immersion solid-phase microextraction. *Anal. Chem.* **2018**, *90*, 11548–11555. [CrossRef] [PubMed]
28. Baimatova, N.; Kenessov, B.; Koziel, J.A.; Carlsen, L.; Bektassov, M.; Demyanenko, O.P. Simple and accurate quantification of BTEX in ambient air by SPME and GC–MS. *Talanta* **2016**, *154*, 46–52. [CrossRef] [PubMed]
29. Cocheo, V.; Boaretto, C.; Sacco, P. High uptake rate radial diffusive sampler suitable for both solvent and thermal desorption. *Ind. Hyg. Assoc.* **1996**, *57*, 897–904. [CrossRef]
30. Chao, K.-P.; Wang, V.-S.; Yang, H.-W.; Wang, C.-I. Estimation of effective diffusion coefficients for benzene and toluene in PDMS for direct solid phase microextraction. *Polym. Test.* **2011**, *30*, 501–508. [CrossRef]

31. Prikryl, P.; Sevcik, J.G.K. Characterization of sorption mechanisms of solid-phase microextraction with volatile organic compounds in air samples using a linear solvation energy relationship approach. *J. Chromatogr. A* **2008**, *1179*, 24–32. [CrossRef] [PubMed]
32. Mackay, D.; Leinonen, P.J. Rate of evaporation of low-solubility contaminants from water bodies to atmosphere. *Environ. Sci. Technol.* **1975**, *9*, 1178–1180. [CrossRef]
33. Mocho, P.; Desauziers, V. Static SPME sampling of VOCs emitted from indoor building materials: Prediction of calibration curves of single compounds for two different emission cells. *Anal. Bioanal. Chem.* **2011**, *400*, 859–870. [CrossRef] [PubMed]
34. Tuduri, L.; Desauziers, V.; Fanlo, J.L. Potential of solid-phase microextraction fibers for the analysis of volatile organic compounds in air. *J. Chromatogr. Sci.* **2001**, *39*, 521–529. [CrossRef] [PubMed]

Sample Availability: Samples of the compounds are not available from the authors.

© 2018 by the authors. Licensee MDPI, Basel, Switzerland. This article is an open access article distributed under the terms and conditions of the Creative Commons Attribution (CC BY) license (http://creativecommons.org/licenses/by/4.0/).

Article

A Dual Ligand Sol–Gel Organic-Silica Hybrid Monolithic Capillary for In-Tube SPME-MS/MS to Determine Amino Acids in Plasma Samples

Luis F. C. Miranda, Rogéria R. Gonçalves and Maria E. C. Queiroz *

Departamento de Química, Faculdade de Filosofia Ciência e Letras de Ribeirão Preto, Universidade de São Paulo, P.O. Box 14040-901 Ribeirão Preto, SP, Brazil; luisfelipe.c22@usp.br (L.F.C.M.); rrgoncalves@ffclrp.usp.br (R.R.G.)
* Correspondence: mariaeqn@ffclrp.usp.br; Tel.: +55-16-3315-9172

Academic Editors: Constantinos K. Zacharis and Paraskevas D. Tzanavaras
Received: 21 March 2019; Accepted: 24 April 2019; Published: 27 April 2019

Abstract: This work describes the direct coupling of the in-tube solid-phase microextraction (in-tube SPME) technique to a tandem mass spectrometry system (MS/MS) to determine amino acids (AA) and neurotransmitters (NT) (alanine, serine, isoleucine, leucine, aspartic acid, glutamic acid, lysine, methionine, tyrosine, and tryptophan) in plasma samples from schizophrenic patients. An innovative organic-silica hybrid monolithic capillary with bifunctional groups (amino and cyano) was developed and evaluated as an extraction device for in-tube SPME. The morphological and structural aspects of the monolithic phase were evaluated by scanning electron microscopy (SEM), Fourier transform infrared spectroscopy (FTIR), nitrogen sorption experiments, X-ray diffraction (XRD) analyses, and adsorption experiments. In-tube SPME-MS/MS conditions were established to remove matrix, enrich analytes (monolithic capillary) and improve the sensitivity of the MS/MS system. The proposed method was linear from 45 to 360 ng mL^{-1} for alanine, from 15 to 300 ng mL^{-1} for leucine and isoleucine, from 12 to 102 ng mL^{-1} for methionine, from 10 to 102 ng mL^{-1} for tyrosine, from 9 to 96 ng mL^{-1} for tryptophan, from 12 to 210 ng mL^{-1} for serine, from 12 to 90 ng mL^{-1} for glutamic acid, from 12 to 102 ng mL^{-1} for lysine, and from 6 to 36 ng mL^{-1} for aspartic acid. The precision of intra-assays and inter-assays presented CV values ranged from 1.6% to 14.0%. The accuracy of intra-assays and inter-assays presented RSE values from −11.0% to 13.8%, with the exception of the lower limit of quantification (LLOQ) values. The in-tube SPME-MS/MS method was successfully applied to determine the target AA and NT in plasma samples from schizophrenic patients.

Keywords: in-tube SPME-MS/MS; dual ligand organic-silica hybrid monolith capillary; amino acids; plasma samples

---

1. Introduction

Schizophrenia is a syndrome of inconclusive etiopathogenesis with a prevalence of about 1% in the general population. Underlying factors include genetic predisposition and impaired neurodevelopment in early life stages [1].

In the past few years, interest in finding out a possible role of amino acids (AA) in schizophrenia pathophysiology has increased [2–9]. Higher glycine, serine, glutamate, and aspartic acid concentrations were reported in plasma samples from schizophrenic patients [2,10]. Levels of other AA have provided inconsistent results [4]. In this context, many researchers have monitored AA and neurotransmitters (NT) in schizophrenic patients [2,6,7,9].

Liquid chromatography tandem–mass spectrometry (LC-MS/MS) has been used to determine AA and NT concentrations in biological samples from schizophrenic patients [2,9,11]. Sample preparation is a significant step in any bioanalytical chromatographic procedure even when powerful analytical instruments are employed.

In-tube solid-phase microextraction (in-tube SPME) is an effective sample preparation technique for biological fluids. It is fast to operate, easy to automate, solvent-free, and requires small sample volume. Capillaries with different stationary phases, including monolithic polymers, have been used in the first dimension of the in-tube SPME-LC system [12–25]. Monolithic materials have binary porous structure (mesopores and macropores). The presence of micron-size macropores ensures fast dynamic transport and low backpressure, leading to high flow rate and analytical speed. Moreover, polymeric monolith presents satisfactory loading capacity (which is superior to open tubular column loading capacity) and favors a convective mass transfer procedure (which is preferable in extraction processes). Organic-inorganic hybrid silica-based monoliths combine the advantages of organic polymers (pH stability and good biocompatibility) and silica-based monoliths (high permeability, high mechanical strength, and good organic solvent tolerance) [26–30].

Organic-inorganic hybrid silica-based monolithic capillaries with cyano [31–33] or amino functionalities [34] have been developed as extraction device for microextraction. To our knowledge, preparation of an organic-inorganic hybrid silica-based monolithic capillary with both of these bifunctional groups (cyano and amino) is an innovation.

Newly developed methods in the mass spectrometry field include direct coupling of sample preparation devices, such as solid-phase microextraction (SPME), to the MS instrumentation [35,36].

In this context, the present article describes direct coupling of the in-tube SPME technique to the MS/MS system to determine alanine, serine, leucine, isoleucine, tryptophan, methionine, tyrosine, lysine, aspartic acid, and glutamic acid in plasma samples from schizophrenic patients. This system employs an organic-inorganic hybrid silica-based monolithic capillary bearing bifunctional groups (amino and cyano).

2. Results

2.1. Hybrid Monolithic Capillary Synthesis

Fused silica capillary pretreatment was important to clean and increase the concentration of silanol groups in the inner surface; these groups can act as chemical binding sites for effective monolith attachment during in situ sol–gel synthesis [37].

In general, the sol–gel reaction encompasses three steps: (a) alkoxysilane precursor hydrolysis; (b) condensation between hydrated silica (Si-OH groups) and non-hydrolyzed alkoxysilane to form siloxane bonds (Si–O–Si); and (c) polycondensation of additional silanol group linkage to form linear or cyclic oligomers and, eventually, a silicate network [38]. The sol–gel matrix properties such as pore size and sorption capacity can be controlled by changing monomer amount and type, water amount, pH, solvent nature, additives, and reaction temperature.

APTES (3-aminopropyl triethoxysilane) is a basic alkoxide precursor that can promote fast TEOS (Tetraethylorthosilicate) hydrolysis (due to the hydroxyl group) and condensation [34,39]. On the other hand, the reaction between CN-TEOS (3-cyanopropyltriethoxysilane) and TEOS is slower in basic medium. Therefore, the main challenge of this synthesis is to elevate CN-TEOS hydrolysis and condensation reaction rates in basic medium to obtain a dual-ligand (cyane and amino) sol–gel organic-inorganic hybrid monolith. To achieve this goal, we used ammonium fluoride as catalyst. Fluoride can increase silicon coordination above four due to the smaller ionic radius of the fluoride anion as compared to the hydroxyl group [40–42]. Thus, ammonium fluoride promotes simultaneous hydrolysis and condensation of both precursors (CN-TEOS and APTES) with TEOS.

Malik et al. have proven that the cyanopropyl moiety in CN-PDMS coatings provides effective extraction of highly and medium polar analytes from aqueous medium [43]; Yan et al. described that amino groups (hybrid silica monoliths) interact with acidic analytes [39]. Moreover, the precursors

CN-TEOS and APTES can establish dipole–dipole, dipole–induced dipole, and charge–transfer interactions [31,43].

The optimization of surfactant and water amounts could help to control pore size and, consequently, permeability [39]. The CTAB surfactant acts as supramolecular template during sol–gel monolith formation and can be easily removed by simple solvent extraction. However, in our experiments, the CTAB amount was slightly different (5 and 7 mg), so it did not significantly influence analyte sorption.

Initially, we investigated different molar ratios of the precursors (TEOS, CN-TEOS, and APTES) (Table 1, procedures 1–4). According to Figure 1a, procedure number 1 presented the highest signal area for majority of the analytes. Thus, the presence of both amino and cyano groups in the monolith structure increased the capillary sorption capacity. All the evaluated synthesis procedures resulted in monolithic phase with adequate permeability and high mechanical strength. The ethanol/water ratio (100:20 v/v) used in procedure number 1 improved capillary performance (Figure 1b). Methanol was also evaluated as a replacement for ethanol, but this change did not modify the sorption capacity (Figure 1c). The aging temperature influence was assessed in procedure number 6 (Table 1). The monolithic phase prepared at 22 °C was not reproducible—the correlation coefficient was higher than 15%. Hence, we selected 60 °C as the aging temperature for subsequent assays.

Table 1. Optimization of the synthesis parameters.

Procedure	TEOS (µL)	APTES (µL)	CN-TEOS (µL)	H₂O/EtOH (µL)	TEOS/APTES/CN-TEOS (µL)	Aging Temperature (°C)
1	56	28	28	100:20	2:1:1	60
2	38	38	38	100:20	1:1:1	60
3	56	56	0	100:20	1:1:0	60
4	56	0	56	100:20	1:0:1	60
5	56	28	28	50:50	2:1:1	60
6	56	28	28	50:50	2:1:1	22

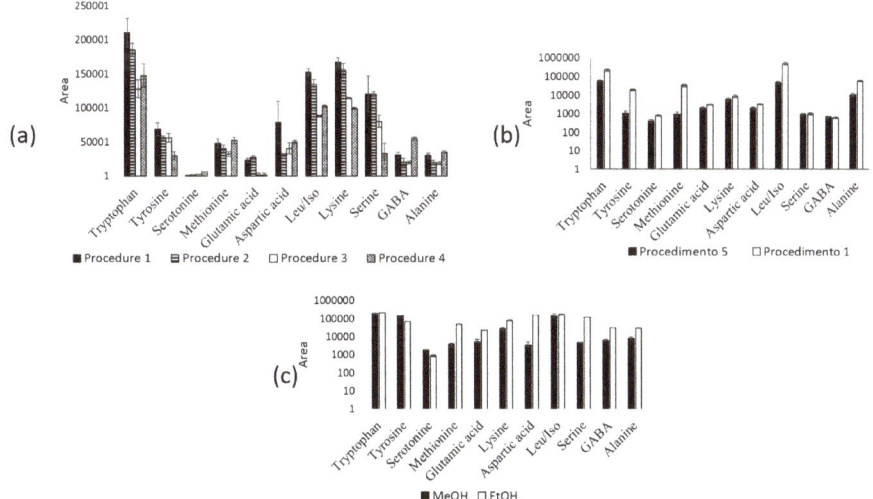

Figure 1. Optimization of the synthesis procedure (Table 1); (a) Molar ratios of the precursors TEOS/APTES/CN-TEOS, procedure 1 (2:1:1 $v/v/v$); procedure 2 (1:1:1 $v/v/v$); procedure 3 (1:1:0 $v/v/v$); and procedure 4 (1:0:1 $v/v/v$); (b) ethanol/water ratios [procedure 1 (107:20 µL) and procedure 5 (50:50 µL)]; (c) methanol or ethanol as solvent. Leu = leucine; Iso = isoleucine; GABA = γ-aminobutiric acid.

Three new different capillaries synthesized by procedure 1 attested that the in-situ polymerization procedure was reproducible. We assayed these capillaries with 100 nmol mL^{-1} AA and NT aqueous solution. The intra-batch and inter-batch assays presented RSD values lower than 15.0%, which demonstrated that the synthesis procedure had good reproducibility.

2.2. Characterization of Hybrid Silica Monoliths

The SEM micrographs in Figure 2 show the morphological features of the monolithic capillary. Because we performed the reaction in the presence of ammonium fluoride as basic catalyst, we expected that the morphological features of the hybrid system would resemble the morphological features of the material obtained from TEOS in the presence of the APTES amino groups (basic precursor).

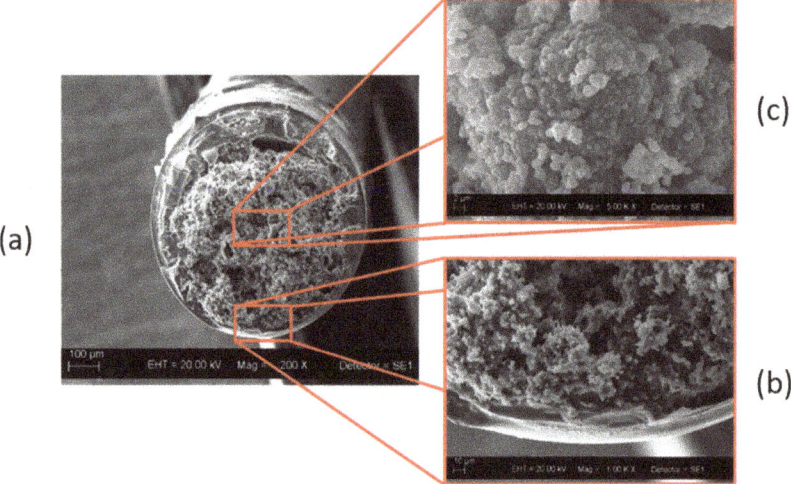

Figure 2. Scanning electron microscopy images of hybrid silica monolith containing cyanopropyl and aminopropyl groups. (**a**) Magnification of 200 X; (**b**) magnification of 1.00 kX; (**c**) magnification of 5.00 kX.

According to Figure 2, the hybrid monolithic capillary did not present any shrinkage and was uniform and regular. The monolith was clearly tightly attached to the capillary inner wall (Figure 2b). Both SEM images evidenced a homogeneous, continuous, and porous skeleton consisting of interconnected particles. The faster condensation kinetics during the ammonium fluoride-catalyzed sol–gel processes generated a highly compacted particulate structure. Morphological features are extremely important to understand microstructure and pore distribution as well as their correlation with sorption efficiency. Nitrogen sorption experiments offered a deeper understanding of porosity: the specific BET surface area and pore volume were 64.12 $m^2\ g^{-1}$ and 0.064 $cm^3\ g^{-1}$, respectively. Compared to the cyanoethyl monolithic sorbent reported by Souza et al. [32], the monolithic sorbent synthesized here had smaller pore volume and larger surface area.

Figure 3a illustrates the XRD pattern of the chemically modified silica. There was a broad band in the 2θ region between 15° and 40°, with maximum at 22°, which corresponded to the amorphous silica-based host. The absence of a peak at higher angles confirmed that the silica was amorphous. However, the diffraction peak at 8.2° suggested that the chemically modified silica had mesoporous structure, which resulted from the use of the CTAB surfactant, as pore template, and functionalized monomers (CN-TEOS and APTES) [44].

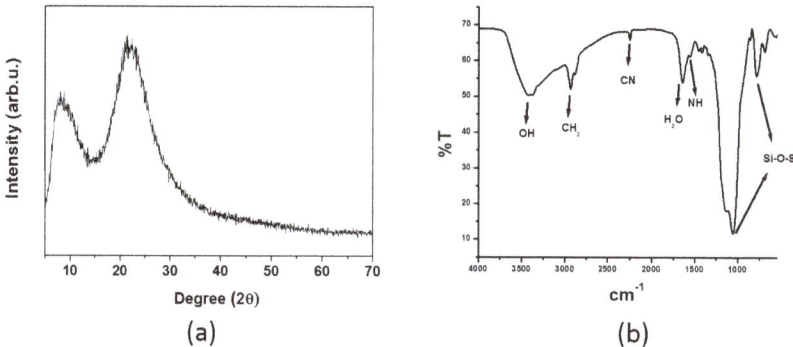

Figure 3. (a) XRD of hybrid silica monolith containing cyanopropyl and aminopropyl groups. (b) FTIR spectrum of hybrid silica monolith containing cyanopropyl and aminopropyl groups.

Figure 3b contains the FTIR absorption spectrum of the hybrid silica monolith functionalized with cyano and amino groups. The peaks at 2950 and 2254 cm^{-1} referred to C–H stretching and –CN stretching vibrational modes, respectively, and corroborated the presence of cyanopropyl groups in the silica network. The bands at 1435 and 1655 cm^{-1} were attributed to C–H bending vibrational modes and molecular water scissor bending vibration, respectively. A broad band at about 3500 cm^{-1} also evidenced the presence of water molecule and silanol groups and was assigned to O–H stretching vibrational modes and significant hydrogen bonding. The shoulder at 1558 cm^{-1} and the overlapped bands at about 3300–3400 cm^{-1} corresponded to NH$_2$ vibrational modes and attested that amino groups were incorporated into the hybrid silica monolith. The stretching band at 795 cm^{-1} indicated that Si–C bonds existed in the prepared hybrid silica monolith. The bands located at 800 and 1100 cm^{-1} were ascribed to Si–O symmetric stretching and to Si–O–Si anti-symmetric stretching, respectively [45–50], which are typical of a silica network.

2.3. In-Tube SPME-MS/MS Optimization

The use of different ion transitions for each analyte favored detection without chromatographic separation (Table 2). Endogenous compounds interferers from plasma samples can suppress ionization of analytes (ESI), decreasing analytical sensitivity. Therefore, we directly coupled the monolithic capillary to the UV detector to optimize the time for analyte sorption and interferers exclusion. On the basis of Figure S1 (Supplementary Materials) and using acetonitrile as mobile phase, we found that approximately two minutes was sufficient to eliminate most plasma macromolecules. No analyte eluted between 2 and 10 min when we used acetonitrile as mobile phase. After 10 min, we changed the mobile phase from acetonitrile to water to elute the analytes from the monolithic capillary to the mass spectrometer (peak at approximately 13 min in Figure S1).

In this work, sample solvent is defined as the solvent that was used to reconstitute the dried extract after protein precipitation. Figure 4 depicts the in-tube SPME-MS/MS optimization. Among the sample solutions evaluated during the pre-concentration step, 50 µL of acetonitrile with 0.1% (v/v) formic acid provided the highest sorption capacity (Figure 4a,b). The aqueous solutions evaluated did not presented adequate sorption due to their hydrophilic nature.

The nature of the mobile phase used to elute the analytes affects the sensitivity of the method (Figure 4c). We selected the mobile phase on the basis of desorption of the analytes from the monolithic capillary and of ESI ionization. Formic acid addition to water improved ionization of the analytes ionization, whereas acetonitrile addition improved the analytical signal due to an increase in desolvation capacity. On the basis of Figure 4c, we selected water as mobile phase. Although formic acid and acetonitrile improve the ESI ionization, the presence of these additives in mobile phase decrease the desorption capacity of the analytes.

Table 2. MS/MS (SRM) ion transitions, cone energy (DP), and collision energy (CE) for each analyte.

Analyte	Precursor Ion	Product Ion (Quantification)	DP (V)	CE (V)	Product Ion (Identification)
Tryptophan	205.2	146.0	20	12	188.1
Methionine	150.0	56.0	20	15	104.0
Methionine d3	153.1	56.0	20	15	107.1
Tyrosine	182.1	136.1	25	15	90.8
Leucine/Isoleucine	132.1	86.0	20	10	44.0
GABA	104.1	87.0	30	15	45.0
Serotonin	177.1	115.1	20	36	104.9
Glutamic acid	148.1	84.0	25	15	102.1
Lysine	147.2	88.0	25	15	107.0
Aspartic acid	134.1	74.0	20	12	88.0
Serine	106.0	60.0	20	10	88.0
Alanine	90.0	44.0	20	10	62.0
Alanine $^{13}C_3^{15}N$	94.2	47.1	20	10	64.8

DP = declustering potential; CE = collision energy.

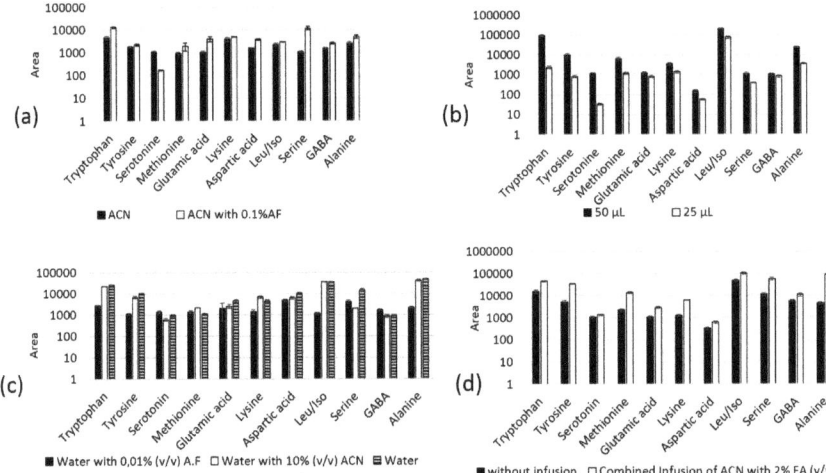

Figure 4. Effect of (**a**) sample solvent (acetonitrile and acetonitrile with 0.1% (*v/v*) formic acid on the pre-concentration step; (**b**) sample solvent volume (acetonitrile with 0.1% formic acid); (**c**) mobile phase for the elution step; and (**d**) post capillary infusion of acetonitrile with 2% formic acid (FA) on the performance of the in-tube SPME-MS/MS procedure.

Post capillary infusion of acetonitrile with 2% formic acid boosted the desolvation capacity and ionization of the analytes, thereby increasing the response (Figure 4d). Acetonitrile infusion reduced the water (mobile phase) dielectric constant and weakened electrostatic interactions between the analytes. The monolithic capillary was reused over 40 times without significant extraction efficiency loss (CV lower than 15%). Table 3 illustrates the optimized in-tube SPME-MS/MS procedure.

Table 3. In-tube SPME-MS/MS procedure.

	MOBILE PHASE A: Water B: Acetonitrile			
Time (min)	% A	% B	Valve Position	Comments
0.0	0	100	1	pre-concentration of analytes and exclusion of plasma macromolecules
2.0	100	0	2	Elution of analytes from monolithic capillary to mass spectrometer
4.0	100	0	2	Post capillary infusion of acetonitrile with 0.1% formic acid
7.0	100	0	1	Final elution step and start of gradient elution to clean up the capillary column

2.4. Adsorption Experiments

Figure 5 shows the sorption isotherms of tryptophan and leucine (representative analytes) and their respective structures. The monolithic capillary presented sorption capacity (binding affinity) of 6.53 µg cm^{-3} and 7.52 µg cm^{-3} for tryptophan and leucine, respectively. The sorption capacities determined for alanine, serine, glutamic acid, isoleucine, methionine, lysine, and aspartic acid were 5.73 µg cm^{-3}, 7.44 µg cm^{-3}, 2.86 µg cm^{-3}, 5.35 µg cm^{-3}, 2.56 µg cm^{-3}, 5.54 µg cm^{-3}, and 3.13 µg cm^{-3}, respectively. Despite structural differences, these compounds have the same functional groups (hydroxyl, carboxyl, and amino), which are responsible for their sorption onto the monolithic capillary. Sorption isotherms for these AA are illustrated in the Supplementary Materials (Figure S2).

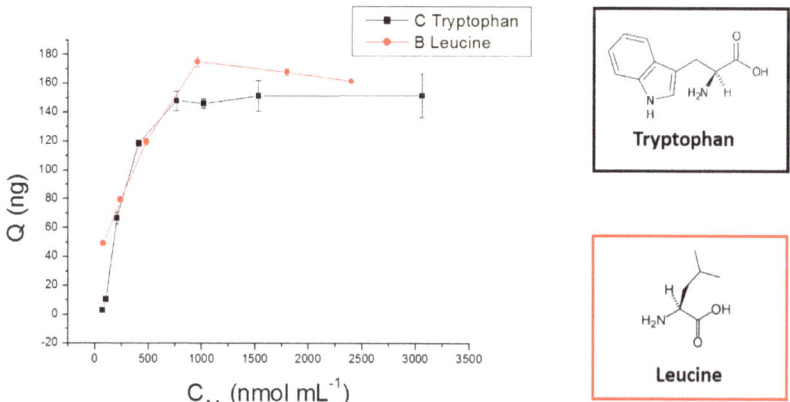

Figure 5. Sorption isotherm of the organic-inorganic hybrid silica-based monolithic capillary for tryptophan and leucine.

The structure of monolithic capillary is amorphous and homogeneous; the skeleton is porous and consists of interconnected particles (Section 2.2). Thus, the adsorption model closest to this application is the "external mass transfer model", which describes mass transfer from de liquid phase (mobile phase) to the solid surface (internal surface of the capillary monolithic phase) into the monolithic capillary. Another model that could describe the sorption of amino acids onto the monolithic phase is the "pseudo first-order model". However, in this work, the sorption of the target analytes onto monolithic capillary is reversible [51–53].

2.5. In-Tube SPME-MS/MS Analytical Validation

Analytical validation of the in-tube SPME-MS/MS method was based on the current FDA (Food and Drug Administration), and EMA (European Medicines Agency) international guidelines for the validation of a bioanalytical method [54–56].

According to the literature, the amino acids in plasma samples are stable for at least 24 h when stored at ambient temperature, over three freeze–thaw cycles, or when stored at −20 °C for six months [57,58]. After pre-treatment (Section 3.3), amino acids were stable for at least 24 h without a decrease in the area obtained in the in-tube SPME-MS/MS method for the sample stored at 10 °C.

Table 4 lists the linear range of the alanine, serine, isoleucine, leucine, aspartic acid, glutamic acid, lysine, methionine, tyrosine, and tryptophan evaluated in plasma samples spiked with standard solutions at different concentrations. Based on the analytical validation, the linear ranges of GABA and serotonin were not adequate. However, the p values of the lack of fit statistical test were higher than 0.05, which confirmed a good fit for all the other analytes. Therefore, the equation obtained during analytical validation can be used to quantify the target analytes [59,60].

Table 4. Analytical curves data, Student's t-test, and lack-of-fit statistical test of the in-tube SPME-MS/MS method to determine AA and NT in plasma samples.

Validation Parameters	Ala	Leu/Iso	Met	Ty	Try	Ser	Glu	Lys	Asp
Linearity (R^2)	0.995	0.993	0.998	0.995	0.997	0.990	0.996	0.991	0.993
Slope	0.0013	0.0110	0.0704	0.0057	0.0251	0.0024	0.0022	0.0046	0.0057
Intercept	0.4277	2.4736	0.6808	0.1844	0.6792	0.3963	0.0746	0.0859	0.0092
LOF (p-value)	0.997	0.803	0.489	0.251	0.876	0.892	0.972	0.251	0.875
Linear range (nmol mL^{-1})	45–360	15–300	12–102	10–102	9–96	12–210	12–90	12–102	6–36
Student's t-test (p-value)	0.520	0.087	0.125	0.907	0.079	0.077	0.219	0.280	0.244

Ala = Alanine; Leu = Leucine; Iso = Isoleucine; Met = Methionine; Ty = Tyrosine; Try = Tryptophan; Ser = Serine; Glu = Glutamic acid; Lys = Lysine; Asp = Aspartic acid; LOF = lack of fit statistical test, (p-value at a significance level of 0.05).

The intra- and inter-assay precision presented CV values ranging from 1.6% to 14.0%. The intra- and inter-assay accuracy presented RSE values spanning from −11.0% to 13.8%, except for the values of lower limit of quantification (LLOQ) concentration, which ranged from −19.2 to 18.0.

The matrix effect was evaluated by comparing the slopes of the calibration curves constructed for the analytes in plasma and aqueous solutions to evaluate parallelism between these analytical curves [55]. The Student's t-test did not reveal any significant difference ($p > 0.05$) between these slopes, confirming parallelism and demonstrating that the matrix effect was not significant, Table 4.

The applicability of the proposed method was evaluated by analyzing plasma samples from six schizophrenic patients undergoing treatment with antipsychotics. Table 5 illustrates the average of these plasma concentrations and the standard deviations. Figure 6 shows a representative in-tube SPME-MS/MS (SRM mode) chromatogram of plasma sample from schizophrenic patient. The AA and NT concentrations determined in plasma from schizophrenic patients agreed with previously published data [9,61].

Table 5. Average values of the AA and NT plasma concentrations (with standard deviation) determined in plasma samples from six schizophrenic patients (n = 6).

Plasma Concentration (nmol mL^{-1})	Ala	Leu/Iso	Met	Ty	Try	Ser	Glu	Lys	Asp
Average values	270.8 ± 60.1	246.1 ± 28.0	18.3 ± 5.1	40.5 ± 10.4	37.1 ± 9.7	143.0 ± 48.6	31.8 ± 11.2	20.6 ± 4.0	11.2 ± 7.5

Ala = Alanine; Leu = Leucine; Iso = Isoleucine; Met = Methionine; Ty = Tyrosine; Try = Tryptophan; Ser = Serine; Glu = Glutamic acid; Lys = Lysine; Asp = Aspartic acid.

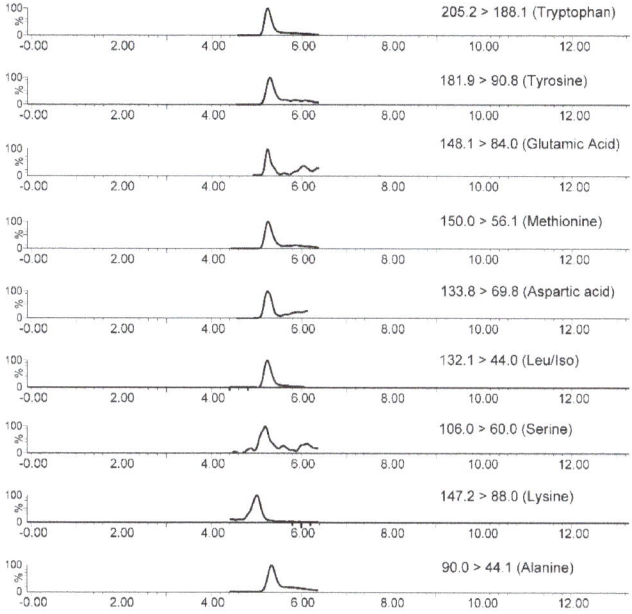

Figure 6. Representative in-tube SPME-MS/MS (SRM mode) chromatogram of plasma sample from a schizophrenic patient. Plasma concentrations: alanine = 243.7 nmol mL^{-1}; serine = 155.1 nmol mL^{-1}; leu/Iso = 199.9 nmol mL^{-1}; aspartic acid = 13.5 nmol mL^{-1}; methionine = 9.6 nmol mL^{-1}; tyrosine = 35.9 nmol mL^{-1}; tryptophan = 32.1 nmol mL^{-1}; glutamic acid = 30.9 nmol mL^{-1}; and lysine = 21.7 nmol mL^{-1}.

2.6. Comparison of the Proposed Method with Literature Methods

We compared the in-tube SPME-MS/MS method with other literature methods for AA and NT determination in plasma samples (see Table 6) [9,62–64].

Table 6. Comparison of the developed method with other methods described in the literature.

Analytes	Sample Preparation	Sample Volume (μL)	Analytical Method	Elution of Analytes (min)	validation Parameters (Intra and Inter Assays)	Ref.
10 amino acids	Protein precipitation	Plasma 50	UHPLC-MS/MS-Ascentis® Express HILIC column (4.6 × 100 mm, 2.7 μm). MP: A = Ammonium acetate solution 10 mM; B = acetonitrile with 0.1% FA	3.2	LLOQ: 9.7–13.3 nmol mL^{-1} Precision: 2–10% (CV) Accuracy: −2.1–9.9% (RSE)	[9]
33 Amino acids	Protein precipitation	Plasma 100	Two columns: 1 - PGC column (Thermo Fisher Scientific, 3 μm Hypercarb, 4.6 mm i.d. × 50 mm), and 2 - fused-core column (Advanced Materials Technology, 2.7 μm Halo C18, 2.1 mm i.d. × 100 mm)	9.4	LLOQ: 0.1–10.0 nmol mL^{-1} Precision: 1.2–9.2% (CV) Accuracy: N.A	[63]
20 amino acids	Protein precipitation	Serum 100	UHPLC-MS/MS CROWNPAK CR-I(+) column (3.0-mm i.d. × 150 mm, 5 μm)	10.1	LLOQ: 0.1–10.0 nmol mL^{-1} Precision: 2.6–10.1% (CV) Accuracy: −12.8–12.4% (RSE)	[62]

Table 6. Cont.

Analytes	Sample Preparation	Sample Volume (µL)	Analytical Method	Elution of Analytes (min)	validation Parameters (Intra and Inter Assays)	Ref.
22 amino acids	Protein precipitation	Plasma 10	HPLC-MS/MS Two Agilent Zorbax SB-C18 columns, (2.1 mm × 50 mm, 1.8 µm)	35.0	LLOQ: 0.01–0.07 nmol mL^{-1} Precision: 1.0–15.0% (CV) Accuracy: −12.8–12.4%	[64]
10 amino acids	In-tube SPME	Plasma 200	In-tube SPME-MS/MS with post capillary infusion	5.2	LLOQ: 6–45 nmol mL^{-1} Precision: 1.1–19.0% (CV) Accuracy: −14.4–19.6% (RSE)	This work

LLOQ = lower limit of quantification; CV = coefficient of variation RSE = relative standard error. N.A = not avaliable.

Compared to recent protocols described in the literature, the in-tube SPME-MS/MS method offered the following advantages: online sample processing, high throughput analysis, and minimal organic solvent consumption (flow at 100 µL min^{-1}) without addition of buffer solution to the mobile phase. The selectivity of both the hybrid silica monolithic capillary and the MS/MS system allowed direct coupling of the in-tube SPME technique to MS/MS. The proposed method did not present the lowest LLOQ values, but the obtained LLOQ values were adequate for the determination of the target analytes in plasma samples. In addition, the other analytical parameters (accuracy and precision) of the in-tube SPME-MS/MS method agree with the values established by the FDA and EMA guidelines.

3. Materials and Methods

3.1. Standards and Reagents

Alanine, serine, isoleucine, leucine, aspartic acid, glutamic acid, lysine, methionine, tyrosine, γ-aminobutiric acid (GABA) and tryptophan standards were purchased from SIGMA Sigma–Aldrich (St. Louis, MO, USA). Acetonitrile (UHPLC grade) was obtained from Sigma–Aldrich (St. Louis, MO, USA). The water used to prepare the solutions had been purified in a Milli-Q system (Millipore, Brazil). Tetraethylorthosilicate (TEOS, 98%), 3 cyanopropyltriethoxysilane (CN-TEOS, 98%), (3-aminopropyl) triethoxysilane (APTES, 98%), and cetyltrimethylammonium bromide (CTAB, 95%) were acquired from Aldrich (São Paulo, SP, Brazil).

3.2. Synthesis of Hybrid Silica-Based Monolithic Capillaries Bearing Amino and Cyano Groups

Organic-silica hybrid monolithic capillaries were synthesized by the sol–gel procedure in one step [39] with some modifications. Initially, the capillary was rinsed with 0.2 mol L^{-1} HCl solution for 30 min and then with water until the pH value of the outlet solution was 7.0. Subsequently, the capillary was flushed with 1 mol L^{-1} NaOH for 2 h, and then with water and methanol for 30 min. Finally, the capillary was purged with nitrogen at 160 °C for 3 h prior to use. After the capillary was pretreated, 5 mg of CTAB was mixed with water/ethanol solution (20 µL/100 µL) in a 1.5 mL Eppendorf vial. Next, 56 µL of TEOS, 28 µL of APTES, 28 µL of CN-TEOS, and 10 µL of ammonium fluoride aqueous solution were added to the initial solution, which was thoroughly vortexed at room temperature for 30 s. The pre-condensation mixture was quickly introduced with a syringe into the pretreated capillary of appropriate length. The capillary ends were sealed with two pieces of rubber and reacted at 40 °C for 15 h. The hybrid gel that emerged within the capillary was rinsed with ethanol to remove CTAB and synthesis residues, washed with water, and dried at 60 °C for 48 h. Different molar ratios of the precursors (TEOS, CN-TEOS, and APTES), aging temperatures (22 °C and 60 °C), CTAB amounts (5 and 7 mg), and ethanol/water ratios (100:20 and 50:50 v/v) were evaluated to optimize the synthesis procedure. Table 1 illustrates the experimental parameters that were assessed in triplicate assays.

3.3. Hybrid Silica Monolithic Capillary Characterization

The monolithic phase had its morphological and structural aspects evaluated by scanning electron microscopy (SEM). To this end, samples were coated with carbon in a Bal-Tec SCD050 Sputter coater instrument (FürstentumLiechtenstein, Cambridge, UK) for 120 s and analyzed under a Zeiss EVO50 scanning electron microscope (Cambridge, UK). Fourier transform infrared spectroscopy (FTIR) was conducted on a Shimadzu-IR Prestige-21 spectrometer (Barueri, Brazil), in KBr pellets, to identify chemical groups. Nitrogen sorption experiments were carried out at 77 K in a Micrometrics ASAP 2020 plus nitrogen sorption porosimeter (São Paulo, Brazil). Specific surface areas were determined by the Brunauer–Emmett–Teller (BET) method. X-ray diffraction (XRD) analyses were accomplished on a Siemens-Bruker D5005-AXS diffractometer (São Paulo, Brazil), with CuKa radiation, graphite monochromator, at $\lambda = 1.5418$ Å and $0.02°$ s^{-1}, in the 5–70° (2θ) range.

3.4. Plasma Samples

Plasma samples were supplied by the Psychiatric Nursing staff of Hospital das Clínicas de Ribeirão Preto, University of São Paulo, Brazil. The plasma samples were collected in agreement with the criteria established by the Ethics Committee of Faculdade de Medicina de Ribeirão Preto, University of São Paulo, Brazil. Blood was collected by venipuncture and placed in tubes containing anticoagulants, EDTA. It was centrifuged immediately after collection, and plasma was stored at −80 °C.

Plasma proteins (200 µL) were precipitated with acetonitrile at a 1:2 (v/v, respectively) ratio. After vortex mixing for 1 min, the mixture was centrifuged at 9000 g (rpm) for 30 min. The supernatant (700 µL) was dried in a vacuum concentrator (Eppendorf, Brazil), and the dried extract was reconstituted with 50 µL of acetonitrile containing 0.1% formic acid (v/v) for the in-tube SPME-MS/MS procedure.

3.5. MS/MS Conditions

In-tube SPME assays were performed in a Waters ACQUITYUPLC H-Class system coupled to the Xevo®TQ-D tandem quadrupole (Waters Corporation, Milford, MA, USA) mass spectrometer equipped with a Z-spray source (electro spray ionization, ESI) operating in the positive mode. Selected reaction monitoring (SRM) transitions and optimal collision energies were optimized for each analyte (Table 2). MS parameters and source were optimized as capillary voltage of 0.50 kV, source temperature of 150 °C, desolvation temperature of 350 °C, desolvation gas flow of 600 L h^{-1} (N_2, 99.9% purity), and cone gas flow of 20 L h^{-1} (N_2, 99.9% purity). Argon (99.9999% purity) was used as collision gas. The dwell time was established for each transition separately, and the inter-scan delay was set at the automatic mode. Data were acquired by using the MassLynx V4.1 software (Waters Corporation, Milford, MA, USA).

3.6. In-Tube SPME Procedure

To optimize the in-tube SPME-MS/MS procedure, 300 µL of plasma sample was used. Parameters were optimized not only to enrich the monolithic capillary with the analytes, but also to improve MS/MS sensitivity.

Different sample solutions (aqueous solution at different pH values (pH 4, 7, and 10), acetonitrile, and acetonitrile with 0.1% (v/v) formic acid), sample solution volumes (25 and 50 µL), and mobile phases to elute the analytes [water, water with 0.01% (v/v) formic acid, and water with 10% (v/v) acetonitrile] were evaluated.

The in-tube SPME-MS/MS system configuration was based on monolithic capillary (10 cm × 530 µm) and not analytical liquid column coupling to the MS/MS valve.

The in-tube SPME procedure comprised three steps (Table 3). In the first step (MS/MS valve in position 1), diluted sample (10 µL) was percolated through the monolithic capillary to pre-concentrate the analytes and to exclude endogenous interferers; acetonitrile was used as mobile phase at a flow rate of 100 µL min^{-1} for 2 min. From 2 to 4 min, the valve was switched to position 2, and the analytes were eluted from the monolithic capillary to the mass spectrometer in tandem by using water as mobile phase at a flow rate of 100 µL min^{-1}. From 4 to 7 min, acetonitrile with 2% (v/v) of formic acid was post-capillary infused to increase MS/MS sensitivity. From 7 to 12 min (third step), the valve was switched back to position 1, and the monolithic capillary was cleaned up with water and acetonitrile with gradient elution from 100% water to 100% acetonitrile at a flow rate of 100 µL min^{-1}. Figure 7 illustrates a schematic diagram of in-tube SPME-MS/MS system configuration.

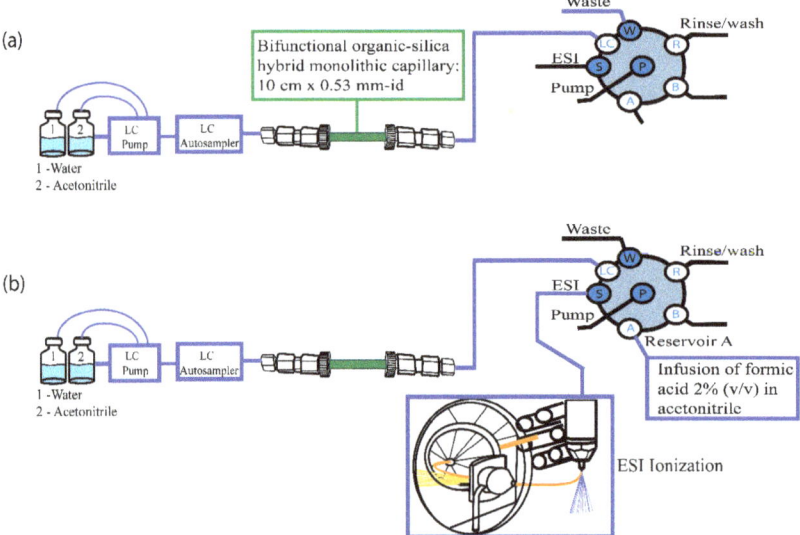

Figure 7. Scheme of in-tube SPME-MS/MS procedure). In the first step (**a**) (MS/MS valve in position 1), diluted sample (10 µL) was percolated through the monolithic capillary to pre-concentrate the analytes. After, (**b**) the valve was switched to position 2 for elution of the analytes From 4 to 7 min, acetonitrile with 2% (v/v) of formic acid was post-capillary infused.

3.7. Adsorption Capacity

Adsorption capacity was evaluated according to reported procedures [55,65]. Considering the target AA and NT chemical structures (cyclic and acyclic), tryptophan and leucine were selected as representative analytes. Standard tryptophan and leucine solutions were prepared in water at different concentrations (CAA, from 0.150 to 3.6 µmol mL^{-1}) to evaluate the monolithic capillary maximum sorption capacity (Q_{max} ng cm^{-3}). These solutions were injected separately into the in-tube SPME-MS/MS system; conditions described in Section 2.6 were employed. Q_{max} was estimated on the basis of the following equation $Q_{max} = Q/V_m$, where Q (ng) is the amount of analyte adsorbed onto the monolithic capillary as determined by calibration curves, and V_m = 22.6 cm^{-3} is the estimated monolithic phase volume immobilized into the capillary. V_m was calculated by using the capillary length (L) and internal radius (r) according to the following equation $V_m = \pi r^2 L$, where π = 3.14, r = 265 µm, and L = 10 cm. Q_{max} was based on the saturation point of the plot of Q (ng) versus CAA.

3.8. Analytical Validation

The linear ranges of the calibration curves were established in agreement with the AA and NT concentrations in plasma samples from schizophrenic patients [9,66]. Calibration curves were constructed by the standard addition method (standard solutions were added directly to the plasma samples): relative peak areas (analyte-to-IS) were plotted as a function of analytes concentration spiked in matrix samples in different ranges.

Accuracy and precision assays were carried out with plasma samples (representative amount of the plasma pool from different voluntaries) spiked with the analytes at five concentrations, namely calibration controls (QC), LLOQ, low (QC), medium QC, high QC, and upper limit of quantitation (ULOQ). Precision assays were evaluated on the same day (intra-assays precision) and on three consecutive days (inter-assays precision) based on the coefficient of variation (CV) values; i.e., within 20%. Intra- and inter-assays accuracy were based on relative standard error (RSE%). RSE values should be within 15% of nominal values for QC samples.

Methionine-d3 and alanine $^{13}C_3^{15}N$ were used as internal standards at concentrations of 30 and 100.0 nmol mL^{-1}, respectively. Leucine and isoleucine present the same transitions because they have the same molecular weights and similar MS/MS characteristics [67]. Thus, these AA could not be separately determined in the triple quadrupole analyzer. These analytes were quantified on the basis of total concentrations; i.e., as a sum of the peak areas of the individual analytes.

The lower limit of quantification (LLOQ) is the lowest analyte concentration in a sample that can be reliably quantified with acceptable accuracy and precision. LLOQ is also the lowest concentration in the calibration curve.

Carryover was assessed by injecting blank aqueous sample after a ULOQ concentration in plasma sample. Carryover in the blank aqueous sample should not be greater than 20% of the LLOQ or 5% for the internal standard.

The matrix effect was examined by comparing the average value of slopes of calibration curves obtained with four aqueous samples spiked with the target analytes at different concentrations with slopes of calibration curves obtained with four plasma samples from different patients [56]. Student's *t*-test (*p*-value at a significance level of 0.05) was applied (Table S2 Supplementary Materials).

4. Conclusions

Optimization of the synthesis procedure (molar ratios of the alkoxysilane precursors, aging temperatures, supramolecular template amounts, and ethanol/water ratios) and the use of ammonium fluoride as catalyst allowed us to develop an innovative organic-inorganic hybrid silica-based monolithic capillary with two bifunctional groups (cyano and amino). This reproducible capillary presented adequate sorption capacity, satisfactory permeability (low pressure), and high mechanical strength, which enabled it to be re-used over forty times without significant changes in extraction reproducibility or system pressure.

Concerning morphology, the hybrid monolithic capillary tightly attached to the capillary inner wall and exhibited a homogeneous, continuous, and porous skeleton. FTIR analyses prove that the cyano and amino groups were incorporated into the monolithic capillary.

The selectivity of the monolithic capillary (amino and cyano groups) and the MS/MS system allowed for direct coupling of in-tube SPME to the MS/MS system without the need for chromatographic separation. In agreement with the analytical validation assays, this automated innovative method is appropriate to determine the target AA and NT in plasma samples from schizophrenic patients for clinical studies involving short analysis time.

Supplementary Materials: The supplementary materials are available online.

Author Contributions: Conceptualization, methodology, software, validation, formal analysis, investigation, resources, data curation, visualization, writing-original draft preparation, writing-review & editing, project administration, L.F.C.M.; writing-review & editing, conceptualization, data curation, R.R.G.; conceptualization, formal analysis, investigation, resources, writing-original draft preparation, writing-review & editing, visualization, supervision, project administration, funding acquisition, M.E.C.Q.

Funding: This research was funded by FAPESP (Fundação de Amparo à Pesquisa do Estado de São Paulo, 2017/02147-0), and INCT-TM (465458/2014-9) (Instituto Nacional de Ciência e Tecnologia Translacional em Medicina) and CAPES (Coordenação de Aperfeiçoamento de Pessoal de Nível Superior).

Conflicts of Interest: The authors declare no conflict of interest.

References

1. McGrath, J.; Saha, S.; Welham, J.; El Saadi, O.; MacCauley, C.; Chant, D. A systematic review of the incidence of schizophrenia: The distribution of rates and the influence of sex, urbanicity, migrant status and methodology. *BMC Med.* **2004**, *2*, 13. [CrossRef]
2. De Luca, V.; Viggiano, E.; Messina, G.; Viggiano, A.; Borlido, C.; Viggiano, A.; Monda, M. Peripheral amino Acid levels in schizophrenia and antipsychotic treatment. *Psychiatry Investig.* **2008**, *5*, 203–208. [CrossRef]
3. Javitt, D.C. Excitatory Amino Acids in Schizophrenia: Both What You Have, and What You Do With Them. *Biol. Psychiatry.* **2018**, *83*, 470–472. [CrossRef]
4. Saleem, S.; Shaukat, F.; Gul, A.; Arooj, M.; Malik, A. Potential role of amino acids in pathogenesis of schizophrenia. *Int. J. Health Sci.* **2017**, *11*, 63.
5. Balu, D.T.; Coyle, J.T. The NMDA receptor 'glycine modulatory site' in schizophrenia: D-serine, glycine, and beyond. *Curr. Opin. Pharmacol.* **2015**, *20*, 109–115. [CrossRef]
6. Cao, B.; Wang, D.; Brietzke, E.; McIntyre, R.S.; Pan, Z.; Cha, D.; Rosenblat, J.D.; Zuckerman, H.; Liu, Y.; Xie, Q.; Wang, J. Characterizing amino-acid biosignatures amongst individuals with schizophrenia: A case–control study. *Amino Acids* **2018**, *50*, 1013–1023. [CrossRef]
7. Panizzutti, R.; Fisher, M.; Garrett, C.; Man, W.H.; Sena, W.; Madeira, C.; Vinogradov, S. Association between increased serum d-serine and cognitive gains induced by intensive cognitive training in schizophrenia. *Schizophr. Res.* **2018**, in press. [CrossRef]
8. MacKay, M.-A.B.; Kravtsenyuk, M.; Thomas, R.; Mitchell, N.D.; Dursun, S.M.; Baker, G.B. D-Serine: Potential Therapeutic Agent and/or Biomarker in Schizophrenia and Depression? *Front. Psychiatry* **2019**, *10*, 25. [CrossRef]
9. Domingues, D.S.; Crevelin, E.J.; Moraes, L.A.B.; Hallak, J.C.E.; Crippa, J.A.S.; Queiroz, M.E.C. Simultaneous determination of amino acids and neurotransmitters in plasma samples from schizophrenic patients by hydrophilic interaction liquid chromatography with tandem mass spectrometry. *J. Sep. Sci.* **2015**, *38*, 780–787. [CrossRef]
10. Altamura, C.A.; Mauri, M.C.; Ferrara, A.; Moro, A.R.; D'andrea, G.; Zamberlan, F.; Meltzer, H.Y. Plasma and platelet excitatory amino acids in psychiatric disorders. *Am. J. Psychiatry* **1993**, *150*, 1731–1733.
11. Cai, H.-L.; Zhu, R.-H. Determination of dansylated monoamine and amino acid neurotransmitters and their metabolites in human plasma by liquid chromatography–electrospray ionization tandem mass spectrometry. *Anal. Biochem.* **2010**, *396*, 103–111. [CrossRef]
12. Queiroz, M.E.C.; Melo, L.P. Selective capillary coating materials for in-tube solid-phase microextraction coupled to liquid chromatography to determine drugs and biomarkers in biological samples: A review. *Anal. Chim. Acta* **2014**, *826*, 1–11. [CrossRef]
13. Saito, A.; Hamano, M.; Kataoka, H. Simultaneous analysis of multiple urinary biomarkers for the evaluation of oxidative stress by automated online in-tube solid-phase microextraction coupled with negative/positive ion-switching mode liquid chromatography–tandem mass spectrometry. *J. Sep. Sci.* **2018**, *41*, 2743–2749. [CrossRef]
14. Lashgari, M.; Yamini, Y. Fiber-in-tube solid-phase microextraction of caffeine as a molecular tracer in wastewater by electrochemically deposited layered double hydroxide. *J. Sep. Sci.* **2018**, *41*, 2393–2400. [CrossRef]
15. Inukai, T.; Kaji, S.; Kataoka, H. Analysis of nicotine and cotinine in hair by on-line in-tube solid-phase microextraction coupled with liquid chromatography-tandem mass spectrometry as biomarkers of exposure to tobacco smoke. *J. Pharm. Biomed. Anal.* **2018**, *156*, 272–277. [CrossRef]
16. Serra-Mora, P.; Jornet-Martinez, N.; Moliner-Martinez, Y.; Campíns-Falcó, P. In tube-solid phase microextraction-nano liquid chromatography: Application to the determination of intact and degraded polar triazines in waters and recovered struvite. *J. Chromatogr. A* **2017**, *1513*, 51–58. [CrossRef]
17. Luo, X.; Li, G.; Hu, Y. In-tube solid-phase microextraction based on NH2-MIL-53 (Al)-polymer monolithic column for online coupling with high-performance liquid chromatography for directly sensitive analysis of estrogens in human urine. *Talanta* **2017**, *165*, 377–383. [CrossRef]
18. Andrade, M.A.; Lanças, F.M. Determination of Ochratoxin A in wine by packed in-tube solid phase microextraction followed by high performance liquid chromatography coupled to tandem mass spectrometry. *J. Chromatogr. A* **2017**, *1493*, 41–48. [CrossRef]

19. Bu, Y.; Feng, J.; Tian, Y.; Wang, X.; Sun, M.; Luo, C. An organically modified silica aerogel for online in-tube solid-phase microextraction. *J. Chromatogr. A* **2017**, *1517*, 203–208. [CrossRef]
20. Wu, F.; Wang, J.; Zhao, Q.; Jiang, N.; Lin, X.; Xie, Z.; Li, J.; Zhang, Q. Detection of trans-fatty acids by high performance liquid chromatography coupled with in-tube solid-phase microextraction using hydrophobic polymeric monolith. *J. Chromatogr. B* **2017**, *1040*, 214–221. [CrossRef]
21. Feng, J.; Wang, X.; Tian, Y.; Bu, Y.; Luo, C.; Sun, M. Electrophoretic deposition of graphene oxide onto carbon fibers for in-tube solid-phase microextraction. *J. Chromatogr. A* **2017**, *1517*, 209–214. [CrossRef]
22. Bu, Y.; Feng, J.; Sun, M.; Zhou, C.; Luo, C. Facile and efficient poly (ethylene terephthalate) fibers-in-tube for online solid-phase microextraction towards polycyclic aromatic hydrocarbons. *Anal. Bioanal. Chem.* **2016**, *408*, 4871–4882. [CrossRef]
23. Bu, Y.; Feng, J.; Wang, X.; Tian, Y.; Sun, M.; Luo, C. In situ hydrothermal growth of polyaniline coating for in-tube solid-phase microextraction towards ultraviolet filters in environmental water samples. *J. Chromatogr. A* **2017**, *1483*, 48–55. [CrossRef]
24. Sun, M.; Feng, J.; Bu, Y.; Luo, C. Ionic liquid coated copper wires and tubes for fiber-in-tube solid-phase microextraction. *J. Chromatogr. A* **2016**, *1458*, 1–8. [CrossRef]
25. Xiang, X.; Shang, B.; Wang, X.; Chen, Q. PEEK tube-based online solid-phase microextraction–high-performance liquid chromatography for the determination of yohimbine in rat plasma and its application in pharmacokinetics study. *Biomed. Chromatogr.* **2017**, *31*, 3866. [CrossRef]
26. Ou, J.; Liu, Z.; Wang, H.; Lin, H.; Dong, J.; Zou, H. Recent development of hybrid organic-silica monolithic columns in CEC and capillary LC. *Electrophoresis* **2015**, *36*, 62–75. [CrossRef]
27. Svec, F. Stellan Hjertén's contribution to the development of monolithic stationary phases. *Electrophoresis* **2008**, *29*, 1593–1603. [CrossRef]
28. Siouffi, A.-M. Silica gel-based monoliths prepared by the sol–gel method: Facts and figures. *J. Chromatogr. A* **2003**, *1000*, 801–818. [CrossRef]
29. Li, Z.; Rodriguez, E.; Azaria, S.; Pekarek, A.; Hage, D.S. Affinity monolith chromatography: A review of general principles and applications. *Electrophoresis* **2017**, *38*, 2837–2850. [CrossRef]
30. Zajickova, Z. Advances in the development and applications of organic–silica hybrid monoliths. *J. Sep. Sci.* **2017**, *40*, 25–48. [CrossRef]
31. Zheng, M.-M.; Wang, S.-T.; Hu, W.-K.; Feng, Y.-Q. In-tube solid-phase microextraction based on hybrid silica monolith coupled to liquid chromatography–mass spectrometry for automated analysis of ten antidepressants in human urine and plasma. *J. Chromatogr. A* **2010**, *1217*, 7493–7501. [CrossRef]
32. de Souza, I.D.; Domingues, D.S.; Queiroz, M.E. Hybrid silica monolith for microextraction by packed sorbent to determine drugs from plasma samples by liquid chromatography–tandem mass spectrometry. *Talanta* **2015**, *140*, 166–175. [CrossRef]
33. Domingues, D.S.; de Souza, I.D.; Queiroz, M.E.C. Analysis of drugs in plasma samples from schizophrenic patients by column-switching liquid chromatography-tandem mass spectrometry with organic–inorganic hybrid cyanopropyl monolithic column. *J. Chromatogr. B* **2015**, *993*, 26–35. [CrossRef]
34. Brothier, F.; Pichon, V. Immobilized antibody on a hybrid organic–inorganic monolith: Capillary immunoextraction coupled on-line to nanoLC-UV for the analysis of microcystin-LR. *Anal. Chim. Acta* **2013**, *792*, 52–58. [CrossRef]
35. Gómez-Ríos, G.A.; Reyes-Garcés, N.; Bojko, B.; Pawliszyn, J. Biocompatible solid-phase microextraction nanoelectrospray ionization: An unexploited tool in bioanalysis. *Anal. Chem.* **2015**, *88*, 1259–1265. [CrossRef]
36. Santos, M.G.; Tavares, I.M.C.; Barbosa, A.F.; Bettini, J.; Figueiredo, E.C. Analysis of tricyclic antidepressants in human plasma using online-restricted access molecularly imprinted solid phase extraction followed by direct mass spectrometry identification/quantification. *Talanta* **2017**, *163*, 8–16. [CrossRef]
37. Li, W.; Fries, D.P.; Malik, A. Sol–gel stationary phases for capillary electrochromatography. *J. Chromatogr. A* **2004**, *1044*, 23–52. [CrossRef]
38. Hench, L.L.; West, J.K. The sol-gel process. *Chem. Rev.* **1990**, *90*, 33–72. [CrossRef]
39. Yan, L.; Zhang, Q.; Zhang, J.; Zhang, L.; Li, T.; Feng, Y.; Zhang, L.; Zhang, W.; Zhang, Y. Hybrid organic–inorganic monolithic stationary phase for acidic compounds separation by capillary electrochromatography. *J. Chromatogr. A* **2004**, *1046*, 255–261. [CrossRef]
40. Rodríguez, R.; Flores, M.; Gómez, J.; Castaño, V.M. Master behaviour for gelation in fluoride-catalyzed gels. *Mater. Lett.* **1992**, *15*, 242–247. [CrossRef]

41. Moreira, J.E.; Cesar, M.L.; Aegerter, M.A. Light scattering of silica particles in solution. *J. Non-Cryst. Solids* **1990**, *121*, 394–396. [CrossRef]
42. Iler, R.K. *Chemistry of Silica: Solubility, Polymerization, Colloid and Surface Properties, and Biochemistry*; John Wiley & Sons: New York, NY, USA, 1979.
43. Kulkarni, S.; Fang, L.; Alhooshani, K.; Malik, A. Sol–gel immobilized cyano-polydimethylsiloxane coating for capillary microextraction of aqueous trace analytes ranging from polycyclic aromatic hydrocarbons to free fatty acids. *J. Chromatogr. A* **2006**, *1124*, 205–216. [CrossRef] [PubMed]
44. Xue, X.; Li, F. Removal of Cu(II) from aqueous solution by adsorption onto functionalized SBA-16 mesoporous silica. *Microporous Mesoporous Mater.* **2008**, *116*, 116–122. [CrossRef]
45. Al-Oweini, R.; El-Rassy, H. Synthesis and characterization by FTIR spectroscopy of silica aerogels prepared using several Si(OR)4 and R″Si(OR′)3 precursors. *J. Mol. Struct.* **2009**, *919*, 140–145. [CrossRef]
46. Innocenzi, P. Infrared spectroscopy of sol–gel derived silica-based films: A spectra-microstructure overview. *J. Non-Cryst. Solids* **2003**, *316*, 309–319. [CrossRef]
47. Wahab, M.A.; Kim, II; Ha, C.-S. Hybrid periodic mesoporous organosilica materials prepared from 1,2-bis(triethoxysilyl)ethane and (3-cyanopropyl)triethoxysilane. *Microporous Mesoporous Mater.* **2004**, *69*, 19–27. [CrossRef]
48. Pasternack, R.M.; Rivillon Amy, S.; Chabal, Y.J. Attachment of 3-(Aminopropyl)triethoxysilane on Silicon Oxide Surfaces: Dependence on Solution Temperature. *Langmuir* **2008**, *24*, 12963–12971. [CrossRef]
49. Peña-Alonso, R.; Rubio, F.; Rubio, J.; Oteo, J.L. Study of the hydrolysis and condensation of γ-Aminopropyltriethoxysilane by FT-IR spectroscopy. *J. Mater. Sci.* **2007**, *42*, 595–603. [CrossRef]
50. Chiang, C.-H.; Ishida, H.; Koenig, J.L. The structure of γ-aminopropyltriethoxysilane on glass surfaces. *J. Colloid Interface Sci.* **1980**, *74*, 396–404. [CrossRef]
51. Qiu, H.; Lv, L.; Pan, B.-C.; Zhang, Q.-J.; Zhang, W.-M.; Zhang, Q.-X. Critical review in adsorption kinetic models. *J. Zhejiang Univ. Sci. A* **2009**, *10*, 716–724. [CrossRef]
52. Largitte, L.; Pasquier, R. A review of the kinetics adsorption models and their application to the adsorption of lead by an activated carbon. *Chem. Eng. Res. Des.* **2016**, *109*, 495–504. [CrossRef]
53. Lagergren, S.K. About the theory of so-called adsorption of soluble substances. *Sven. Vetenskapsakad. Handingarl* **1898**, *24*, 1–39.
54. FDA. Guidance for Industry: Bioanalytical Method Validation. Available online: https://www.fda.gov/downloads/Drugs/Guidances/ucm070107.pdf (accessed on 17 April 2019).
55. EMA. Guideline on Bioanalytical Method Validation. Available online: https://www.ema.europa.eu/en/documents/scientific-guideline/guideline-bioanalytical-method-validation_en.pdf (accessed on 17 April 2019).
56. Kollipara, S.; Bende, G.; Agarwal, N.; Varshney, B.; Paliwal, J. International Guidelines for Bioanalytical Method Validation: A Comparison and Discussion on Current Scenario. *Chromatographia* **2011**, *73*, 201–217. [CrossRef]
57. Schaefer, A.; Piquard, F.; Haberey, P. Plasma amino-acids analysis: Effects of delayed samples preparation and of storage. *Clin. Chim. Acta* **1987**, *164*, 163–169. [CrossRef]
58. González, O.; Blanco, M.E.; Iriarte, G.; Bartolomé, L.; Maguregui, M.I.; Alonso, R.M. Bioanalytical chromatographic method validation according to current regulations, with a special focus on the non-well defined parameters limit of quantification, robustness and matrix effect. *J. Chromatogr. A* **2014**, *1353*, 10–27. [CrossRef]
59. Zhang, L.; Gionfriddo, E.; Acquaro, V.; Pawliszyn, J. Direct Immersion Solid-Phase Microextraction Analysis of Multi-class Contaminants in Edible Seaweeds by Gas Chromatography-Mass Spectrometry. *Anal. Chim. Acta* **2018**, *1031*, 83–97. [CrossRef]
60. Otto, M. *Chemometrics: Statistics and Computer Application in Analytical Chemistry*; John Wiley & Sons: New York, NY, USA, 2016.
61. Li, Q.Z.; Huang, Q.X.; Li, S.C.; Yang, M.Z.; Rao, B. Simultaneous Determination of Glutamate, Glycine, and Alanine in Human Plasma Using Precolumn Derivatization with 6-Aminoquinolyl-*N*-hydroxysuccinimidyl Carbamate and High-Performance Liquid Chromatography. *Korean J Physiol. Pharmacol.* **2012**, *16*, 355–360. [CrossRef]

62. Han, M.; Xie, M.; Han, J.; Yuan, D.; Yang, T.; Xie, Y. Development and validation of a rapid, selective, and sensitive LC-MS/MS method for simultaneous determination of D-and L-amino acids in human serum: Application to the study of hepatocellular carcinoma. *Anal. Bioanal. Chem.* **2018**, *410*, 2517–2531. [CrossRef]
63. Le, A.; Ng, A.; Kwan, T.; Cusmano-Ozog, K.; Cowan, T.M. (LC-MS/MS). *J. Chromatogr. B* **2014**, *944*, 166–174. [CrossRef]
64. Harder, U.; Koletzko, B.; Peissner, W. Quantification of 22 plasma amino acids combining derivatization and ion-pair LC-MS/MS. *J. Chromatogr. B* **2011**, *879*, 495–504. [CrossRef]
65. De Gisi, S.; Lofrano, G.; Grassi, M.; Notarnicola, M. Characteristics and adsorption capacities of low-cost sorbents for wastewater treatment: A review. *Sustain. Mater. Technol.* **2016**, *9*, 10–40. [CrossRef]
66. Goldstein, D.S.; Eisenhofer, G.; Kopin, I.J. Sources and Significance of Plasma Levels of Catechols and Their Metabolites in Humans. *J. Pharmacol. Exp. Ther.* **2003**, *305*, 800–811. [CrossRef] [PubMed]
67. Guo, S.; Duan, J.-a.; Qian, D.; Tang, Y.; Qian, Y.; Wu, D.; Su, S.; Shang, E. Rapid determination of amino acids in fruits of Ziziphus jujuba by hydrophilic interaction ultra-high-performance liquid chromatography coupled with triple-quadrupole mass spectrometry. *J. Agric. Food Chem.* **2013**, *61*, 2709–2719. [CrossRef] [PubMed]

Sample Availability: Samples of the compounds are not available from the authors.

© 2019 by the authors. Licensee MDPI, Basel, Switzerland. This article is an open access article distributed under the terms and conditions of the Creative Commons Attribution (CC BY) license (http://creativecommons.org/licenses/by/4.0/).

Article

High-Throughput Analysis of Selected Urinary Hydroxy Polycyclic Aromatic Hydrocarbons by an Innovative Automated Solid-Phase Microextraction

Stefano Dugheri [1,*], Alessandro Bonari [2], Matteo Gentili [3], Giovanni Cappelli [2], Ilenia Pompilio [2], Costanza Bossi [2], Giulio Arcangeli [2], Marcello Campagna [4] and Nicola Mucci [2]

1. Laboratorio di Igiene e Tossicologia Industriale, Azienda Ospedaliero-Universitaria Careggi, Largo P. Palagi 1, 50139 Firenze, Italy
2. Dipartimento di Medicina Sperimentale e Clinica, Università degli Studi di Firenze, Largo G.A. Brambilla 3, 50139 Firenze, Italy; alessandro.bonari@unifi.it (A.B.); giovanni.cappelli@unifi.it (G.C.); ilenia.pompilio@unifi.it (I.P.); costanza.bossi@unifi.it (C.B.); giulio.arcangeli@unifi.it (G.A.); nicola.mucci@unifi.it (N.M.)
3. Giotto Biotech Srl, Via Madonna del Piano 6, 50019 Sesto Fiorentino (Firenze), Italy; gentili@giottobiotech.com
4. Dipartimento di Scienze Mediche e Sanità Pubblica, Università di Cagliari, Cittadella Universitaria di Monserrato, SS 554 bivio Sestu, 09042 Monserrato (Cagliari), Italy; mam.campagna@gmail.com
* Correspondence: stefano.dugheri@unifi.it; Tel.: +39-055-794-8296

Academic Editors: Constantinos K. Zacharis and Paraskevas D. Tzanavaras
Received: 22 June 2018; Accepted: 26 July 2018; Published: 26 July 2018

Abstract: High-throughput screening of samples is the strategy of choice to detect occupational exposure biomarkers, yet it requires a user-friendly apparatus that gives relatively prompt results while ensuring high degrees of selectivity, precision, accuracy and automation, particularly in the preparation process. Miniaturization has attracted much attention in analytical chemistry and has driven solvent and sample savings as easier automation, the latter thanks to the introduction on the market of the three axis autosampler. In light of the above, this contribution describes a novel user-friendly solid-phase microextraction (SPME) off- and on-line platform coupled with gas chromatography and triple quadrupole-mass spectrometry to determine urinary metabolites of polycyclic aromatic hydrocarbons 1- and 2-hydroxy-naphthalene, 9-hydroxy-phenanthrene, 1-hydroxy-pyrene, 3- and 9-hydroxy-benzoantracene, and 3-hydroxy-benzo[a]pyrene. In this new procedure, chromatography's sensitivity is combined with the user-friendliness of *N-tert*-butyldimethylsilyl-*N*-methyltrifluoroacetamide on-fiber SPME derivatization using direct immersion sampling; moreover, specific isotope-labelled internal standards provide quantitative accuracy. The detection limits for the seven OH-PAHs ranged from 0.25 to 4.52 ng/L. Intra-(from 2.5 to 3.0%) and inter-session (from 2.4 to 3.9%) repeatability was also evaluated. This method serves to identify suitable risk-control strategies for occupational hygiene conservation programs.

Keywords: SPME; OH-PAHs; gas-chromatography; MTBSTFA

1. Introduction

Polycyclic Aromatic Hydrocarbons (PAHs) are ubiquitous, despite their danger to humans. PAHs can be found in both gaseous and particulate forms. The latter are considered very hazardous to human health; many of the studies on the effects of air pollution have correlated solid aerosols with PAHs with cancer [1]. Outdoor air pollution in both cities and rural areas was estimated to cause 4.2 million premature deaths worldwide in 2016 [2], mainly attributable to airborne particulate matter (PM) [3,4].

Gherardi et al. indicates that 80% of the suspended PM is represented by PAHs [5]. Benzo[a]pyrene (B[a]P) is often used as a marker for total exposure to carcinogenic PAHs, and Ohura et al. [6] reported that B[a]P contribution to the total carcinogenic potential was in the 51–64% range. The Institute of Occupational Medicine [7] estimated that in 2006 in the EU there were 234,000 workers who were potentially exposed to high levels of B[a]P and about seven million to low levels. Recently, Stec et al. revealed that cancer incidence appears to be higher amongst firefighters compared to the general population [8].

Urinary hydroxylated-PAHs (OH-PAHs) have been used as biomarkers to assess total human exposure to PAHs, with 1-hydroxy-pyrene (1-OH-P) as the most commonly used indicator in biomonitoring studies [9]. The Center for Disease Control and Prevention (CDC) developed a OH-PAHs method used for analyzing urinary samples from the National Health and Nutrition Examination Survey, a comprehensive survey that CDC performs annually to assess exposure of the U.S. general population to PAHs [10]. For many years the American Conference of Governmental Industrial Hygienist (ACGIH) had recommended the determination of urinary 1-OH-P as a biomarker to occupational exposure to PAH mixtures, but without any indication of a limit value. Then in 2017, the ACGIH introduced a Biological Exposure Indices (BEI) value of 2.5 µg/L for 1-OH-P, while proposing urinary 3-OH-B[a]P and the sum of 1- and 2-naphthols as non-quantitative markers [11].

The development of analytical methods to identify suitable risk-control strategies for occupational hygiene conservation programs have aroused interest in the scientific community. Most analytical methods have been published to measure urinary OH-PAHs [12–25], which have two to three benzene rings, and only eleven studies [26–36] have considered the determination of OH-PAHs with more than three benzene rings, particularly 3-OH-B[a]P.

The use of MS techniques, particularly gas chromatography (GC) and liquid chromatography (LC), are indispensable tools in metabolomic science owing to their high sensitivity and specificity. In assessing B[a]P exposure—the only PAH classified as category 1 by the International Agency for Research on Cancer—urinary 3-OH-B[a]P determination plays a fundamental role; however the hyphenated chromatographic MS procedures proposed for its analysis are based on age-old methodologies resulting in many manual operations, these have several drawbacks such as uncertainty of the determination and higher overall costs [26,27,29,33–36]. The use of liquid/liquid extraction (LLE) or SPE with evaporation to dryness of the analyte solution followed by reconstitution in a suitable solvent for injection into the chromatographic system—with or without derivatization—are the typical sequences for monohydroxy PAHs in urine. Currently, four GC methods using N-methyl-N-(trimethylsilyl) trifluoroacetamide (BSTFA) as a derivatizing agent via extraction with hexane or pentane and related analysis by single [27], QpQ [34], or high-resolution [26,35] MS were proposed. Regarding the LC-triple quadrupole analyses, Raponi et al. [29] and Zangh et al. [36] reported using SPE, while Luo et al. [33] included also reaction with dansyl chloride (DNS). These existing assays have limitations, namely, their complexity, their use of solvents, and the need for clean-up steps to extract and eliminate interfering compounds from the urine; all of those involved lengthy manual operations, bigger costs, uncertainty in the determination analysis, and the possible loss of analytes. For these reasons, simultaneous and more sensitive assay methods were clearly needed.

Miniaturization is increasingly applied in analytical chemistry, resulting in savings in both time and costs throughout the sampling process, typically the most time-consuming and error-prone stage. Specifically, the microextraction via solid phase technique was initially distributed by Supelco (Bellefonte, PA, USA) under the name of Solid Phase MicroExtraction (SPME), [37–41]. At the same time, other companies were also working on devices using the same concept: SPME Arrows [42], MicroExctraction by Packed Sorbent (MEPS) [43], Stir Bar Sorptive Extraction (Twister, SBSE) [44], Solid Phase Dynamic Extraction (Magic Needle, SPDE) [45], In-Tube Extraction (ITEX) [46], HiSorb Sorptive Extraction [47], and Monolithic Material Sorptive Extraction (MonoTrap) [48].

SPME analysis is considered one of the major breakthroughs in 20th-century analytical chemistry; it was the first powerful miniaturized sampling technique developed for GC. SPME, which was invented by Pawliszyn in 1989 [49], integrates sampling, extraction, concentration, and sample introduction into a single step and the extraction requires no polluting organic solvent. Through this, the principles of green chemistry are applied to not only chemical engineering and synthesis, but also increasingly analytical chemistry [10,21,50,51]. Since 2009, significant progress has been achieved by market introduction of the Fast Fit Fiber Assemblies (FFA) [52]. This new generation of SPME fiber was developed by Chromline (Prato, Italy), in cooperation with Supelco, expanding the applicability of SPME; the product line is centered around the SPME FFA barcodes that can be automatically exchanged on a three axis autosampler equipped with the Multi Fiber EXchanger (MFX) system[53].

Therefore, we sought to simplify sample treatment by using the SPME technique in off- and on-line modes for seven OH-PAHs, namely, 1-hydroxy-naphthalene (1-OH-Nap), 2-hydroxy-napthalene (2-OH-Nap), 9-hydroxy-phenanthrene (9-OH-Phen), 1-OH-P, 3-hydroxy-benzoanthracene (3-OH-B[a]A), 9-hydroxybenzo-anthracene (9-OH-B[a]A), and 3-OH-B[a]P. In order to do this, we fully automated the on-line mode, and in the off-line mode we reduced human intervention before GC/triple quadrupole-mass spectrometry (QpQ-MS) analysis. These innovations form a new user-friendly SPME platform which provides relatively prompt results with a high degree of selectivity, precision, and accuracy.

2. Results and Discussion

Several OH-PAHs have been suggested as urinary biomarkers to estimate PAH exposure. For instance, 1-OH-P has long been used due to its relatively high levels, even if pyrene is not carcinogenic. However, the profile of PAHs depends on their emission source, and therefore extrapolation of its presence would be imprecise. Conversely, analyzing the hydroxy metabolites of B[a]P and B[a]A would more accurately assess exposure to PAHs. In addition, Gundel et al. [54] proposed 3-OH-B[a]A as an indicator for this, also due to the fact that it is excreted in relatively high concentrations in urine. However, since smoking is a significant source of PAHs it represents a confounding factor. The largest difference between smokers and non-smokers in PAH metabolite concentrations regard 2-OH-Nap, 1-OH-P and OH-phenanthrenes [55–57]. Moreover, several authors have shown that urinary OH-fluorene levels are positively correlated with smoking status, particularly 1-OH-fluorene [22,57].

An evaluation of the processing steps of the previously-proposed MS methods including 3-OH-B[a]P [26,27,29,33–36] revealed five critical phases. First off, phenolic compounds are susceptible to oxidation with consequent loss of OH-PAHs; adding gallic acid (50 µg/mL) prior to the evaporation and derivatization steps effectively inhibits this loss, according to Jacob et al. [9]. Otherwise, Woudneh et al. [35] controlled oxidation by employing 2-mercaptoethanol in a nitrogen atmosphere. Secondly, since OH-PAHs are photodegradable [58], this should be avoided by using amber glassware. A third critical area is the wearing down of the injectors, columns, and GC detectors due to the high amounts of BSTFA injected with conventional sample preparation methods. The fourth point we evaluated is a more rapid and less-solvent-consuming derivatization step by using 1.2-dimethylimidazole-4-sulfonyl (DMISC) instead of DNS; this not only reduced the retention time (RT) by three times, but maintained good quality, as shown in our previous work [59]. Lastly, the seven methods do not allow for full automation.

Accordingly, we developed a method in which the fiber SPME derivatization technique was applied after direct immersion (DI), and then coupled with quantitative determination via GC/QpQ-MS. Here three key aspects behind this choice: SPME extraction, derivatization, and automation.

2.1. SPME Extraction

The 85-μm polyacrylate (PA) absortive coating was chosen for sampling such a complex matrix as urine, because the analytes do not compete with each other. Because of the liquid coating's properties, the extraction obeys the rules of liquid-liquid equilibrium:

$$n = C_0 V_1 V_2 K / K V_1 + V_2, \quad (1)$$

where K is the partition constant SPME fiber liquid polymeric coating/sample, C_0 is the initial concentration of the analyte in the aqueous solution, V_1 and V_2 are the volumes of the coating and the solution in the equilibrium concentration of the analyte in the urine. Although SPME is an equilibrium extraction, it is not exhaustive.

The equilibrium and kinetics of the OH-PAHs versus SPME fiber with liquid coating were calculated. Table 1 illustrates the physicochemical constants of the seven OH-PAHs obtained by Performs Automated Reasoning in Chemistry (ARChem, Danielsville, GA, USA)—a physicochemical calculator that uses computational algorithms based on their basic chemical structures to foresee a wide variety of reactivity parameters, to anticipate trends in sampling extraction.

Pacenti et al. [60] indicated K_{ow} as a good estimator of K, whose value is often very close to the gas phase partition coefficient/aqueous matrix partition coefficient ($K_2 = K_H/RT$) and to the SPME coating/gas phase partition coefficient (K_1). $K = K_2 \cdot K_1$; consequently, knowing K_1 and K_2 values allows you to know in advance whether or not the SPME method is advantageous. The K_H value of the OH-PAHs ranges between 8.39×10^{-8} atm m^3/mol for OH-Nap and 4.56×10^{-10} atm m^3/mol for OH-B[a]P, as indicated by Pacenti et al. [60]. DI-SPME is efficient for compounds with Henry's constant lower than 1.75×10^{-7} atm m^3/mol when GC-MS/MS is used.

An excellent SPME extraction sensitivity for the urinary OH-PAHs was achieved by immerging the PA fiber in diluted urine; dilution of the urine with distilled water reduces the sensitivity of the method, but increases the precision and the fiber lifetime. The best results were obtained with DI times up to 30 min with temperature-controlled agitation (60 °C and 500 rpm). To remove any liquid sample remaining on the SPME PA fiber after DI extraction, the fiber was placed for 45 s into an SPME fiber conditioning station set at 100 °C. Moreover, reduction in vial diameter by a factor of 3 resulted in an order of magnitude decrease in extraction time, since t, the average time of diffusion through the aqueous layer is proportional to the square of the migration distance, x, and inversely proportional to D_{water} [61],

$$t = x^2 / 2 D_{water}, \quad (2)$$

Hence for high-concentration samples, 2 mL filled to the top, using the same dilution ratio can be used instead of 10 mL amber vials.

Table 1. Physical properties and partition coefficients of OH-PAHs evaluated using SPARC. (M.W. = molecular weight; T_{eb} = boiling point; D_{water} = diffusion coefficient of the analyte in water; K_H = Henry's constant; K_{ow} = octanol-water partition coefficient; P_{vap} = vapour pressure).

SMILES strings	M.W. (g/mol)	T_{eb} (°C)	D_{water} (cm²/s)	K_H (atm/(mol/m³))	K_{OW} (Log)	P_{vap} (log(atm))
OC1=CC=CC2=CC=CC=C21	144	269.7	8.08×10^{-6}	8.39×10^{-8}	3.04	−6.0
OC1=CC2=CC=CC=C2C=C1	144	269.8	8.08×10^{-6}	9.10×10^{-8}	3.11	−6.14
OC1=CC2=C(C3=C1C=CC=C3)C=CC=C2	194	378.9	6.92×10^{-6}	6.93×10^{-9}	4.49	−8.67
OC1=CC=C((C=C2)C3=C1C=CC4=CC=CC2=C34	218	454.6	6.41×10^{-6}	6.37×10^{-9}	5.01	−8.9
OC(C=C1)=CC2=C1C3=CC4=CC=CC=C4C=C3C=C2	244	537.2	6.15×10^{-6}	3.24×10^{-10}	5.71	−11.86
OC1=CC=C2C(C=C(C=CC3=C4C=CC=C3)C4=C2)=C1	244	537.2	6.15×10^{-6}	3.21×10^{-10}	5.71	−11.86
OC1=CC=C((C=C2)C3=C1C=CC4=CC5=CC=CC=C5C2=C43	268	564.5	5.74×10^{-6}	4.56×10^{-10}	6.28	−11.81

2.2. SPME Derivatization

N-*tert*-butyldimethylsilyl-N-methyltrifluoroacetamide (MTBSTFA) as a TBDMS derivatizing agents was used in GC analysis of amino acids and in GC-MS analysis of hydroxylated fluorenes, and it was shown that TBDMS derivatives were thermally stable and had favorable fragmentations upon electron impact (EI) ionization [22,62] (Figure 1).

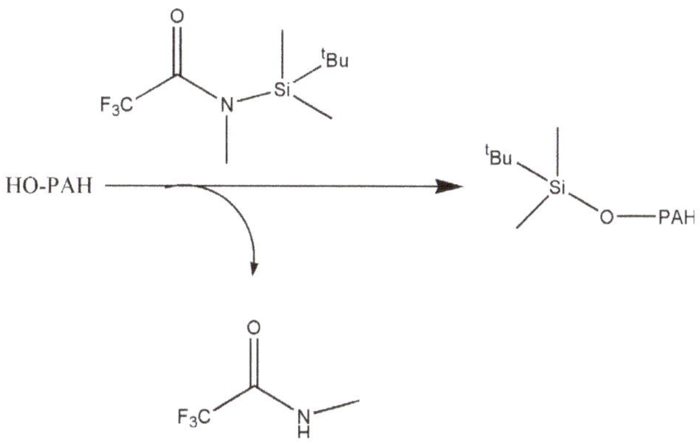

Figure 1. Derivatization of hydroxylated polycyclic aromatic hydrocarbons (OH-PAHs) with N-*tert*-butyldimethylsilyl-N-methyltrifluoroacetamide (MTBSTFA).

The MS spectra show two intense fragment ion peaks corresponding to the TBDMS-derivate molecular ion produced by adding 115 Da to the original mass of the compound, and its loss of a tert-butyl fragment of 57 Da. We found that the intensity due to the base peaks of OH-PAHs-TBDMS was about five times higher than that of OH-PAHs-TMS even if the TBDMS derivatives had longer eluting times than the TMS ones.

Next, the effects of time, temperature, and volume of urine and derivatization reagent in automated analysis were evaluated. For on-fiber derivatization low and high values for these three variables (15 and 100 µL MTBSTFA, 25 and 60 °C, and 10 and 60 min) were selected on the basis of previously reported results [63]. The volume of MTBSTFA, the derivatization time, and temperature were fixed at 15 µL, 30 min, and 60 °C, respectively. In order to avoid contamination problems between consecutive samples, on-fiber derivatization was performed in an argon atmosphere in 2 mL silanized amber vials, placed in a 98-position vial tray set to +4 °C.

2.3. Automation of the SPME Procedure

New full automation of the procedure was achieved using an *xyz* robotic autosampler coupled with FFA-SPME fibers. In off-line SPME sampling mode by a Multi Off-Line Sampler, the fibers—after the extraction and derivatization steps are manually performed—are placed into the *xyz* autosampler and transported from the MFX 45-position tray to the injector by a SPME holder equipped with a plunger/magnetic system; at the end of the analysis each desorbed fiber is moved back to the tray and the cycle is repeated with a new loaded SPME fiber. Instead, in fully automated on-line SPME mode, the FFA fiber is transported from the vials—containing urine or derivatization agents—to the injector. Figure 2 clearly shows the advantages of using an SPME-FFA Multi Off-Line Sampler calculating a urine extraction time of 30 min, followed by 30 min of MTBSTFA derivatizating reaction plus an analysis time of 20 min; the results are excellent, reducing total analysis time by 2200 min for 60 samples, compared to SPME on-line analysis. The initial economic commitment for the purchase of SPME fibers, as well as for the manual transport steps for extracting and derivatizing, is thus overridden by the off-line method's high-throughput approach.

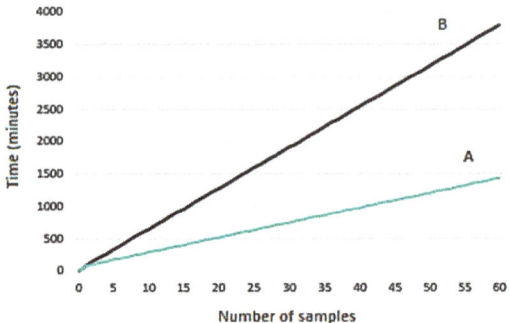

Figure 2. Comparison between solid-phase microextraction fast fit fiber assemblies (SPME-FFA) Multi Off-Line Sampler (**A**) and SPME on-line (**B**) for the analysis of 60 urinary OH-PAHs samples.

To show the above findings the authors present the final results in Table 2.

Table 2. Limits of detection (LOD), limit of quantification (LOQ), accuracy and precision for each OH-PAH measured in urine samples.

		Response factor Plot and Limit of Detection and Quantification						
		1-OH-Nap	2-OH-Nap	9-OH-Phen	1-OH-P	3-OH-B[a]A	9-OH-B[a]A	3-OH-B[a]P
Least-squares linear regression parameters	m	1.0924	1.0922	1.1125	1.1126	1.1244	1.1240	1.1245
	b	0.1893	0.2204	0.0769	0.0828	0.0455	0.0453	0.0368
Coefficient of Correlation		0.99	0.99	1.00	0.99	0.99	1.00	1.00
LOD (ng L^{-1})		4.52	4.03	2.53	1.89	0.31	0.33	0.25
LOQ (ng L^{-1})		14.91	13.20	8.34	6.23	1.02	1.08	0.82
		Accuracy and precision (%)						
Within-session accuracy		10.0	9.3	10.3	9.8	10.5	10.2	10.7
Within-session repeatability		2.7	2.7	3.0	2.7	2.5	2.6	2.5
Inter-session repeatability		3.0	3.0	3.6	3.9	3.4	2.4	3.4

The resulting calibration curves were linear, in the investigated range for all the OH-PAHs considered, with correlation coefficients >0.99. The precision of the assay (reported as a coefficient of variation, CV%), estimated both as within-session and as inter-session repeatability, resulted in the 2.5–3.0% and 2.4–3.9% range, respectively. Accuracy was within 15% of the theoretical concentration, in line with requirements of the US Food and Drug Administrations for bioanalytical method validation. To demonstrate the applicability of the method to urinary samples, the content of these compounds in non-occupationally-exposed humans, 19 smokers and 21 non-smokers, was analyzed and indicated in Table 3.

Table 3. OH-PAHs in human urine of smokers and non-smokers. (SD = standard deviation).

	Non-Smokers	Smokers
	Average (ng/L) ± SD (min-max Value)	Average (ng/L) ± SD (min-max Value)
1-OH-Nap	1040.6 ± 340.7 (150.3–1500.2)	2966.6 ± 904.3 (240.1–3500.6)
2-OH-Nap	1879.2 ± 402.4 (201.6–2001.3)	4297.5 ± 1151.2 (2898.3–8214.4)
9-OH-Phen	<LOD ± 0.54 (<LOD–3.2)	<LOD ± 0.66 (<LOD–3.6)
1-OH-P	59.3 ± 27.4 (25.1–166.7)	291.4 ± 89.3 (178.0–647.2)
3-OH-B[a]A	0.43 ± 0.21 (<LOD–1.2)	0.60 ± 0.23 (<LOD–1.6)
9-OH-B[a]A	<LOD ± 0.25 (<LOD–1.41)	1.44 ± 0.59 (<LOD–2.3)
3-OH-B[a]P	<LOD ± 0.17 (<LOD–0.91)	0.98 ± 0.14 (<LOD–1.32)

3. Materials and Methods

3.1. Hydrolysis of Conjugated OH-PAHs

Sample processing was conducted in a dark room with limited yellow light. Three mL of urine were spiked with 5 µL of β-Glucuronidase from Helix pomatia (Sigma-Aldrich, Saint Louis, MO, USA, cat. no. G7017-5ML) in 10 mL amber vial (Sigma-Aldrich, Saint Louis, MO, USA, cat. no. 27389). The headspace (HS) over each sample was purged with argon, sealed with screwed caps (Agilent Technologies, St. Clara, CA, USA, cat. no. 8010-1039) and incubated in the dark at 37 °C. After 17 h the samples were diluted with 7 mL of water (10 mL total volume) and doped by deuterated internal standards (ISs) for on- or off-line analysis.

3.2. On-Line DI-SPME and xyz Robotic Apparatus

Automated DI-SPME and on-fiber derivatization experiments were carried out by a Flex Autosampler (EST Analytical, Fairfield, CT, USA). The *xyz* robotic system was assembled with devices developed by Chromline (Prato, Italy): a 32-position tray for 10 mL vials, a 98-position tray for 2 mL vials, a Peltier cooler tray (set to 4 °C), a MFX 6-position SPME system, a SPME fiber conditioning station, and agitator. The 10 mL amber vial containing standards/sample was taken automatically from the 32-position tray and was inserted into the agitator, heated (60 °C), and agitated (pulsed agitation, 2 s at 500 rpm and off 4 s). During that period, the FFA-SPME 85-µm PA fiber (Supelco, Bellefonte, PA, USA, cat. no. FFA 57294-U) was immersed directly into the sample solution. After SPME extraction, the fiber was placed for 45 s into an SPME fiber conditioning station set at 100 °C. Subsequently, the SPME on-fiber HS derivatization was performed in the agitator for 30 min at 60 °C, exposing the SPME fiber in 2 mL amber silanized vials (Thermo Fisher Scientific, Waltham, MA, USA, cat. no. MSCERT 5000-S41W) fit with screw thread caps for magnetic transport (Thermo Fisher Scientific, Waltham, MA, USA, cat. no. 9-MSC(BG)-ST101) and containing 15 µL of MTBSTFA (Sigma-Aldrich Saint Louis, MO, USA, cat. no. 394882-10X1ML). Finally, the fiber was inserted into the GC injector equipped with Merlin Microseals (Sigma-Aldrich, Saint Louis, MO, USA, cat. no. 24817-U) for the thermal desorption of the analytes.

3.3. Off-Line DI-SPME and xyz Robotic Apparatus

The SPME Multi Off-Line Sampler (Chromline, Prato, Italy) is a holder designed (Figure 3) to be used with FFA SPME fibers; in our case PA 85-µm SPME FFA were used. The holder acts as a support when exposing the SPME fibers in the 10 mL amber vials (60 °C for 30 min), after which they are placed on 15-position magnetic stirrer plates (Chromline, Prato, Italy). After extraction, the FFAs are removed from the Multi Off-Line Sampler and placed for 45 s into an SPME fiber conditioning station set at 100 °C. Subsequently, the SPME on-fiber HS derivatization (30 min at 60 °C) was performed in 2 mL amber silanized vials, placed into the SPME Multi Off-Line Sampler. For desorption the fiber was put into a MFX 45-position SPME system installed on the Flex autosampler coupled with GC instrumentation.

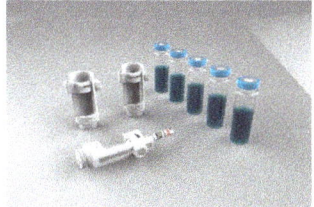

Figure 3. The SPME Multi Off-Line Sampler.

3.4. GC/QpQ-MS

Analyses were performed with a Varian 3900 GC equipped with electronic flow control and a Varian 320-QpQ-MS (Agilent Technologies, St. Clara, CA, USA) detector (Table 4).

A VF-5 ms + 10 m EZ-Guard fused silica capillary column (internal diameter 0.25 mm, length 30 m and film thickness 0.25 μm) (Agilent Technologies, St. Clara, CA, USA, cat. no. CP9013) was used (Figure 4). For desorbing the analytes, the SPME fiber was introduced into the 1177 Varian GC injector port. A connection with the Laboratory Information Management System (Bika Lab System, Pringle Bay, South Africa) provides a user-programmable suite of options.

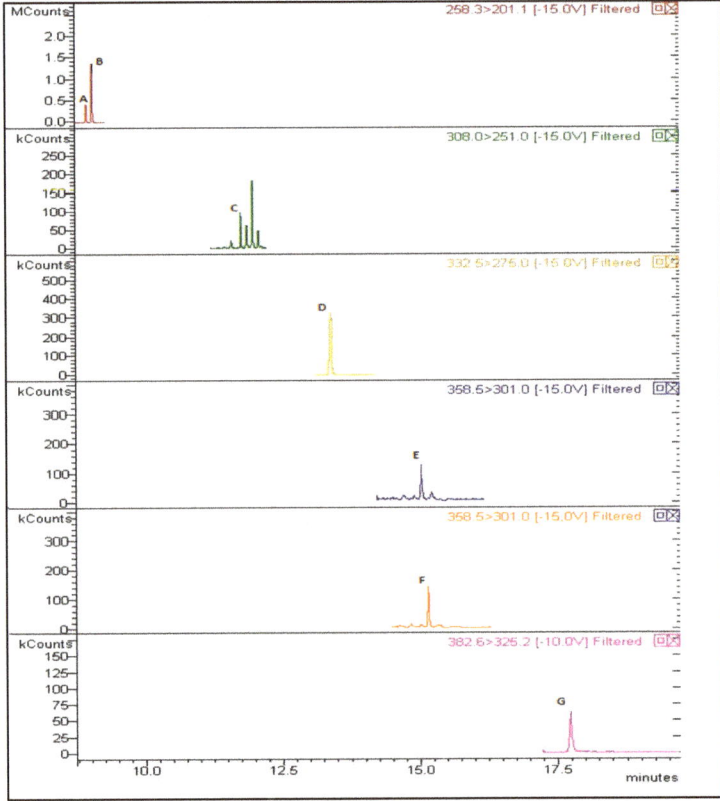

Figure 4. Chromatogram of OH-PAHs by gas chromatography (GC)/triple quadrupole-mass spectrometry (QpQ-MS) in spiked artificial urine. A = 1-OH-Nap (200 ng/L); B = 2-OH-Nap (600 ng/L); C = 9-OH-Phen (50 ng/L); D = 1-OH-P (400 ng/L); E = 3-OH-B[a]A (1 ng/L); F = 9-OH-B[a]A (1 ng/L); G = 3-OH-B[p]A (5 ng/L).

Table 4. GC/QpQ-MS method parameters.

GC Conditions	
Injection	300 °C, 20:1 split mode. Liner 0.75 mm i.d.
Oven	80 °C (1 min) increased to 20 °C/min end to 320 °C (5 min)
Column flow	Helium (99.999%) at a flow rate of 1.2 mL/min
Retention time	1-OH-Nap (8.90 min); 2-OH-Nap (9.05 min); 9-OH-Phen (11.96 min); 1-OH-P (13.38 min); 3-OH-B[a]A (14.93 min); 9-OH-B[a]A (14.93 min); 3-OH-B[a]P (17.72 min)
GC interface	280 °C
MS Parameters	
Mode	EI
Filament	Electron energy, 70 eV. Filament current 50 µA
Source	Temperature, 200 °C. Pressure, 8 Torr.
Collision gas	CID gas, Argon. CID gas pressure, 2.00 mTorr
Collision energy	1-OH-Nap 15 eV; 2-OH-Nap 15 eV; 9-OH-Phen 15 eV; 1-OH-P 15 eV; 3-OH-B[a]A 15 eV; OH-9-B[a]A 15 eV; 3-OH-B[a]P 10 eV.
SRM Transition	
1-OH-Nap	Fragment Q1 > Q3 Quantification m/z 258.5→201.2 Confirmation m/z 201.4→185.0
2-OH-Nap	Q1 > Q3 258.5→201.2 201.4→185.0
9-OH-Phen	Q1 > Q3 308.5→251.2 251.4→235.0
1-OH-P	Q1 > Q3 332.5→275.0 275.4→259.0
3-OH-B[a]A	Q1 > Q3 358.5→301.1 301.5→285.0
9-OH-B[a]A	Q1 > Q3 358.5→301.1 301.5→285.0
3-OH-B[a]P	Q1 > Q3 382.6→325.2 382.6→309.6

3.5. Synthesis

1-OH-Nap (cat. no. N1000), 2-OH-Nap (cat. no. 185507), 9-OH-Phen (cat. no. 211281), and 1-OH-P (cat. no. 361518) were purchased from Sigma-Aldrich (Saint Louis, MO, USA). As described by Xu et al, 3-OH-B[a]P was synthesized [64], while 3-OH-B[a]A, and 9-OH-B[a]A were prepared following McCourt's procedure [65]. The deuterated compounds 1-hydroxy-naphthalene-D7, 2-hydroxy-napthalene-D7, 9-hydroxy-phenanthrene-D9,1-hydroxypyrene-D9, 3-hydroxy-benzoanthracene-D11, 9-hydroxybenzo-anthracene-D11, and 3-hydroxy-benzo[a]pyrene-D11 were prepared by perdeuteration of the unlabeled starting material under the conditions described by Duttwyler et al. [66]; in all cases two reaction cycles were enough to reach a deuteration of above 98%.

3.6. Calibration and Method Validation

Spiked artificial urine [12] (1, 2, 4, 8, 16, 32 ng/L) was analyzed and five replicates of each sample were performed to calculate their limits of detection (LOD). A linear regression plot was generated; the LOD is reported as [(YB + 3SB)/m], where Y B is the intercept, SB is its standard deviation, and m is the plot slope. The limit of quantification (LOQ) was then calculated in the same fashion, using 10SB, which corresponds to 3.3 LOD. Samples were processed by calibration curves set as follow: (a) 100, 200, 400, 800, 1600, 3200, 6400 ng/L for 1-OH-Nap and 2-OH-Nap (b) 10, 20, 40, 80, 160, 320, 640 ng/L for 1-OH-P (c) 0.5, 1.0, 2.0, 4.0, 8.0, 16.0 ng/L for 3-OH-B[a]A, 9-OH-B[a]A and 3-OH-B[a]P. The precision of the assay (as a coefficient of variation, CV%) was based on both within-session and inter-session repeatability. Accuracy was evaluated by the recoveries (calculated from the percentage ratio between the measured and the nominal concentration solutions) at all concentrations used for the calibration plot and from certified analytical standards for 1-OH-P (Chromsystems Instruments & Chemicals GmbH, Gräfelfing, Germany, cat. no. 53003). Values of accuracy were then compared with the requirements of the US Food and Drug Administration for analytical method validation. Level quality control samples were prepared and processed in every analytical session from a fresh solution with the IPA with ISs to ensure the precision validity of reported results.

3.7. Evaluation of Processing Steps of Previously Proposed MS Methods Including 3-OH-B[a]P

We found five critical processing steps in the seven above-indicated methods [26,27,29,33–36]: (i) susceptibility to oxidation during evaporation and derivatization, (ii) photodegradation, (iii) BSTFA contamination, (iv) excessive analysis time using DNS, and (v) undue handling time.

4. Conclusions

Occupational studies indicate that there is a correlation between PAH exposure and cancer incidence for various human tissues such as lungs, skin, and bladder. As a result, a regular control of the concentrations in the workplace and in life environments trough measuring their metabolites has become mandatory. PAH metabolites in human urine can be used as biomarkers of internal dose to assess recent exposure to PAHs. In previous studies, the oft-reported use of solvent and/or clean-up steps were necessary to extract and eliminate most of the interfering compounds from the urine. But these laboratories used techniques based on age-old methodologies with a low level of automation in which each step represents additional time and potential source of error. Instead, a straightforward, optimized sample preparation strategy minimizes the number of steps.

Our data suggests that automated SPME extraction coupled with GC/QpQ-MS is a viable alternative for OH-PAH analyses. Customized and automated MS systems for high-throughput screening are not only user-friendly, but they reduce the costs of monitoring occupational health hazards. New sample preparation techniques are currently being increasingly explored because of the considerable need for the automating of sample preparation, and for integrating data management into the analytical process.

Author Contributions: S.D. and N.M. conceived of the presented idea, the research methodology and wrote the present manuscript; A.B., I.P. and C.B. collected materials to carry out and performed the experiment; M.G. contributes in the organic chemical synthesis; G.C. contributes in preparation of references; G.A. and M.C. contributed to the interpretation of the results.

Funding: This research received no external funding.

Conflicts of Interest: The authors declare no conflict of interest.

References

1. Grant, W.B. Air pollution in relation to U.S. cancer mortality rates: An ecological study; likely role of carbonaceous aerosols and polycyclic aromatic hydrocarbons. *Anticancer Res.* **2009**, *29*, 3537–3545. Available online: http://ar.iiarjournals.org/content/29/9/3537.full (accessed on 21 June 2018). [PubMed]
2. Lelieveld, J.; Evans, J.S.; Fnais, M.; Giannadaki, D.; Pozzer, A. The contribution of outdoor air pollution sources to premature mortality on a global scale. *Nature* **2015**, *525*, 367–371. [CrossRef] [PubMed]
3. Lim, S.S.; Vos, T.; Flaxman, A.D.; Danaei, G.; Shibuya, K.; Adair-Rohani, H.; AlMazroa, M.A.; Amann, K.; Anderson, H.R.; Andrews, G.K.; et al. A comparative risk assessment of burden of disease and injury attributable to 67 risk factors and risk factor clusters in 21 regions, 1990–2010: A systematic analysis for the Global Burden of Disease Study 2010. *Lancet* **2012**, *380*, 2224–2260. [CrossRef]
4. Pacenti, M.; Lofrumento, C.; Dugheri, S.; Zoppi, A.; Borsi, I.; Speranza, A.; Boccalon, P.; Arcangeli, G.; Antoniucci, A.; Castellucci, E.M.; et al. Physicochemical characterization of exhaust particulates from gasoline and diesel engines by solid-phase micro extraction sampling and combined raman microspectroscopic/fast gas-chromotography mass spectrometry analysis. *Eur. J. Inflamm.* **2009**, *7*, 25–37. [CrossRef]
5. Gherardi, M.; Gatto, M.P.; Gordani, A.; L'Episcopo, N.; L'Episcopo, A. Caratterizzazione e confronto dei profili di Idrocarburi Policiclici Aromatici su particolato indoor e outdoor in uffici di un centro ricerca in un'area suburbana di Roma. In Proceedings of the 35° Congresso Nazionale di Igiene Industriale e Ambientale, Centro Internazionale di Formazione ITC-ILO, Torino, Italy, 13–15 June 2018.
6. Ohura, T.; Amagai, T.; Fusaya, M.; Matsushita, H. Polycyclic Aromatic Hydrocarbons in Indoor and Outdoor Environments and Factors Affecting Their Concentrations. *Environ. Sci. Technol. Sci. Technol.* **2004**, *38*, 77–83. [CrossRef]

7. Health, Socio-Economic and Environmental Aspects of Possible Amendments to the EU Directive on the Protection of Workers from the Risks Related to Exposure to Carcinogens and Mutagens at Work Hexavalent Chromium. 2011, p. 101. Available online: Ec.europa.eu/social/BlobServlet?docId=10165&langId=en (accessed on 21 June 2018).
8. Stec, A.A.; Dickens, K.E.; Salden, M.; Hewitt, F.E.; Watts, D.P.; Houldsworth, P.E.; Martin, F.L. Occupational Exposure to Polycyclic Aromatic Hydrocarbons and Elevated Cancer Incidence in Firefighters. *Sci. Rep.* **2018**, *8*, 4–11. [CrossRef] [PubMed]
9. Jacob, J.; Seidel, A. Biomonitoring of polycyclic aromatic hydrocarbons in human urine. *J. Chromatogr. B* **2002**, *778*, 31–47. [CrossRef]
10. Grainger, J.; Huang, W.; Li, Z.; Edwards, S.; Walcott, C.; Smith, C.; Turner, W.; Wang, R.; Patterson, D. Polycyclic aromatic hydrocarbon reference range levels in the U.S. population by measurement of urinary monohydroxy metabolites. *Polycycl. Aromat. Compd.* **2004**, *24*, 385–404. [CrossRef]
11. Limit, T.; Tlvs, V. *Annual Reports for the Year 2016: And Biological Exposure Indices (BEIs®)*; BEIs: London, UK; pp. 1–22.
12. Jacob, P.; Wilson, M.; Benowitz, N.L. Determination of phenolic metabolites of polycyclic aromatic hydrocarbons in human urine as their pentafluorobenzyl ether derivatives using liquid chromatography-tandem mass spectrometry. *Anal. Chem.* **2007**, *79*, 587–598. [CrossRef] [PubMed]
13. Smith, C.J.; Walcott, C.J.; Huang, W.; Maggio, V.; Grainger, J.; Patterson, D.G., Jr. Determination of selected monohydroxy metabolites of 2-, 3- and 4-ring polycyclic aromatic hydrocarbons in urine by solid-phase microextraction and isotope dilution gas chromatography-mass spectrometry. *J. Chromatogr. B Anal. Technol. Biomed. Life Sci.* **2002**, *778*, 157–164. [CrossRef]
14. Smith, C.J.; Huang, W.; Walcott, C.J.; Turner, W.; Grainger, J.; Patterson, D.G. Quantification of monohydroxy-PAH metabolites in urine by solid-phase extraction with isotope dilution-GC–MS. *Anal. Bioanal. Chem.* **2002**, *372*, 216–220. [CrossRef] [PubMed]
15. Ramsauer, B.; Sterz, K.; Hagedorn, H.W.; Engl, J.; Scherer, G.; McEwan, M.; Errington, G.; Shepperd, J.; Cheung, F. A liquid chromatography/tandem mass spectrometry (LC-MS/MS) method for the determination of phenolic polycyclic aromatic hydrocarbons (OH-PAH) in urine of non-smokers and smokers. *Anal. Bioanal. Chem.* **2011**, *399*, 877–889. [CrossRef] [PubMed]
16. Chetiyanukornkul, T.; Toriba, A.; Kameda, T.; Tang, N.; Hayakawa, K. Simultaneous determination of urinary hydroxylated metabolites of naphthalene, fluorene, phenanthrene, fluoranthene and pyrene as multiple biomarkers of exposure to polycyclic aromatic hydrocarbons. *Anal. Bioanal. Chem.* **2006**, *386*, 712–718. [CrossRef] [PubMed]
17. Lintelmann, J.; Wu, X.; Kuhn, E.; Ritter, S.; Schmidt, C.; Zimmermann, R. Detection of monohydroxylated polycyclic aromatic hydrocarbons in urine and particulate matter using LC separations coupled with integrated SPE and fluorescence detection or coupled with high-resolution time-of-flight mass spectrometry. *Biomed. Chromatogr.* **2018**, *34*, e4183. [CrossRef] [PubMed]
18. Desmet, K.; Tienpont, B.; Sandra, P. Analysis of 1-hydroxypyrene in urine as PAH exposure marker using in-situ derivatisation stir bar sorptive extraction-thermal desorption-Capillary gas chromatography-Mass spectrometry. *Chromatographia* **2003**, *57*, 681–685. Available online: http://hdl.handle.net/1854/LU-208904 (accessed on 21 June 2018). [CrossRef]
19. Itoh, N.; Tao, H.; Ibusuki, T. Optimization of aqueous acetylation for determination of hydroxy polycyclic aromatic hydrocarbons in water by stir bar sorptive extraction and thermal desorption-gas chromatography-mass spectrometry. *Anal. Chim. Acta* **2005**, *535*, 243–250. [CrossRef]
20. Shin, H.S.; Lim, H.H. Simultaneous determination of 2-naphthol and 1-hydroxy pyrene in urine by gas chromatography-mass spectrometry. *J. Chromatogr. B Anal. Technol. Biomed. Life Sci.* **2011**, *879*, 489–494. [CrossRef] [PubMed]
21. Romanoff, L.C.; Li, Z.; Young, K.J.; Blakely, N.C., III; Patterson, D.G., Jr.; Sandau, C.D. Automated solid-phase extraction method for measuring urinary polycyclic aromatic hydrocarbon metabolites in human biomonitoring using isotope-dilution gas chromatography high-resolution mass spectrometry. *J. Chromatogr. B Anal. Technol. Biomed. Life Sci.* **2006**, *835*, 47–54. [CrossRef] [PubMed]
22. Gmeiner, G.; Gärtner, P.; Krassnig, C.; Tausch, H. Identification of various urinary metabolites of fluorene using derivatization solid-phase microextraction. *J. Chromatogr. B Anal. Technol. Biomed. Life Sci.* **2002**, *766*, 209–218. [CrossRef]

23. Jongeneelen, F.J.; Anzion, R.B.M.; Henderson, P.T. Determination of hydroxylated metabolites of polycyclic aromatic hydrocarbons in urine. *J. Chromatogr. B Biomed. Sci. Appl.* **1987**, *413*, 227–232. [CrossRef]
24. Pigini, D.; Cialdella, A.M.; Faranda, P.; Tranfo, G. Comparison between external and internal standard calibration in the validation of an analytical method for 1-hydroxypyrene in human urine by high-performance liquid chromatography/tandem mass spectrometry. *Rapid Commun. Mass Spectrom.* **2006**, *20*, 1013–1018. [CrossRef] [PubMed]
25. Cahours, X.; Blanchet, M.; Rey, M. Fast and simple method for the determination of urinary 1-hydroxypyrene. *J. Sep. Sci.* **2009**, *32*, 3403–3410. [CrossRef] [PubMed]
26. Li, Z.; Romanoff, L.C.; Trinidad, D.A.; Hussain, N.; Jones, R.S.; Porter, E.N.; Patterson, D.G., Jr.; Sjödin, A. Measurement of urinary monohydroxy polycyclic aromatic hydrocarbons using automated liquid-liquid extraction and gas chromatography/isotope dilution high-resolution mass spectrometry. *Anal. Chem.* **2006**, *78*, 5744–5751. [CrossRef] [PubMed]
27. Campo, L.; Rossella, F.; Fustinoni, S. Development of a gas chromatography/mass spectrometry method to quantify several urinary monohydroxy metabolites of polycyclic aromatic hydrocarbons in occupationally exposed subjects. *J. Chromatogr. B Anal. Technol. Biomed. Life Sci.* **2008**, *875*, 531–540. [CrossRef] [PubMed]
28. Fan, R.; Dong, Y.; Zhang, W.; Wang, Y.; Yu, Z.; Sheng, G. Fast simultaneous determination of urinary 1-hydroxypyrene and 3-hydroxybenzo[a]pyrene by liquid chromatography-tandem mass spectrometry. *J. Chromatogr. B Anal. Technol. Biomed. Life Sci.* **2006**, *836*, 92–97. [CrossRef] [PubMed]
29. Raponi, F.; Bauleo, L.; Ancona, C.; Forastiere, F.; Paci, E.; Pigini, D.; Tranfo, G. Quantification of 1-hydroxypyrene, 1- and 2-hydroxynaphthalene, 3-hydroxybenzo[a]pyrene and 6-hydroxynitropyrene by HPLC-MS/MS in human urine as exposure biomarkers for environmental and occupational surveys. *Biomarkers* **2017**, *22*, 575–583. [CrossRef] [PubMed]
30. Hollender, J.; Koch, B.; Dott, W. Biomonitoring of environmental polycyclic aromatic hydrocarbon exposure by simultaneous measurement of urinary phenanthrene, pyrene and benzo[a]pyrene hydroxides. *J. Chromatogr. B Biomed. Sci. Appl.* **2000**, *739*, 225–229. [CrossRef]
31. Barbeau, D.; Persoons, R.; Marques, M.; Hervé, C.; Laffitte-Rigaud, G.; Maitre, A. Relevance of urinary 3-hydroxybenzo(a)pyrene and 1-hydroxypyrene to assess exposure to carcinogenic polycyclic aromatic hydrocarbon mixtures in metallurgy workers. *Ann. Occup. Hyg.* **2014**, *58*, 579–590. [CrossRef] [PubMed]
32. Strickland, P.; Kang, D.; Sithisarankul, P. Polycyclic aromatic hydrocarbon metabolites in urine as biomarkers of exposure and effect. *Environ. Health Perspect.* **1996**, *104*, 927–932. [CrossRef] [PubMed]
33. Luo, K.; Gao, Q.; Hu, J. Derivatization method for sensitive determination of 3-hydroxybenzo[a]pyrene in 9human urine by liquid chromatography-electrospray tandem mass spectrometry. *J. Chromatogr. A* **2015**, *1379*, 51–55. [CrossRef] [PubMed]
34. Gaudreau, É.; Bienvenu, J.F.; Bérubé, R.; Daigle, É.; Chouinard, S.; Kim, M. Using the Agilent 7000B Triple Quadrupole GC/MS for Parts per Trillion Detection of PAH Metabolites in Human Urine. 2012. Available online: http://hpst.cz/sites/default/files/attachments/5991-0991en-using-agilent-7000b-triple-quadrupole-gc-ms-ms-parts-trillion-detection-pah-metabolites.pdf (accessed on 21 June 2018).
35. Woudneh, M.B.; Benskin, J.P.; Grace, R.; Hamilton, M.C.; Magee, B.H.; Hoeger, G.C.; Forsberg, N.D.; Cosgrove, J.R. Quantitative determination of hydroxy polycyclic aromatic hydrocarbons as a biomarker of exposure to carcinogenic polycyclic aromatic hydrocarbons. *J. Chromatogr A* **2016**, *1454*, 93–100. [CrossRef] [PubMed]
36. Zhang, X.; Hou, H.; Xiong, W.; Hu, Q. Development of a method to detect three monohydroxylated polycyclic aromatic hydrocarbons in human urine by liquid chromatographic tandem mass spectrometry. *J. Anal. Methods Chem.* **2015**. [CrossRef] [PubMed]
37. Bianchi, F.; Bisceglie, F.; Dugheri, S.; Arcangeli, G.; Cupelli, V.; del Borrello, E.; Sidisky, L.; Careri, M. Ionic liquid-based solid phase microextraction necklaces for the environmental monitoring of ketamine. *J. Chromatogr. A* **2014**, *1331*, 1–9. [CrossRef] [PubMed]
38. Pan, L.; Adams, M.; Pawliszyn, J. Determination of Fatty Acids Using Solid-Phase Microextraction. *Anal. Chem.* **1995**, *67*, 4396–4403. [CrossRef]
39. Bartelt, R.J. Calibration of a Commercial Solid-Phase Microextraction Device for Measuring Headspace Concentrations of Organic Volatiles. *Anal. Chem.* **1997**, *69*, 364–372. [CrossRef] [PubMed]

40. Marini, F.; Bellugi, I.; Gambi, D.; Pacenti, M.; Dugheri, S.; Focardi, L.; Tulli, G. Compound A, formaldehyde and methanol concentrations during low-flow sevoflurane anaesthesia: Comparison of three carbon dioxide absorbers. *Acta Anaesthesiol. Scand.* **2007**, *51*, 625–632. [CrossRef] [PubMed]
41. Pacenti, M.; Dugheri, S.; Gagliano-Candela, R.; Strisciullo, G.; Franchi, E.; Degli Esposti, F.; Perchiazzi, N.; Boccalon, P.; Arcangeli, G.; Cupelli, V. Analysis of 2-Chloroacetophenone in air by multi-fiber solid-phase microextraction and fast gas chromatography-mass spectrometry. *Acta Chromatogr.* **2009**, *21*, 379–397. [CrossRef]
42. Kremser, A.; Jochmann, M.A.; Schmidt, T.C. PAL SPME Arrow-Evaluation of a novel solid-phase microextraction device for freely dissolved PAHs in water. *Anal. Bioanal. Chem.* **2016**, *408*, 943–952. [CrossRef] [PubMed]
43. Asgari, S.; Bagheri, H.; Es-haghi, A.; AminiTabrizi, R. An imprinted interpenetrating polymer network for microextraction in packed syringe of carbamazepine. *J. Chromatogr. A* **2017**, *1491*, 1–8. [CrossRef] [PubMed]
44. David, F.; Sandra, P. Stir bar sorptive extraction for trace analysis. *J. Chromatogr. A* **2007**, *1152*, 54–69. [CrossRef] [PubMed]
45. Rossbach, B.; Kegel, P.; Letzel, S. Application of headspace solid phase dynamic extraction gas chromatography/mass spectrometry (HS-SPDE-GC/MS) for biomonitoring of n-heptane and its metabolites in blood. *Toxicol. Lett.* **2012**, *210*, 232–239. [CrossRef] [PubMed]
46. Laaks, J.; Jochmann, M.A.; Schilling, B.; Schmidt, T.C. Optimization strategies of in-tube extraction (ITEX) methods. *Anal. Bioanal. Chem.* **2015**, *407*, 6827–6838. [CrossRef] [PubMed]
47. Application Note 120 Flavour Profiling of Milk Using HiSorb Sorptive Extraction and TD-GC-MS. 2016. 44. Available online: http://kinesis-australia.com.au/media/wysiwyg/knowledebase/pdf/Flavour_profiling_of_various_drinks_using_HiSorb_sorptive_extraction_and_TD_GC_MS.pdf (accessed on 21 June 2018).
48. Ma, W.; Gao, P.; Fan, J.; Hashi, Y.; Chen, Z. Determination of breath gas composition of lung cancer patients using gas chromatography/mass spectrometry with monolithic material sorptive extraction. *Biomed. Chromatogr.* **2015**, *29*, 961–965. [CrossRef] [PubMed]
49. Belardi, R.P.; Pawliszyn, J.B. The application of chemically modified fused silica fibers in the extraction of organics from water matrix samples and their rapid transfer to capillary columns. *Water Pollut. Res. J. Can.* **1989**, *24*, 79–191. Available online: http://digital.library.mcgill.ca/wqrj/pdfs/WQRJ_Vol_24_No_1_Art_09.pdf (accessed on 21 June 2018).
50. Ranawat, K.K.; Singh, S.; Singh, G.P. A Green Microwave Assisted Synthesis Of New (Anthracene-9-Yl) Methylamines as an Environmentally Friendly Alternatives. *Rasayan J. Chem.* **2014**, *7*, 343–345. Available online: http://rasayanjournal.co.in/vol-7/issue_4/7_%20Vol.7_4_,%20343-345,%202014,%20RJC-1156.pdf (accessed on 21 June 2018).
51. Tobiszewski, M.; Mechlińska, A.; Namieśnik, J. Green analytical chemistry-theory and practice. *Chem. Soc. Rev.* **2010**, *39*, 2869–2878. [CrossRef] [PubMed]
52. Dugheri, S.; Bonari, A.; Pompilio, I.; Mucci, N.; Montalti, M.; Arcangeli, G. Development of New Gas Chromatography/Mass Spectrometry Procedure for the Determination of Hexahydrophthalic Anhydride in Unsaturated Polyester Resins. *Rasayan J. Chem.* **2016**, *9*, 657–666. Available online: http://www.rasayanjournal.co.in/admin/php/upload/76_pdf.pdf (accessed on 21 June 2018).
53. Pacenti, M.; Dugheri, S.; Traldi, P.; Degli Esposti, F.; Perchiazzi, N.; Franchi, E.; Calamante, M.; Kikic, I.; Alessi, P.; Bonacchi, A.; et al. New automated and high-throughput quantitative analysis of urinary ketones by multifiber exchange-solid phase microextraction coupled to fast gas chromatography/negative Chemical-Electron Ionization/Mass Spectrometry. *J. Autom. Methods Manag. Chem.* **2010**, *2010*, 972926. [CrossRef] [PubMed]
54. Gündel, J.; Schaller, K.H.; Angerer, J. Occupational exposure to polycyclic aromatic hydrocarbons in a fireproof stone producing plant: Biological monitoring of 1-hydroxypyrene, 1-, 2-, 3- and 4-hydroxyphenanthrene, 3-hydroxybenz (a) anthracene and 3-hydroxybenzo (a) pyrene. *J. Int. Arch. Occup. Environ Health* **2000**, *73*, 270–274. [CrossRef]
55. Jeng, H.A.; Pan, C.H.; Chang-Chien, G.P.; Diawara, N.; Peng, C.Y.; Wu, M.T. Repeated measurements for assessment of urinary 2-naphthol levels in individuals exposed to polycyclic aromatic hydrocarbons. *J. Environ. Sci. Health A Tox. Hazard Subst. Environ. Eng.* **2011**, *46*, 865–873. [CrossRef] [PubMed]

56. Kim, H.; Cho, S.H.; Kang, J.W.; Kim, Y.D.; Nan, H.M.; Lee, C.H.; Lee, H.; Kawamoto, T. Urinary 1-hydroxypyrene and 2-naphthol concentrations in male Koreans. *Int. Arch. Occup. Environ. Health* **2000**, *74*, 59–62. [CrossRef]
57. St Helen, G.; Goniewicz, M.L.; Dempsey, D.; Wilson, M.; Jacob, P., III; Benowitz, N.L. Exposure and kinetics of polycyclic aromatic hydrocarbons (PAHs) in cigarette smokers. *Chem. Res. Toxicol.* **2012**, *25*, 952–964. [CrossRef] [PubMed]
58. Ge, L.; Li, J.; Na, G.; Chen, C.E.; Huo, C.; Zhang, P.; Yao, Z. Photochemical degradation of hydroxy PAHs in ice: Implications for the polar areas. *Chemosphere* **2016**, *155*, 375–379. [CrossRef] [PubMed]
59. Dugheri, S.; Palli, L.; Bossi, C.; Bonari, A.; Mucci, N.; Santianni, D.; Arcangeli, G.; Sirini, P.; Gori, R. Developmnet of an automated LC-MS/MS method for the determination of eight pharmaceutical compounds in wastewater. *Fresenius Environ. Bull.* **2018**, in press.
60. Pacenti, M.; Dugheri, S.; Villanelli, F.; Bartolucci, G.; Calamai, L.; Boccalon, P.; Arcangeli, G.; Vecchione, F.; Alessi, P.; Kikic, I.; et al. Determination of organic acids in urine by solid-phase microextraction and gas chromatography–ion trap tandem mass spectrometry previous 'in sample' derivatization with trimethyloxonium tetrafluoroborate. *Biomed. Chromatogr.* **2008**, *22*, 1155–1163. [CrossRef] [PubMed]
61. Louch, D.; Motlagh, S.; Pawliszyn, J. Dynamics of organic compound extraction from water using liquid-coated fused silica fibers. *Anal. Chem.* **1992**, *64*, 1187–1199. [CrossRef]
62. Moreau, N.M.; Goupry, S.M.; Antignac, J.P.; Monteau, F.J.; Le Bizec, B.J.; Champ, M.M.; Martin, L.J.; Dumon, H.J. Simultaneous measurement of plasma concentrations and 13C-enrichment of short-chain fatty acids, lactic acid and ketone bodies by gas chromatography coupled to mass spectrometry. *J. Chromatogr. B Anal. Technol. Biomed. Life Sci.* **2003**, *784*, 395–403. [CrossRef]
63. Canosa, P.; Rodriguez, I.; Rubí, E.; Cela, R. Optimization of solid-phase microextraction conditions for the determination of triclosan and possible related compounds in water samples. *J. Chromatogr. A* **2005**, *1072*, 107–115. [CrossRef]
64. Xu, D.; Penning, T.M.; Blair, I.A.; Harvey, R.G. Synthesis of phenol and quinone metabolites of benzo[a]pyrene, a carcinogenic component of tobacco smoke implicated in lung cancer. *J. Org. Chem.* **2009**, *74*, 597–604. [CrossRef] [PubMed]
65. McCourt, D.W.; Roller, P.P.; Gelboin, H.V. Tetrabutylammonium hydroxide: A reagent for the base-catalyzed dehydration of vicinal dihydro diols of aromatic hydrocarbons. Implications to ion-pair chromatography. *J. Org. Chem.* **1981**, *46*, 4157–4161. [CrossRef]
66. Duttwyler, S.; Butterfield, A.M.; Siegel, J.S. Arenium Acid-Catalyzed Deuteration of Aromatic Hydrocarbons. *J. Org. Chem.* **2013**, *78*, 2134–2138. [CrossRef] [PubMed]

Sample Availability: Samples of the compounds are all available from the authors.

© 2018 by the authors. Licensee MDPI, Basel, Switzerland. This article is an open access article distributed under the terms and conditions of the Creative Commons Attribution (CC BY) license (http://creativecommons.org/licenses/by/4.0/).

Article

Analysis of Contact Traces of Cannabis by In-Tube Solid-Phase Microextraction Coupled to Nanoliquid Chromatography

Neus Jornet-Martínez, Adrián Ortega-Sierra, Jorge Verdú-Andrés, Rosa Herráez-Hernández * and Pilar Campíns-Falcó

MINTOTA Research Group, Department of Analytical Chemistry, Faculty of Chemistry, University of Valencia, Dr. Moliner 50, 46100 Burjassot, Valencia, Spain; neus.jornet@uv.es (N.J.-M.); aorsie@alumni.uv.es (A.O.-S.); jorge.verdu@uv.es (J.V.-A.); pilar.campins@uv.es (P.C.-F.)
* Correspondence: rosa.herraez@uv.es; Tel.: +34-96-354-4978

Received: 12 August 2018; Accepted: 13 September 2018; Published: 15 September 2018

Abstract: Because of its inherent qualities, in-tube solid-phase microextraction (IT-SPME) coupled on-line to nanoliquid chromatography (nanoLC) can be a very powerful tool to address the new challenges of analytical laboratories such as the analysis of traces of complex samples. This is the case of the detection of contact traces of drugs, especially cannabis. The main difficulties encountered in the analysis of traces of cannabis plants on surfaces are the low amount of sample available (typically < 1 mg), the complexity of the matrix, and the low percentages of cannabinoic compounds in the samples. In this work, a procedure is described for the detection of residues of cannabis on different surfaces based on the responses obtained by IT-SPME coupled to nanoLC with UV diode array detection (DAD) for the cannabinoids Δ^9-tetrahydrocannabinol (THC), cannabidiol (CBD), and cannabinol (CBN); the proposed conditions can also be applied for quantitative purposes through the measurement of the percentage of THC, the most abundant cannabinoid in plants. The method is based on collecting the suspected drug samples with cotton swabs, followed by the extraction of the target compounds by ultrasound assisted extraction. The extracts are then separated and processed by IT-SPME-nanoLC. The proposed approach has been applied to the detection of traces of cannabis in different kind of items (plastic bags, office paper, aluminum foil, cotton cloths, and hand skin). Sample amounts as low as 0.08 mg have been collected and analysed for THC. The selectivity and effect of the storage conditions on the levels of THC have also been evaluated. The percentages of THC in the samples typically ranged from 0.6% to 2.8%, which means that amounts of this compound as low as 1–2 μg were adequately detected and quantified. For the first time, the reliability of IT-SPME-nanoLC for the analysis of complex matrices such as cannabis plant extracts has been demonstrated.

Keywords: in-tube solid-phase microextraction (IT-SPME); nanoliquid chromatography (nanoLC); contact trace analysis; cannabis; THC

---

1. Introduction

The detection and characterization of contact traces of some substances, such as illicit drugs or explosives, is a challenging task that needs to be addressed in some investigations. For example, the presence of drug traces on clothes, packaging, skin, or vehicle interiors may be due to simple contact with the bulk drug during its production, transport, or consumption and may persist for relatively long periods of time. Thus, positive identification and characterization of drugs in these kinds of samples may play a significant role in criminal investigations, particularly in those related to drug trafficking [1–3]. The main difficulty in the analysis of contact traces of drugs is the low amount of available sample (<1 mg) that is often only visible through microscopy. Furthermore, unlike other illicit drug samples (e.g., amphetamine-derived drug street samples), a large number of constituents are present in cannabis plants, and the percentages of the cannabinoic compounds in them are usually low. Cannabis is one of the most important products in the illicit drug market. For example, in Spain, although the private possession of small amounts of cannabis for consumption is allowed, the lucrative sale of this product is illegal and it is the predominant material seized in the context of drug trafficking control activities [4]. Therefore, high sensitivity and selective techniques for the analysis of traces of cannabis are required.

Some spectroscopic techniques have been proposed for the direct detection of residues of drugs on different items. For example, the detection of traces of cocaine on dealers' clothes and cars has been reported using ion mobility spectrometry [1]; Raman spectroscopy and ambient mass spectrometry have been applied to detect cocaine, amphetamine, ketamine, and N-methyl-3,4-methylenedioxy methamphetamine (MDMA) in a variety of fabrics [5–7]. These techniques have proved to be useful for samples with few components, mostly one or two active drugs and some adulterants. However, for the analysis of more complex samples such as cannabis, chromatographic techniques are predominant because the psychoactive ingredients used for identification and quantification have to be separated from the other plant constituents. The samples are generally collected with swabs or wipes that are subsequently treated to extract the target compounds. Finally, the extracts are analysed, typically by liquid chromatography (LC) coupled on-line to mass spectrometry (MS). This has been the approach used for the analysis of traces of cannabis on police station work surfaces as well as in the clothes and hands of workers involved in the custody and/or destruction of drug seizures for assessing the levels of exposure to THC [8–11]. Multiple extraction and recombination of the extracts followed by solvent evaporation is necessary in order to reach the required sensitivity.

The miniaturization of the chromatographic systems (capillary LC and nanoLC) is one of the options available to improve the sensitivity of chromatographic analysis [12]. The sensitivity can be further increased if an on-line preconcentration technique such as in-tube solid-phase microextraction (IT-SPME) is used [13,14]. In this form of microextraction, the analytes are concentrated in an extractive capillary packed or coated with a proper sorbent. Although there are different modalities to achieve the extraction, the employment of an extractive capillary as the loop of the injection valve of the liquid chromatograph (in-valve IT-SPME) is one of the most attractive options for organic analytes. This is because the target compounds are retained in the capillary during sample loading and sent to the analytical column with the mobile phase when changing the valve position. In such a way, relatively large volumes of the sample can be loaded into the system until the required amount of the target compound is retained in the capillary. The potential of IT-SPME coupled to capillary LC has been extensively documented for a variety of organic compounds [13–15], and more recently, IT-SPME coupled to nanoLC has been applied to the analysis of some pollutants in water samples [16].

In the present work, we describe a new method for the detection and quantification of contact traces of cannabis in different kinds of surfaces using in-valve IT-SPME coupled on-line to nanoLC. The method is based on the employment of cotton swabs for collecting the suspected sample, the extraction of the cannabinoids by treating the cotton swabs in an ultrasonic bath, and the direct processing of the extracts by IT-SPEM-nanoLC. Cotton swabs were selected for sampling, as it is the methodology commonly used to collect residues of drugs from surfaces [8–11]. Δ^9-tetrahydrocannabinol (THC), cannabidiol (CBD), and cannabinol (CBN) were the target analytes selected; for quantitative purposes, THC was the only compound used because it was the predominant cannabinoid in plants. The IT-SPME and chromatographic conditions were previously optimized using extracts obtained from cannabis plants. The proposed approach has been successfully used for the detection of residues of cannabis plants (<1 mg). The selectivity towards other plants (infusion herbs, tobacco) and the effects of storage conditions and item type have been studied.

2. Results

2.1. Chromatographic Analysis

Conditions used to effect IT-SPME of the target cannabinoids were selected according to the results obtained in our previous works. A polydimethylsiloxane-based coated capillary (TRB5) was used because this phase provided satisfactory results in the extraction of compounds of similar polarity [15]; the sample volume and the capillary length were selected taking into account the dimensions of the chromatographic column and the mobile-phase flowrate [16]. Initially, different experiments were carried out in order to find chromatographic conditions suitable for the separation of THC, CBD, and CBN from other plant constituents. The selected elution conditions allowed a satisfactory separation of CBD (retention time, t_r = 13.3 min), CBN (t_r = 15.1 min), and THC (t_r = 16.2 min) from the rest of the components extracted from the plants, as most of them were expected to elute at shorter retention times [17–19].

The quantitative performance of IT-SPME-nanoLC was studied by processing aqueous solutions of the analytes. The results obtained are listed in Table 1. As it can be observed, satisfactory linearity was found for the three compounds studied up to concentrations of 100 ng/mL. The instrumental limits of detection (LODs), established for each analyte as the concentration that provided a signal-to-noise ratio (S/N) of 3, were 2 ng/mL for THC and 5 ng/mL for the other compounds, and the limits of quantification (LOQs) (established for an S/N of 10) were 8–15 ng/mL.

Table 1. Quantitative performance of the proposed method (values established from aqueous standard solutions of the target compounds processed by in-tube solid-phase microextraction coupled on-line to liquid chromatography (IT-SPME-nanoLC)).

Compound	Linearity [1], y = ax + b			n	LOD (ng/mL)	LOQ (ng/mL)
	$a \pm s_a$ (mL/ng)	$b \pm s_b$	r^2			
THC	24 ± 1	−270 ± 80	0.98	7	2	8
CBD	28 ± 2	−300 ± 100	0.990	7	5	15
CBN	7.8 ± 0.5	−10 ± 30	0.990	6	5	15

[1] Tested up to concentrations of 100 ng/mL of each compound.

The analytical performance of the proposed conditions was considered satisfactory and, therefore, applied to evaluate the presence and concentrations of THC, CBD, and CBN in the extracts obtained from cannabis plants.

2.2. Analysis of Extracts of Cannabis Plants

Four cannabis samples (M1–M4) were used throughout the study. These samples were analysed in order to estimate the content in cannabinoids for subsequent comparison with the results of the studies with residues. Samples were roughly homogenized manually, and portions of 10 mg were subjected to extraction with 3 mL of a mixture of methanol and chloroform (9:1 (v/v)). Extractions were performed in an ultrasonic bath for 15 min as proposed in [16]. The liquid phase was then separated and filtrated (<0.20 µm), and portions of 10 µL were analysed by IT-SPME-nanoLC. As the amounts of cannabinoids vary with the storage conditions, portions of the samples that were dried at 50 °C for 6 days were also analysed.

The direct injection of the collected extracts saturated the detector, particularly at retention times of 9–13 min, where most matrix components eluted. Moreover, the peak areas obtained for THC were much higher than those obtained for standards at concentrations within the linear range (see Table 1). It has to be noted that for in-valve IT-SPME and for a given analyte and extractive phase, the extraction efficiency is mainly determined by the solvent sample composition [13,14]. For the extraction of low-medium polarity compounds, such as the cannabinoids used in the present study with apolar coatings (such as TRB 5), the analytes must be loaded in the capillary in a water-rich eluent [20], otherwise, they do not interact with the extractive coating and are excluded from the capillary during sample loading. For these reasons, in the present study, the extracts obtained from the plants were diluted with ultrapure water before being processed by IT-SPME.

A dilution factor of 1:100 led to suitable chromatographic profiles and adequate peak areas for the predominant cannabinoid (THC). It was observed that the peak of THC increased after drying at 50 °C, most probably due to the loss of humidity. In all the samples assayed, the peaks corresponding to the other two cannabinoids, CBD and CBN, were much lower than that of THC, and in some of the samples, the concentration of CBN was below its LOQ (see Table 1). Consequently, THC was the only compound used for quantitative studies.

The percentages of THC in the plants were estimated from the peak areas measured for this compound in the collected extracts and the calibration equation of Table 1, taking into account the dilution factor. As conditions for the extraction were those proposed in a previously validated method [18], it was assumed that extraction of cannabinoids from the plants was quantitative. The results obtained for plant extracts after applying different dilution factors (1:100–1:200) are listed in Table 2.

Table 2. Percentages of Δ^9-tetrahydrocannabinol (THC) in cannabis plants stored at ambient conditions and at 50 °C.

Storage Conditions	Percentage of THC (%) [1]			
	M1	M2	M3	M4
Ambient	0.4, 0.8, 0.8	0.4, 0.5	0.4, 0.8	0.6, 0.7
50 °C	1.0, 1.0, 0.8	3.2, 2.8	4.8, 4.7	3.2, 2.6

[1] Values obtained from three independent assays for sample M1 and for two replicates of the same plant extract in samples M2–M4.

As observed in this table, the percentages of THC found for samples exposed to ambient conditions were <1%. The values obtained for three independent assays for one of the samples (M1) were 0.4%, 0.4%, and 0.8%. This variability (relative standard deviation, RDS = 23%) can be explained by the heterogeneity of the sample. It has to be noted that all samples were processed as they were expected to be consumed by users, that is, without being homogenized with lab equipment such as mortars or mills. The consecutive analysis of three aliquots of the same extract led to peak areas of THC with a relative standard deviation of 4%. In samples dried at 50 °C, the percentages of THC slightly increased (up to 4.8%) most probably due to the loss of humidity. These values are about the same order as those reported by other authors [17–19]. As for the samples exposed to ambient conditions, the precision was evaluated by performing three independent analyses of the same sample (M1), with the resulting relative standard deviation (RSD) of 12%. In another set of experiments, one of the extracts was spiked with standard solutions of the analytes (added concentration, 50 ng/mL). Then, the increments on the peak areas were used to estimate the added concentration from the calibration equations of Table 1. The calculated concentration to added concentration ratios were used to calculate the recoveries, and the mean values were 91%, 88%, and 128% for CBN, CBD, and THC, respectively. It was concluded that the analyte responses were not substantially affected by the matrix and, therefore, the values presented in Table 1 were valid for the analysis of cannabis plant extracts.

2.3. Analysis of Residues of Cannabis on Surfaces

2.3.1. Collection and Extraction Procedure

In order to develop a protocol for the analysis of contact traces of cannabis, different studies were carried out using plastic as a model surface, more specifically, polyethylene bags (6 × 4 cm). The bags were previously put into contact with cannabis by placing about 1.0 g of plant inside them and pressing it. Then, the bags were emptied by shaking them repeatedly, so that most parts of the plants were removed (only small particles could be visually detected). Next, the inner surface of the bag was wiped with a cotton swab in order to collect possible residues of the plant. The amount of residue collected was calculated by the difference of mass of the swabs before and after the wiping step. As an illustrative example, Figure 1 displays images of one of the bags (a) and the swab obtained after wiping the bag as well as an unused swab (blank) (b). The mass of the residues collected during the study ranged from 0.08 to 0.87 mg. It has to be noted that, unlike other procedures, dry swabs were used to collect the traces of cannabis in order to avoid error during weight operations due to evaporation of the solvent.

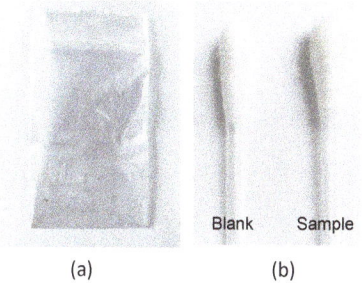

Figure 1. Images obtained in the studies for the analysis of residues of cannabis: (**a**) a plastic bag after the removal of the plant; (**b**) swabs after collecting the residue; and an unused swab (blank).

Next, the cotton tips of the swabs were introduced into 2-mL glass vials, and after adding 1 mL of the extracting solvent (the cotton tip of the swab was completely soaked), the vials were introduced into an ultrasonic bath for 15 min. Finally, the liquid phase was removed and filtered for further processing.

As explained above, the high sensitivity attainable by IT-SPME-nanoLC allowed the detection and quantification of the main cannabinoids in only 10 mg of plants, making the dilution of the collected extracts necessary. However, in the analysis of traces of cannabinoids, much lower amounts of samples are expected to be available. Therefore, dilution factors as low as possible should be applied. In an attempt to eliminate intermediate dilutions and since water-rich media are necessary for IT-SPME, water and different water–methanol mixtures were tested for the extraction of THC from the residues of cannabis collected on the swabs. The results were compared with those obtained by using methanol and chloroform (9:1, v/v).

No THC was detected in the chromatograms obtained when using water for extraction. According to previous works, a water-methanol mixture (5:1, v/v) is suitable for IT-SPME with a PDMS-based coating, such as that used in the present study [20]. However, with this solvent, the extraction of THC was also unacceptably low, which can be explained by its high hydrophobicity (log $k_{octanol/water}$ = 6.97) [21]. Much better results (higher peak areas for the analytes) were observed when the extraction was carried out with methanol, followed by the dilution of the extracts with water. On the other hand, no significant increments of the peak areas were observed when methanol:chloroform was used for extraction. For simplicity, methanol was selected for extraction in further assays.

As an example, Figure 2 shows the chromatograms obtained for one of the samples (M1). The amount of sample collected with the swab was 0.82 mg, and the methanolic extract was diluted with ultrapure water, with a methanol:water proportion of 1:5 (v/v). This figure also shows the good concordance between the UV spectra recorded for the peaks of the suspected analytes and those obtained for standard solutions of the analytes. Therefore, the presence of cannabis in the collected trace sample could be properly confirmed.

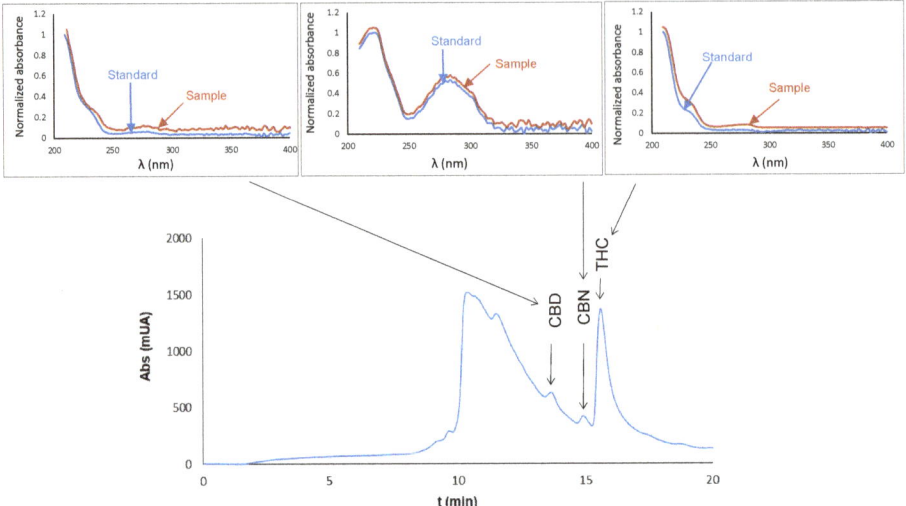

Figure 2. Chromatogram obtained in the analysis of residues of cannabis (sample M1, 0.82 mg), and comparison of the spectra obtained for the peaks of the suspected cannabidiol (CBD), cannabinol (CBN), and THC and those of standard solutions. The methanolic extract collected after processing the swab was diluted 1:5 with ultrapure water before the IT-SPME-nanoLC step. Conditions used for IT-SPME and chromatographic analysis were identical to those indicated in Sections 4.2 and 4.3, respectively.

2.3.2. Type of Surface

The proposed procedure was applied to detect residues of cannabis on other surfaces, namely, aluminum foil (7 × 10 cm), office paper (5 × 5 cm), a piece of cotton cloth (2 × 2 cm), and skin (hand). In the assays with the office paper, a piece of cloth and hand small particles of cannabis were visually detected after the removal of the drug; no traces were detected by naked eye on the aluminum, which was the less porous material (see Figure 3).

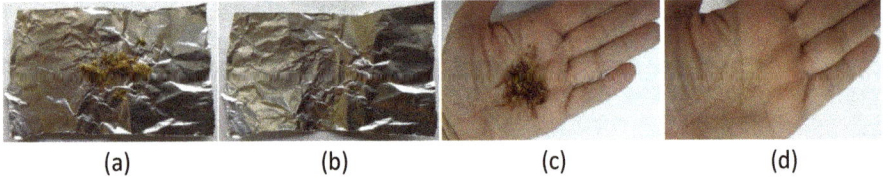

(a)   (b)   (c)   (d)

Figure 3. Images obtained in the studies for the analysis of residues of cannabis on: aluminum foil in contact with cannabis (**a**) and after the removal of the plant (**b**); hand of a volunteer in contact with cannabis (**c**); and after the removal of the plant (**d**).

Possible residues of cannabis were collected with swabs, and the swabs were treated with 1 mL of methanol as described above. For comparison purposes, a piece of the cotton fabric (2 × 2 cm) was also directly immersed in 2 mL of methanol, and the extract was treated the same as the solutions obtained for the swabs.

The amounts of samples collected were of about the same order as those collected in the assays with plastic bags, but the amount of cannabis that remained bound to the piece of cloth was much higher (4.10 mg). The three cannabinoids could be detected and identified, although the concentration of CBN in some of the samples was below its LOQ. Examples of the chromatograms obtained are given in Figure 4.

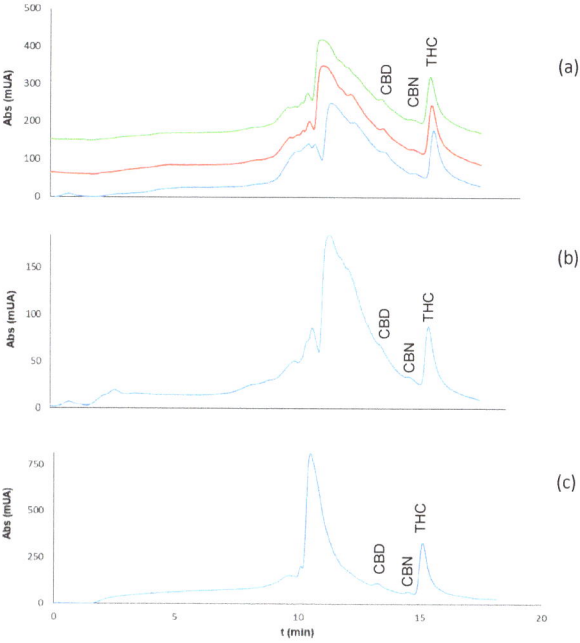

Figure 4. Chromatograms obtained for the analysis of residues of cannabis (sample M1) in: (**a**) aluminum foil, (**b**) cellulose paper, and (**c**) and skin. The sample amounts collected with the swabs were 0.27, 0.10, and 0.25 mg in (**a**–**c**), respectively; the extracts were diluted 1:20 with ultrapure water before the IT-SPME–nanoLC step in (**a**,**b**) and 1:5 in (**c**). Conditions for sampling and extraction were identical to those indicated in Section 2.3.1; conditions used for IT-SPME and chromatographic analysis are indicated in Sections 4.2 and 4.3, respectively.

2.3.3. Quantitation of THC

The percentage of THC in the collected residues was established from the peak areas of the chromatograms obtained. The dilution factor was selected according to the amount of residue collected with the swabs; dilution factors of 1:20–1:100 were adequate in most of the samples to adjust the concentrations of THC to the linear concentration interval (Table 1). However, due to the high amount of cannabis collected in the assay with the piece of cloth, a dilution factor of 1:250 had to be applied.

The results obtained in quantitative assays are summarized in Table 3. As observed in this table, the concentrations of THC measured were in the 41–99 µg/mL concentration interval. These concentrations correspond to percentages of THC in the residues ranging from 0.6% to 2.8%. These values were similar or slightly higher than those found for the plants, which indicates that the parts of the plant with a higher tendency to adhere to surfaces are those with higher THC contents.

The intraday precision of the method was estimated for the entire procedure from three independent consecutive analyses of the same cannabis sample (M1). The RSD obtained was 17%. The interday precision was calculated from the analysis of residues of the sample M1 on three different days, resulting in RSD 27% (n = 3). The reproducibility was also tested for the consecutive injection of aliquots of the same extract. The RSD values obtained were ≤18% for bags, 4% for aluminum foil (see also Figure 3a), 6% for cellulose paper, 21% for cloth, and 7% for skin. As for the assays with plants, a slight increment in the percentage of THC was found for samples previously dried at 50 °C.

Table 3. Mean concentrations of THC found in the analysis with residues of THC (*n* = 3), and their equivalence in amount of THC collected and percentage of THC in the samples.

Item	Cannabis Sample	Amount of Sample Collected (mg)	Dilution Factor	Concentration of THC Measured (ng/mL)	Estimated Amount of THC Collected (µg)	Percentage of THC in the Sample (%)
Plastic bag	M1	0.82	1:100	79	8.2	1.0
Plastic bag	M2	0.42	1:50	78	3.8	0.9
Plastic bag	M3	0.87	1:400	61	24.4	2.8
Plastic bag	M4	0.38	1:100	99	9.9	2.6
Plastic bag	Mixture of plants	0.10	1:20	58	1.2	1.2
Plastic bag	Mixture of plants	0.08	1:100	41	4.1	5.1
Aluminum foil	M1	0.27	1:20	82	1.6	0.6
Office paper	M1	0.10	1:20	75	1.5	1.5
Piece of cotton cloth	M1	4.28	1:250	71	34.2	0.4
Skin	M1	0.20	1:20	94	1.9	0.9

2.3.4. Selectivity

The proposed method was applied to other kinds of samples in order to evaluate the selectivity; the products tested were three herbal infusions (HI1–HI3) and tobacco (see Figure 5a). A mixture of tobacco:cannabis in a proportion of masses of 4:1 was also tested. This study was carried out with plastic bags.

The chromatograms obtained for all the herbal infusions and tobacco tested were free of peaks in the region where CND, CBN, and THC eluted. Examples of the chromatograms obtained are depicted in Figure 5b. As observed in this figure, in the tobacco and cannabis mixture, the three analytes were detected, with their respective peak areas being about a quarter of those found when the assay was carried out with cannabis (chromatogram also shown in Figure 5b).

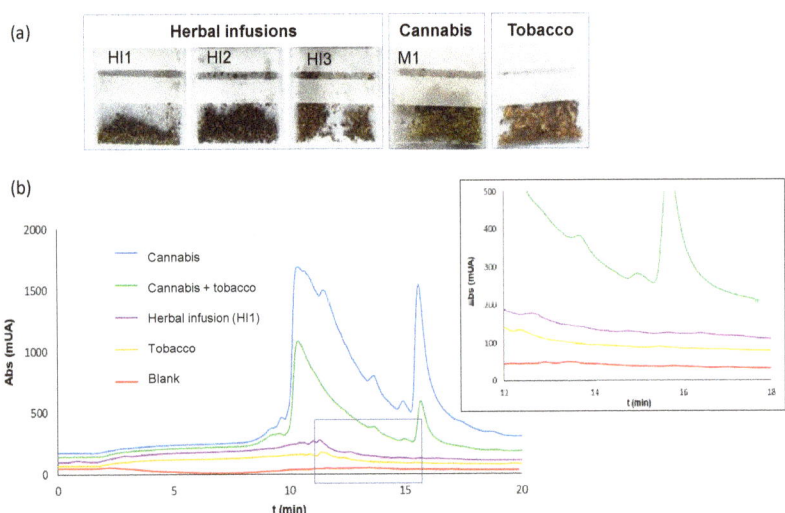

Figure 5. (**a**) Images of the bags with the different products used in the selectivity study: herbal infusions (HI1–HI3), tobacco, and cannabis (M1). (**b**) Chromatograms obtained for the analysis of residues collected from some of the bags (**a**) after the removal of the products: cannabis (M1), tobacco, a mixture of M1 and tobacco (1:4, m/m), herbal infusion (HI1), and a blank. The extracts were diluted 1:20 with ultrapure water before being processed by IT-SPME-nanoLC. Conditions for sampling and extraction were identical to those indicated in Section 2.3.1; conditions used for IT-SPME and chromatographic analysis were identical to those indicated in Sections 4.2 and 4.3, respectively.

In view of these results, it was concluded that the proposed approach was suitable for the differentiation of residues of cannabis from noncannabinoid plants.

3. Discussion

The analysis of cannabis has been undertaken from different perspectives such as the characterization of the varieties of plants or the quantification of the active components [17–19], typically using LC-DAD or LC-MS(/MS). Chromatographic techniques have been also used for confirmatory analysis of samples that give positive results in screening tests for drugs, for example, in colorimetric tests [22]. Those methods are not suitable for the analysis of contact traces of cannabis because the amounts of sample required are much higher, typically 100–300 mg [15–17]. Only a few procedures have been described for the detection and quantification of THC in surfaces so far and have been generally addressed to the evaluation of the exposure to cannabis on work surfaces [9–12]. Those methods allow the detection and quantification of THC (LOQ, 5 ng/mL) with analyte detectability similar to that of the present approach, but they involve wiping larger areas (which may not always be possible) and multiple extraction/evaporation of the extracts (3 × 8 mL of methanol followed by evaporation to dryness).

In the present study, we have illustrated the potential of IT-SPME coupled to nanoLC for the detection and characterization of contact traces of complex materials, such as cannabis plants. For the first time, it has been demonstrated that IT-SPME-nanoLC can be used for the analysis of extracts obtained for such samples with adequate selectivity. The system stability was also suitable; more than 100 extracts were analysed by the proposed method with the same capillary without observing deterioration in its retention properties or in the system background pressure. The proposed approach uses green extraction techniques, such as ultrasound assisted extraction, and only 1 mL of methanol per sample is necessary. It has been applied to the detection and quantification of amounts 1–2 µg of THC from areas typically lower than 50 cm^2, but lower amounts could be detected. As an example, considering the instrumental detection limit (2 ng/mL), a percentage of THC of 1% in the sample, and a dilution factor 1:20, the method could detect THC in residues of cannabis of only 4 µg.

Throughout our study, we observed that most parts of the residues were collected with swabs. However, in more porous surfaces, incomplete removal of the sample could occur, and thus, the amount of residue collected would be affected by parameters such as pressure or time of sampling. In such cases, the proposed procedure could not be applied to estimate the mass of residue, although the positive identification of cannabis would be still possible. If required for a more complete characterization of the sample, the proposed approach could be applied to the quantification of CBD and CBN by processing the undiluted extracts.

Because of the low amount of sample necessary, the protocol can be considered minimally invasive. Moreover, as only 10 µL of diluted extracts is necessary, most part of the original extract can be stored and used in further studies, if required. The concordance of the percentages of THC found in the study with plants (Table 2) and with residues of the same plants (Table 3) confirms the reliability of the proposed approach.

In summary, this work illustrates the potential utility of new integrated techniques such as IT-SPME coupled on-line to nanoLC for contact trace analysis of complex samples [23].

4. Materials and Methods

4.1. Chemicals and Solutions

All reagents were of analytical grade. Standards of Δ^9-trans-tetrahidrocannabinol, cannabidiol, and cannabinol (solutions in methanol) were obtained from Sigma-Aldrich (St. Louis, MO, USA). Acetonitrile, chloroform, and methanol were of HPLC grade (VWR, Radnor, PA, USA).

Intermediate stock standard solutions of the analytes (1 mg/mL) were prepared by diluting the commercial standards with acetonitrile and kept at $-20\,°C$ until use. Working solutions were prepared by diluting the stock solutions with ultrapure water.

4.2. Chromatographic Conditions

Chromatographic analyses were performed using an Agilent 1260 Infinity nanoLC chromatograph equipped with a quaternary nanopump, a six-port microscale manual injector (Rheodyne, Rohnert Park, CA, USA), and a UV-Vis diode array detector with an 80-nL nanoflow cell (Agilent, Waldbronn, Germany). The detector was coupled to a data system (Agilent, ChemStation) for data acquisition and treatment. The analytical signal was recorded between 190 and 400 nm and monitored at 210 nm. A Zorbax 300SB C18 (50 × 0.075 mm id, 3.5 µm particle size) column (Agilent) was used for separation.

The mobile phase was a mixture of water and acetonitrile in gradient elution mode. The percentage of acetonitrile in the mobile phase was linearly increased from 55% at 0–5 min to 75% at 10 min, which was then kept constant until 12 min; finally, the acetonitrile percentage was linearly decreased to reach a percentage of 55% at 15 min and remained constant until the end of the run. The flow rate was 0.5 µL/min. Solvents were filtered through 0.22-µm nylon membranes before use (Teknokroma, Barcelona, Spain).

4.3. IT-SPME Conditions

For the IT-SPME a segment of a TBR-5 column (95% polydimethyl siloxane, 5% polydiphenylsiloxane), 15 cm length, 0.320 mm id, and 3 µm coating thickness (Teknokroma, Barcelona, Spain), was used. For connecting the extractive capillary to the valve, a 2.5-cm sleeve of 1/6 i.n. polyether ether ketone (PEEK) tubing (1/6 i.n. PEEK nuts and ferrules (Teknokroma)) was used.

The volume of sample loaded into the IT-SPME device was 10 µL.

4.4. Analysis of Cannabis Plants

Four cannabis samples (M1–M4) collected from different plants within two to three months before the study were used. The samples, intended for self-consumption, were voluntarily donated by users who were previously informed of the aim of the study. The samples were stored in plastic bags in the dark until use.

For the extraction of cannabinoids, accurately weighted portions of the samples (10 mg) were placed in 10-mL glass vials, and then 3 mL of a mixture of methanol and chloroform (9:1, v/v) was added. The vials were closed with a screw cap and placed into an ultrasonic bath (300 W, 40 kHz, Sonitech, Guarnizo, Spain) for 15 min. Next, a portion of about 1 mL of the extraction solvent was removed and filtered through 0.22-µm nylon membranes (Teknokroma). Unless otherwise stated, the extracts were properly diluted with ultrapure water and processed by IT-SPEM-nanoLC. Each sample was assayed in duplicate.

4.5. Analysis of Traces of Cannabis on Surfaces

The presence of traces of cannabis on the surface of different types of materials was evaluated: plastic (polyethylene), cellulose office paper, aluminum foam, cotton cloth, and hand skin. For this purpose, the items were put into contact with portions of cannabis of about 1 g by exerting pressure with hands, and then the plant was removed by shaking the item repeatedly. Next, the surface of the item was wiped with dry swabs for 20–30 s in different directions in order to collect possible traces of cannabis; the surface areas wiped were of about 20–50 cm^2. The swabs were accurately weighted before and after sampling, so the amount of sample collected was calculated as the difference of masses. After sample collecting, the swabs were placed in glass vials and the target compounds were extracted in 1 mL of methanol (unless otherwise stated) in an ultrasonic bath for 15 min. The extracts were properly filtered and diluted with ultrapure water and then processed by IT-SPEM-nanoLC. Unless otherwise stated, each extract was processed in duplicate.

Cotton swabs purchased from local supermarkets were used for collecting the samples.

For the selectivity studies, three different herbal infusions—lime-orange (HI1), pennyroyal (HI2), and valerian (HI3)—were used. These products were purchased from local supermarkets.

Author Contributions: Data curation, N.J.-M., A.O.-S., J.V.-A., R.H.-H., and P.C.-F.; Formal analysis, N.J.-M., A.O.-S., J.V.-A., R.H.-H., and P.C.-F.; Funding acquisition, N.J.-M., J.V.-A., R.H.-H., and P.C.-F.; Investigation, N.J.-M., A.O.-S., J.V.-A., R.H.-H., and P.C.-F.; Methodology, N.J.-M., A.O.-S., J.V.-A., R.H.-H., and P.C.-F.; Validation, N.J.-M., A.O.-S., J.V.-A., R.H.-H., and P.C.-F.; Writing—original draft, N.J.-M., A.O.-S., J.V.-A., R.H.-H., and P.C.-F.

Funding: The authors are grateful to EU FEDER and the Spanish AEI (project CTQ2017-90082-P) and the Generalitat Valenciana (PROMETEO 2016/109) for the financial support received. N.J.-M. expresses her gratitude to the Generalitat Valenciana and the EU for her postdoctoral grant (APOSTD2016/113).

Conflicts of Interest: The authors declare no conflict of interest. The funders had no role in the design of the study; in the collection, analyses, or interpretation of data; in the writing of the manuscript; or in the decision to publish the results.

References

1. Keller, T.; Keller, A.; Tutsch-Bauer, E.; Monticelli, F. Application of ion mobility spectrometry in cases of forensic interest. *Forensic Sci. Int.* **2006**, *161*, 130–140. [CrossRef] [PubMed]
2. Castillo-Peinado, L.S.; de Castro, M.D.L. An overview on forensic analysis devoted to analytical chemists. *Talanta* **2017**, *167*, 181–192. [CrossRef] [PubMed]
3. Pawlowski, W.; Matyjasek, L.; Karpińska, M. Detection of contact traces of powdery substances. *J. Forensic Sci.* **2017**, *62*, 1028–1032. [CrossRef] [PubMed]
4. Klimer, B. New Developments in Cannabis Regulations. Available online: http://www.emcdda.europa.eu/document-library/new-developments-cannabis-regulation_en (accessed on 1 September 2018).
5. Ali, E.M.A.; Edwards, G.M.; Hargreaves, M.D.; Scowen, I.J. In-situ detection of drugs-of-abuse on clothing using confocal Raman microscopy. *Anal. Chim. Acta* **2008**, *615*, 63–72. [CrossRef] [PubMed]
6. West, M.J.; Went, M.J. The spectroscopic detection of drugs of abuse on textile fibers after recovery with adhesive lifters. *Forensic Sci. Int.* **2009**, *189*, 100–103. [CrossRef] [PubMed]
7. Talaty, N.; Mulligan, C.C.; Justes, D.R.; Jackson, A.U.; Noll, R.J.; Cooks, R.G. Fabric analysis by ambient mass spectrometry for explosive and drugs. *Analyst* **2008**, *133*, 1532–1540. [CrossRef] [PubMed]
8. Martyny, J.W.; Serrano, K.A.; Schaeffer, J.W.; Joshua, W.; Van Dyke, M.V. Potential exposures associated with indoor marijuana growing operations. *J. Occup. Environ. Hyg.* **2013**, *10*, 622–639. [CrossRef] [PubMed]
9. Doran, G.S.; Deans, R.; De Filippis, C.; Kostakis, C.; Howitt, J.A. Work place drug testing of police officers after THC exposure during large volume cannabis seizures. *Forensic Sci. Int.* **2017**, *275*, 224–233. [CrossRef] [PubMed]
10. Doran, G.S.; Deans, R.; De Filippis, C.; Kostakis, C.; Howitt, J.A. The presence of licit and illicit drugs in police stations and their implications for workplace drug testing. *Forensic Sci. Int.* **2017**, *278*, 125–136. [CrossRef] [PubMed]
11. Doran, G.S.; Deans, R.; De Filippis, C.; Kostakis, C.; Howitt, J.A. Quantification of licit and illicit drugs on typical police station work surface using LC-MS/MS. *Anal. Methods* **2017**, *9*, 198–210. [CrossRef]
12. Nazario, C.E.D.; Silva, M.R.; Franco, M.S.; Lanças, F.M. Evolution in miniaturized column liquid chromatography instrumentation and applications: An overview. *J. Chromatogr. A* **2015**, *1421*, 18–37. [CrossRef] [PubMed]
13. Moliner-Martinez, Y.; Herráez-Hernández, R.; Verdú-Andrés, J.; Molins-Legua, C.; Campíns-Falcó, P. Recent advances of in-tube solid-phase microextraction. *Trends Anal. Chem.* **2015**, *71*, 205–213. [CrossRef]
14. Serra-Mora, P.; Moliner-Martínez, Y.; Molins-Legua, C.; Herráez-Hernández, R.; Verdú-Andrés, J.; Campíns-Falcó, P. Trends in online in-tube solid phase microextraction. *Compr. Anal. Chem.* **2017**, *76*, 427–461. [CrossRef]
15. Serra-Mora, P.; Rodríguez-Palma, C.E.; Verdú-Andrés, J.; Herráez-Hernández, R.; Campíns-Falcó, P. Improving the on-line extraction of polar compounds by IT-SPME with silica nanoparticles modified phases. *Separations* **2018**, *5*, 10. [CrossRef]

16. Serra-Mora, P.; Jornet-Martínez, N.; Moliner-Martínez, Y.; Campíns-Falcó, P. In tube-solid phase microextraction-nano liquid chromatography: Application to the determination of intact and degraded polar triazines in waters and recovered struvite. *J. Chromatogr. A* **2017**, *1513*, 51–58. [CrossRef] [PubMed]
17. Ambach, L.; Penitschka, F.; Broillet, A.; König, S.; Weinmann, W.; Bernhard, W. Simultaneous quantification of delta-9-THC, THC-acid, CBN and CBD in seized drugs using HPLC-DAD. *Forensic Sci. Int.* **2014**, *243*, 107–111. [CrossRef] [PubMed]
18. Gul, W.; Gul, S.W.; Radwan, M.M.; Wanas, A.S.; Mehmedic, Z.; Khan, I.I.; Sharaf, M.H.M.; ElSohly, M.A. Determination of 11 cannabinoids in biomass and extracts of different varieties of cannabis using high-performance liquid chromatography. *J. AOAC Int.* **2015**, *98*, 1523–1528. [CrossRef] [PubMed]
19. Patel, B.; Wene, D.; Fan, Z. Qualitative and quantitative measurement of cannabinoids in cannabis using modified HPLC/DAD method. *J. Pharm. Biomed. Anal.* **2017**, *146*, 15–23. [CrossRef] [PubMed]
20. Campíns-Falcó, P.; Verdú-Andrés, J.; Sevillano-Cabeza, A.; Molins-Legua, C.; Herráez-Hernández, R. New micromethod combining miniaturized matrix solid-phase dispersion and in-tube-in-valve solid phase microextraction for estimating polycyclic aromatic hydrocarbons in bivalves. *J. Chromatogr. A* **2008**, *1211*, 13–21. [CrossRef] [PubMed]
21. Molnar, A.; Lewis, J.; Fu, S. Recovery of spiked Δ9-tetrahydrocannabinol in oral fluid from polypropylene containers. *Forensic Sci. Int.* **2013**, *227*, 69–73. [CrossRef] [PubMed]
22. Philp, M.; Fu, S. A review of chemical 'spot' tests: A presumptive illicit drug identification technique. *Drug Test. Anal.* **2017**, 1–14. [CrossRef] [PubMed]
23. Stoney, D.A.; Stoney, P.L. Critical review of forensic trace evidence analysis and the need for a new approach. *Forensic Sci. Int.* **2015**, *251*, 159–170. [CrossRef] [PubMed]

Sample Availability: Sample Availability: Not available.

© 2018 by the authors. Licensee MDPI, Basel, Switzerland. This article is an open access article distributed under the terms and conditions of the Creative Commons Attribution (CC BY) license (http://creativecommons.org/licenses/by/4.0/).

Article

Skin Permeation of Solutes from Metalworking Fluids to Build Prediction Models and Test A Partition Theory

Jacqueline M. Hughes-Oliver [1], Guangning Xu [2] and Ronald E. Baynes [3,*]

1. Department of Statistics, North Carolina State University, Raleigh, NC 27695-8203, USA; hughesol@ncsu.edu
2. Wells Fargo and Company, Charlotte, NC 28202-0901, USA; gxu@ncsu.edu
3. Center for Chemical Toxicology Research & Pharmacokinetics, Department of Population Health and Pathobiology, College of Veterinary Medicine, North Carolina State University, 1060 William Moore Dr., Raleigh, NC 27607, USA
* Correspondence: ronald_baynes@ncsu.edu; Tel.: +1-919-513-6261

Academic Editors: Constantinos K. Zacharis and Paraskevas D. Tzanavaras
Received: 27 October 2018; Accepted: 23 November 2018; Published: 24 November 2018

Abstract: Permeation of chemical solutes through skin can create major health issues. Using the membrane-coated fiber (MCF) as a solid phase membrane extraction (SPME) approach to simulate skin permeation, we obtained partition coefficients for 37 solutes under 90 treatment combinations that could broadly represent formulations that could be associated with occupational skin exposure. These formulations were designed to mimic fluids in the metalworking process, and they are defined in this manuscript using: one of mineral oil, polyethylene glycol-200, soluble oil, synthetic oil, or semi-synthetic oil; at a concentration of 0.05 or 0.5 or 5 percent; with solute concentration of 0.01, 0.05, 0.1, 0.5, 1, or 5 ppm. A single linear free-energy relationship (LFER) model was shown to be inadequate, but extensions that account for experimental conditions provide important improvements in estimating solute partitioning from selected formulations into the MCF. The benefit of the Expanded Nested-Solute-Concentration LFER model over the Expanded Crossed-Factors LFER model is only revealed through a careful leave-one-solute-out cross-validation that properly addresses the existence of replicates to avoid an overly optimistic view of predictive power. Finally, the partition theory that accompanies the MCF approach is thoroughly tested and found to not be supported under complex experimental settings that mimic occupational exposure in the metalworking industry.

Keywords: leave-one-solute-out (LOSO) cross-validation; leave-one-out (LOO) cross-validation; linear free-energy relationship (LFER) model; membrane-coated fiber (MCF) approach; partition coefficient; quantitative structure-activity relationship (QSAR); metalworking fluid

1. Introduction

The assessment of skin permeation of chemical solutes can be used to inform scientific research and regulatory agencies in the risk management of chemical solutes that may be of concern especially for occupational exposures [1–3]. For example, in the metalworking industry, certain performance enhancing solutes such as corrosive inhibitors, emulsifiers, and biocides/preservatives are often added to the metalworking fluids (MWF). Contact with these industrial fluids containing some or all of these performance additives could sometimes cause skin irritation or even more harmful consequences [4–7]. Thus, it is of interest to study the permeation capability of the added solutes through skin, in the hopes of finding less permeable solutes that can be used in metalworking fluids.

Unfortunately, conducting skin absorption studies of the many industrial chemicals and many formulations can be very expensive, and many efforts have been made to mimic the skin using synthetic membranes [8–13]. Xia et al. [14] proposed an intriguing technique, called the membrane-coated fiber (MCF) assay approach, to simulate the different molecular interactions in skin permeation by different types of materials. In this approach, an MCF is used as the absorption membrane to determine partition coefficients, namely the ratio of the concentration of solute partitioning to the MCF relative to the concentration of solute not partitioning to the MCF. The partition coefficient is a measurement of the strength of molecular interaction that governs percutaneous absorption processes. Assuming that the MCF adequately represents skin absorption, larger values of partition coefficients suggest greater levels of absorption of the solute into skin, translating to possible health implications during the metalworking processes.

To relate the dermal permeability of a solute to the solute's chemical structure or properties, it is very common practice to develop and study a relevant quantitative structure-activity relationship (QSAR) model as classically demonstrated by [15] and [16], and also demonstrated more recently in studies more relevant to this paper ([17–19]). Many commonly used QSAR models are linear regression models that use the biological activity (partition coefficients, permeation coefficients, etc.) as the response variable and the molecular descriptors as predictors. The linear free-energy relationship (LFER) model of [20] is a particular type of QSAR model that is widely used in modeling results from dermal permeability studies. The LFER model is easy to use and interpret, however, when experimental conditions are complex, a simple LFER model may not be able to appropriately account for the observed variability, leading to a model with poor fit statistics and low predictive power. Xu et al. [19] expanded the LFER model to account for the heterogeneity introduced by experimental factors, in which one set of partial slopes are defined for each experimental condition. This model proved to be useful, improving both the model fit statistics and predictive power. This article pursues extensions of the LFER model that are in the spirit of [19], but we are able to obtain further improvements in model performance by incorporating additional features observed in the current study. The critical role played by model assessment criterion Q^2_{LOSO} is also reviewed. The resulting model provides interpretations that are useful for identifying solutes whose chemical structures are consistent with low predicted levels of skin permeability.

An attractive feature of the MCF approach of [14] is their proposed partition theory, namely that the partition coefficient of a solute from a formulation is not affected by the starting concentration of that solute in the formulation. This theory, if realized, can lead to simplified analysis even in the most complex of experimental conditions. By applying an expanded LFER model, we are able to test this theory that could not otherwise be tested.

Earlier efforts by Xia et al. 2007 [13] demonstrated the use of a MCF array to simulate skin permeability in simple binary mixtures. However the present paper utilizes the MCF and molecular structure parameters within an LFER model described above to now better estimate the effects of several real world formulations at various concentrations on the partitioning behavior of 37 solutes at different concentrations in an effort to estimate solute partitioning into MCF which serves as a surrogate for skin permeability

2. Results and Discussion

2.1. Data Summaries

Formulations are designed to mimic fluids used in the metalworking process. For this article, a *formulation* refers to: a particular metalworking fluid (MWF), at a particular MWF concentration, spiked with a solute at a particular concentration. Formulations are spiked with trace levels of solutes in such a way that the chemistry of the MWF is not altered.

In this study, we considered 37 solutes (see Table 1) and five solvatochromic descriptors believed to be most relevant to the solvation process during permeation [16,20]. These descriptors represent different characteristics of compounds involved in the solvation process, specified as follows. E is the solute excess molar refractivity, S is the solute dipolarity/polarizability, A is the overall hydrogen bond acidity, B is the overall hydrogen bond basicity, and V is the McGowan characteristic volume. For most solutes, V can be calculated directly, E can be obtained from experiment or calculated, but A, B, and S must be experimentally derived.

Table 1. Set of 37 solutes, and their descriptor values, used in this study.

Solute	Solute Name	E	S	A	B	V
1	Toluene	0.60	0.52	0	0.14	0.8573
2	Chloro-benzene	0.72	0.65	0	0.07	0.8388
3	Ethylbenzene	0.61	0.51	0	0.15	0.9982
4	p-Xylene	0.61	0.52	0	0.16	0.9982
5	Bromo-benzene	0.88	0.73	0	0.09	0.8914
6	Propyl-benzene	0.60	0.50	0	0.15	1.1391
7	1-Chloro-4-methyl-benzene	0.71	0.74	0	0.05	0.9797
8	Phenol	0.81	0.89	0.60	0.30	0.7751
9	Benzonitrile	0.74	1.11	0	0.33	0.8711
10	4-Fluoro-phenol	0.67	0.97	0.63	0.23	0.7927
11	Benzyl alcohol	0.80	0.87	0.39	0.56	0.9160
12	Iodo-benzene	1.19	0.82	0	0.12	0.9746
13	Phenyl ester acetic acid	0.66	1.13	0	0.54	1.0726
14	2-Chloro-acetophenone	1.02	1.59	0	0.41	1.1363
15	Phenol, 4-methyl-	0.82	0.87	0.57	0.31	0.9160
16	Nitro-Benzene	0.87	1.11	0	0.28	0.8906
17	Methyl ester benzoic acid	0.73	0.85	0	0.46	1.0726
18	1-chloro-4-methoxy-benzene	0.84	0.86	0	0.24	1.0384
19	Phenylethyl alcohol	0.81	0.86	0.31	0.65	1.0569
20	3-Methylbenzyl alcohol	0.82	0.90	0.39	0.59	1.0569
21	4-Ethyl-phenol	0.80	0.90	0.55	0.36	1.0569
22	3,5-Dimethyl-phenol	0.82	0.84	0.57	0.36	1.0569
23	Ethyl ester benzoic acid	0.69	0.85	0	0.46	1.2135
24	2-Methyl-methyl ester benzoic acid	0.77	0.87	0	0.43	1.2135
25	Naphthalene	1.34	0.92	0	0.20	1.0854
26	3-Chloro-phenol	0.91	1.06	0.69	0.15	0.8975
27	p-Chloroaniline	1.06	1.13	0.30	0.31	0.9386
28	1-methyl-4-nitro-benzene	0.87	1.11	0	0.28	1.0315
29	1-(4-Chlorophenyl)-ethanone	0.96	1.09	0	0.44	1.1363
30	3-Bromo-phenol	1.06	1.13	0.70	0.16	0.9501
31	4-Chloro-3-methyl-phenol	0.92	1.02	0.67	0.22	1.0384
32	1-Methyl-naphthalene	1.34	0.92	0	0.20	1.2263
33	Biphenyl	1.36	0.99	0	0.26	1.3242
34	Chloroxylenol	0.93	0.96	0.64	0.21	1.1793
35	4-(1,1-Dimethylpropyl)-phenol	0.79	0.80	0.50	0.44	1.4796
36	o-Hydroxybiphenyl	1.55	1.40	0.56	0.49	1.3829
37	Clorophene	1.53	1.42	0.67	0.47	1.6462

We varied the three other factors to create a formulation: the MWF, MWF concentration, and solute concentration. Five MWFs were considered: mineral oil (MO), polyethylene glycol-200 (PEG), soluble oil (SO), synthetic oil (SYN), and semi-synthetic oil (SSYN). MWF concentrations were at three levels: 0.05 percent, 0.5 percent, and 5 percent. Six solute concentrations were considered: 0.01, 0.05, 0.1, 0.5, 1, and 5 ppm. As a result, there were 5 × 3 × 6 = 90 treatment combinations, as displayed in Table A1 in Appendix A.

The study was designed to obtain partition coefficients, $K_{MCF/mix}$, for all 37 solutes, under each of the 90 treatment combinations, using three replicates. Unfortunately, due to a variety of reasons (e.g., lack of detection in gas chromatography, records outside the calibration range, etc.), not all replicates were recordable, with some treatment combinations even ending in no replicates for a particular solute. Fitting the QSAR model does not require replicates because of the structure provided by the model, and all collected data informs the fitting process. Having replicates would likely result in smaller measures of variability and hence greater power to make inference beyond what could done here, but the lack of replicates has not impeded the ability to conduct statistical analysis and model building. Of the maximum possible 37 × 90 × 3 = 9990 observations, we actually generated 4646 partition coefficients.

Summary statistics are displayed in Table 2 for all variables, based on the complete dataset of 4646 observations. Partition coefficients range from 0.015 to 1279 (−1.820 to 3.107 on the base 10 logarithm scale). To get a more detailed view of the range of values for partition coefficients, Figure 1 shows boxplots of log $K_{MCF/mix}$ grouped by solute concentration. It is somewhat surprising that the smallest partition coefficients are associated with higher concentrations of solute present in the formulation; we return to this observation later in the article.

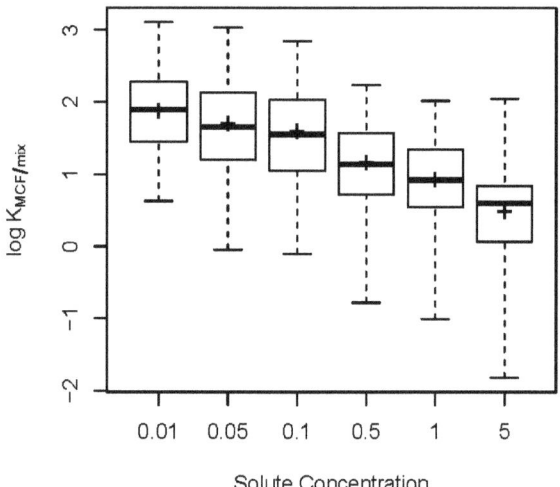

Figure 1. Boxplots of log $K_{MCF/mix}$ across different solute concentrations. Thick horizontal lines are the medians, the means are shown as +, boxes contain the middle half of the data, and dotted lines extend to the minimum and maximum.

Table 2. Summary statistics for all variables, based on the complete dataset of 4646 observations.

Variable	Minimum	Lower Quartile	Mean	Median	Upper Quartile	Maximum	Std Dev
log $K_{MCF/mix}$	−1.820	0.841	1.329	1.380	1.879	3.107	0.719
E	0.600	0.710	0.862	0.800	0.960	1.550	0.225
S	0.500	0.800	0.928	0.900	1.110	1.590	0.266
A	0.000	0.000	0.120	0.000	0.000	0.700	0.232
B	0.050	0.150	0.293	0.280	0.440	0.650	0.146
V	0.775	0.939	1.058	1.038	1.136	1.646	0.170

2.2. Insufficiency of the LFER Model

Abraham and Martins [20] proposed the general linear free-energy relationship (LFER) model to study dermal absorption:

$$SP = \beta_0 + \beta_1 E + \beta_2 S + \beta_3 A + \beta_4 B + \beta_5 V,$$

where SP is the property of interest for the solutes (such as log K_p, log P, etc.). Given data, the coefficients in the LFER model are determined by multiple linear regression. These coefficients are also commonly denoted as c, e, s, a, b, and v; we used β_0, β_1, β_2, β_3, β_4, and β_5 as this is more common in the literature of multiple linear regression. In this article, logarithm of the partition coefficient, log $K_{MCF/mix}$, is the property of interest. The resulting LFER model is shown in Equation (1):

$$\log K_{MCF/mix} = \beta_0 + \beta_1 E + \beta_2 S + \beta_3 A + \beta_4 B + \beta_5 V. \quad (1)$$

While the LFER model in Equation (1) is simple and easy to interpret, it is not always sufficient, especially for large datasets under complicated experimental conditions. Equation (1) suggests that the expected value of log $K_{MCF/mix}$ is a function of only E, S, A, B, and V. However, as is clearly demonstrated in Figure 1, log $K_{MCF/mix}$ decreases as solute concentration increases, suggesting that solute concentration should likely be included as a predictor in Equation (1); we return to this observation below.

Focusing for the moment on the LFER model, Equation (1) was separately applied to data from each of the 90 treatment combinations, resulting in 90 separate estimated models. If all 90 estimated models essentially coincide, then the LFER model that only accounts for E, S, A, B, and V, and does not adjust for experimental conditions, is sufficient. To investigate this, Table 3 presents details on three of the 90 estimated models; details include estimated coefficients, their standard errors, and associated 95 percent confidence intervals. Estimated models are shown for: treatment combination 5, with mineral oil at 0.05 percent and solute concentration 1 ppm; treatment combination 17, with mineral oil at five percent and solute concentration 1 ppm; and treatment combination 52, with soluble oil at five percent and solute concentration 0.5 ppm.

Table 3. Results from fitting separate LFER models (Equation (1)) for each of three treatment combinations (T).

T		β_0 (Intercept)	β_1 (for E)	β_2 (for S)	β_3 (for A)	β_4 (for B)	β_5 (for V)
5	est(se)	0.21(0.45)	1.77(0.43)	−1.59(0.27)	−1.87(0.18)	−0.50(0.42)	1.61(0.32)
	ci	(−0.71, 1.12)	(0.89, 2.65)	(−2.14, −1.03)	(−2.23, −1.51)	(−1.36, 0.36)	(0.97, 2.25)
17	est(se)	−0.61(0.27)	−0.91(0.22)	0.42(0.18)	−1.14(0.12)	−2.00(0.24)	2.47(0.25)
	ci	(−1.15, −0.07)	(−1.35, −0.48)	(0.07, 0.77)	(−1.37, −0.91)	(−2.48, −1.53)	(1.97, 2.97)
52	est(se)	1.50(0.31)	−0.03(0.23)	−0.36(0.25)	−0.42(0.20)	−0.98(0.48)	−0.14(0.34)
	ci	(0.89, 2.11)	(−0.50, 0.44)	(−0.87, 0.15)	(−0.83, −0.02)	(−1.94, −0.02)	(−0.81, 0.53)

For each of the intercept, E, S, A, B, and V, the table provides: the estimated coefficient (est), the associated standard error (se), and a 95 percent confidence interval (ci) for the coefficient. With large differences in estimated coefficients for different treatment combinations, these estimated models indicate a clear dependency on treatment combinations.

The estimated models in Table 3 did not coincide. Consider, for example, the coefficient β_1 corresponding to E. For treatment combination 5, the 95 percent confidence interval consists of only positive values (0.89 to 2.65), suggesting that log $K_{MCF/mix}$ is expected to increase as excess molar refractivity increases. On the other hand, the 95 percent confidence interval consists of only negative values (−1.35 to −0.48) for treatment combination 17, suggesting that log $K_{MCF/mix}$ is expected to decrease as excess molar refractivity increases. These conflicting interpretations are not isolated. Figure 2 graphs the 95 percent confidence intervals for coefficient β_1 corresponding to E from all 90 treatment combinations, and these intervals clearly do not coincide. Moreover, similar results hold for all coefficients, as demonstrated in Table 3.

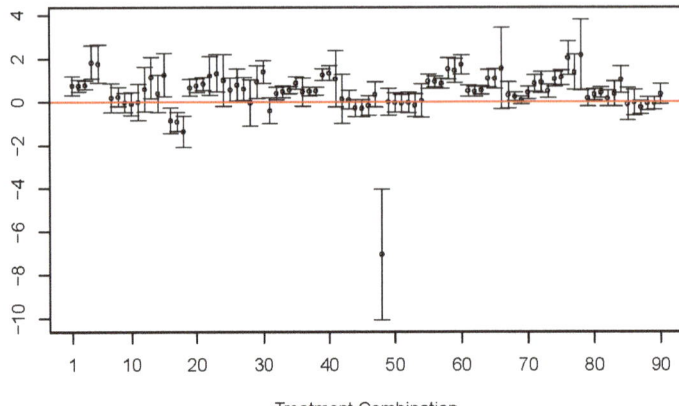

Figure 2. Estimated β_1 coefficients (circles) for molecular descriptor E from fitting separate LFER models across the 90 treatment combinations. Ninety-five percent (95%) confidence intervals are also shown, as vertical lines with two bars at the ends.

2.3. Improvement by Expanded LFER Models

Xu et al. [19] demonstrate insufficiency of the LFER model for accounting for experimental conditions defined by four MWFs. They extend the LFER model by allowing for different sets of estimated coefficients for each of the four MWFs, all while using a single model. They obtained substantial improvements in predictive power of the Extended LFER model compared to the (single) LFER model. Hoping to achieve similar levels of improvement as [19], we also fitted an Extended LFER model that allows for different sets of estimated coefficients for each of the 90 treatment combinations, while using a single model, as follows:

$$\begin{aligned}
\log K_{MCF/mix,ijkl} &= F_{1jkl}C_{i1kl}W_{ij1l}(\beta_{0111} + \beta_{1111}E_l + \beta_{2111}S_l + \beta_{3111}A_l + \beta_{4111}B_l \\
&\quad + \beta_{5111}V_l) \\
&\quad + F_{1jkl}C_{i1kl}W_{ij2l}(\beta_{0112} + \beta_{1112}E_l + \beta_{2112}S_l + \beta_{3112}A_l + \beta_{4112}B_l \\
&\quad + \beta_{5112}V_l) + \cdots \\
&\quad + F_{5jkl}C_{i3kl}W_{ij6l}(\beta_{0536} + \beta_{1536}E_l + \beta_{2536}S_l + \beta_{3536}A_l + \beta_{4536} \\
&\quad + \beta_{5536}V_l),
\end{aligned}$$

where $\log K_{MCF/mix,ijkl}$ is the lth observation from MWF i ($i = 1$ for MO, $i = 2$ for PEG, $i = 3$ for SO, $i = 4$ for SYN, and $i = 5$ for SSYN), MWF concentration j ($j = 1$ for 0.05, $j = 2$ for 0.5, and $j = 3$ for 5 percent), and solute concentration k ($k = 1$ for 0.01, $k = 2$ for 0.05, $k = 3$ for 0.1, $k = 4$ for 0.5, $k = 5$ for 1, and $k = 6$ for 5 ppm). In Equation (2), β_{dijk} denotes the coefficient for descriptor d (with $d = 0$ for the intercept, $d = 1$ for E, $d = 2$ for S, $d = 3$ for A, $d = 4$ for B, and $d = 5$ for V) corresponding to MWF i, MWF concentration j, and solute concentration k. For example, β_{1111} is the partial slope for descriptor E under treatment combination 1, with mineral oil at 0.05 percent and solute concentration 0.01 ppm. Three "dummy variables" F_{ijkl}, C_{ijkl}, and W_{ijkl} are defined to indicate treatment combinations; these variables take value zero or one according to the levels of MWF, MWF concentration, and solute concentration. $F_{ijkl} = 1$ if the observation comes from MWF i, otherwise $F_{ijkl} = 0$; $C_{ijkl} = 1$ if the observation comes from MWF concentration j, otherwise $C_{ijkl} = 0$; and $W_{ijkl} = 1$ if the observation comes from solute concentration k, otherwise $W_{ijkl} = 0$.

The model in Equation (2) is quite large, having a maximum of 90 intercepts (one for each treatment combination) and $5 \times 90 = 450$ partial slopes (slopes corresponding to each of E, S, A, B, and V for each treatment combination). For any given observation, Equation (2) activates only a single set of coefficients because the product $F_{ijkl}C_{ijkl}W_{ijkl}$ will only be nonzero for a single treatment combination. For example, if the observation is in treatment combination 2 (mineral oil at concentration 0.05 percent with solute concentration 0.05 ppm), then $F_{1jkl}C_{i1kl}W_{ij2l} = 1$ and all other $F_{ijkl}C_{ijkl}W_{ijkl} = 0$, thus activating only $\beta_{0112} + \beta_{1112}E_l + \beta_{2112}S_l + \beta_{3112}A_l + \beta_{4112}B_l + \beta_{5112}V_l$ in Equation (2). Since Equation (2) is based on multiplying the dummy variables, we refer to it as the Expanded Crossed-Factors LFER model.

Table 4 shows regression statistics of fitting the Expanded Crossed-Factors LFER model of Equation (2). Regression statistics are also shown for the (single) LFER model of Equation (1), and another model to be described later. The improvements in r^2, Adj-r^2, Q^2_{LOO}, and Q^2_{LOSO} are quite noticeable in favor of the Expanded Crossed-Factors LFER model over the LFER model. While r^2 and Adj-r^2 are widely known, Q^2_{LOO}, and Q^2_{LOSO} may be less familiar. Both Q^2_{LOO} and Q^2_{LOSO} are designed to measure predictive ability of a model, but [19] demonstrate the advantage of Q^2_{LOSO} over Q^2_{LOO} for the current context. Leave-one-out (LOO) cross-validation is employed in both, meaning models are fit after reducing the dataset, then the resulting fit is used to make prediction on the portion of the data that was left out. The difference is that Q^2_{LOSO} leaves out an entire solute at a time, whereas Q^2_{LOO} omits a single row from the dataset. If only a single row is removed from the dataset, we are left with the possibility that a single replicate of a solute in a particular formulation may be removed, but the other two replicates remain in the dataset. The result is that the model is fit with almost full knowledge of the solute in question, and the consequence is that we are misled about the quality of the model for fitting "new, unseen" solutes. By removing every instance of a solute, Q^2_{LOSO} provides a better assessment of the quality of the model for predicting new, unseen solutes. Large values are desirable for both Q^2_{LOO} and Q^2_{LOSO}, but the extra demands placed on Q^2_{LOSO} usually result in smaller values of Q^2_{LOSO} compared to Q^2_{LOO}, in much the same way that Adj-r^2 is often smaller than r^2. (It is important to note that Q^2_{LOSO} in this article is equivalent to $Q^2_{LOO-adj}$ in [19]. We prefer the simpler "LOSO" as it more clearly explains the difference from "LOO".)

Table 4. Fit statistics of a single LFER model (1), the Expanded Crossed-Factors LFER model (2) and the Expanded Nested-Solute-Concentration LFER model (5).

Regression Statistics	LFER Model (1)	Expanded Crossed-Factors LFER Model (2)	Expanded Nested-Solute-Concentration LFER Model (5)
r^2	0.60	0.90	0.88
Adj-r^2	0.60	0.89	0.87
Q^2_{LOO}	0.60	0.87	0.87
Q^2_{LOSO}	0.57	0.68	0.80

Q^2_{LOO} is calculated as

$$Q^2_{LOO} = 1 - \frac{\sum_{l=1}^{n}(y_l - \hat{y}_{l,-l})^2}{\sum_{l=1}^{n}(y_l - \bar{y})^2}, \tag{3}$$

where y_l is the lth observed response of log $K_{MCF/mix}$, $\hat{y}_{l,-l}$ is the leave-one-out prediction of the lth observation based on the model fit without the lth observation, and $\bar{y}$ is the average of all the observed responses. Q^2_{LOSO}, designed by [19] to handle pseudo or real replicates in leave-one-out cross-validation for proper assessment of predictive power, is defined as:

$$Q^2_{LOSO} = 1 - \frac{\sum_{s=1}^{37} \sum_{l=1}^{n_s}(y_{sl} - \hat{y}_{sl,-s})^2}{\sum_{s=1}^{37} \sum_{l=1}^{n_s}(y_{sl} - \bar{y})^2}, \tag{4}$$

where y_{sl} is the lth observation of the sth solute, $\bar{y}$ is the average of all the observed responses, and $\hat{y}_{sl,-s}$ is the predicted value of y_{sl} based on the model fit from leaving out all the observations belonging to the sth solute.

While Q^2_{LOSO} showed improvement of the Expanded Crossed-Factors LFER model over the LFER model, the value of 0.68 is not impressive and indicates some deficiency of the model. One possible reason may be overfitting. With so many regression parameters, this model seems to fit the data too closely, thus the idiosyncrasies of the data are captured instead of the general trends. The problem of overfitting is that when the model is applied to a new dataset, it cannot predict the new data well, as indicated by the weak value of Q^2_{LOSO}. This motivates us to look for an alternative model, which not only accounts for the heterogeneity introduced by different experimental conditions, but is also simpler and more predictive. The LFER model may be expanded in a variety of ways that accommodate experimental conditions, and the goal is to identify the simplest adequate expansion. As previously mentioned, the Expanded Crossed-Factors LFER model of Equation (2) is quite large, and we wondered whether it could be simplified.

Figure 1 tells us that partition coefficients decrease as the solute concentration increases. This suggests that there may be a quantifiable relationship between $\log K_{MCF/mix}$ and solute concentration. However, Figure 1 is the overall effect of solute concentration, not accounting for the effect of MWF or MWF concentration. Thus, a more detailed visualization is desired. Figure 3 depicts the trend of $\log K_{MCF/mix}$ over solute concentration in all 15 combinations of MWF and MWF concentration. It shows a similar trend as in Figure 1, for each of the 15 combinations of MWF and MWF concentration. Figure 3 suggests that instead of viewing solute concentration as a third factor crossed with MWF and MWF concentration, we can take it as a (numerically) nested factor within each of the combinations of MWF and MWF concentration. In other words, for each combination of MWF and MWF concentration, allow a different partial slope for solute concentration. By doing this, we place a structure within each MWF x MWF concentration condition, and may be able to see how $\log K_{MCF/mix}$ changes as a function of solute concentration.

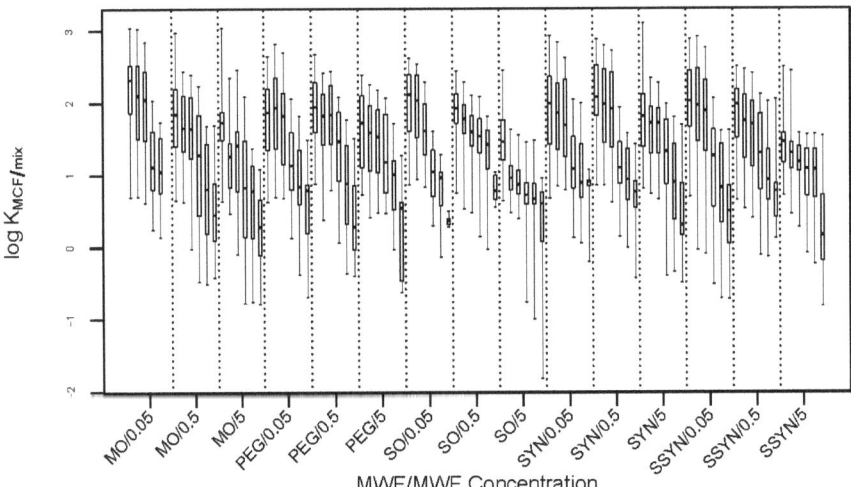

Figure 3. Boxplots of $\log K_{MCF/mix}$ across different solute concentrations in each of the 15 combinations of MWF and MWF concentration. Within each of the 15 panels, boxplots are shown for solute concentrations of 0.01, 0.05, 0.1, 0.5, 1, and 5 ppm. The 15 combinations (MWF/MWF concentration), from left to right, are: MO/0.05, MO/0.5, MO/5, PEG/0.05, PEG/0.5, PEG/5, SO/0.05, SO/0.5, SO/5, SYN/0.05, SYN/0.5, SYN/5, SSYN/0.05, SSYN/0.5, and SSYN/5.

We propose a new Expanded Nested-Solute-Concentration LFER model as in Equation (5):

$$\log K_{MCF/mix,ijl} = F_{1jl}C_{i1l}(\beta_{011} + \beta_{111}E_l + \beta_{211}S_l + \beta_{311}A_l + \beta_{411}B_l + \beta_{511}V_l + \beta_{611}t_l) \\ + F_{1jl}C_{i2l}(\beta_{012} + \beta_{112}E_l + \beta_{212}S_l + \beta_{312}A_l + \beta_{412}B_l + \beta_{512}V_l + \beta_{612}t_l) + \cdots \\ + F_{5jl}C_{i3l}(\beta_{053} + \beta_{153}E_l + \beta_{253}S_l + \beta_{353}A_l + \beta_{453}B_l + \beta_{553}V_l + \beta_{653}t_l), \quad (5)$$

where $\log K_{MCF/mix,ijl}$ is the lth observation from MWF i, MWF concentration j, t_l is the logarithm (base 10) of solute concentration of the lth observation, β_{dij} is the regression coefficient of descriptor d ($d = 0$ for intercept, $d = 1$ for E, $d = 2$ for S, $d = 3$ for A, $d = 4$ for B, $d = 5$ for V, and $d = 6$ for logarithm of solute concentration), for MWF i and MWF concentration j. We take the logarithm of solute concentration as it is common practice and it linearizes the relationship. This model is relatively small, with a maximum of 15 × 7 = 105 coefficients to be estimated, compared to a maximum of 540 for the model in Equation (2).

Regression statistics are shown in Table 4, and it is clear that the Expanded Nested-Solute-Concentration LFER model of Equation (5) is at least as good as the Expanded Crossed-Factors LFER model of Equation (2), because it has comparable or larger values for all regression statistics. However, the Expanded Nested-Solute-Concentration LFER model of Equation (5) has a tremendous advantage in that: (1) it is much smaller, and so more amenable to interpretation; and (2) it is more predictive as indicated by a much larger value for Q^2_{LOSO}.

Figure 4 plots observed versus predicted $\log K_{MCF/mix}$ values for both the LFER and Expanded Nested-Solute-Concentration LFER models. The tighter grouping around the line for the latter model is yet another demonstration of that model's better predictive power.

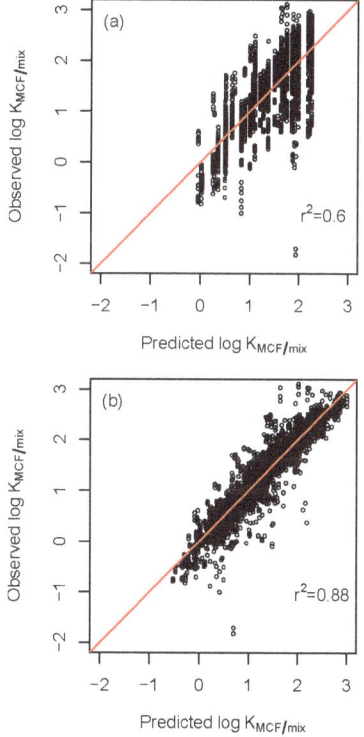

Figure 4. Observed versus predicted $\log K_{MCF/mix}$ for (**a**) the LFER model of Equation (1) and (**b**) the Expanded Nested-Solute-Concentration LFER model of Equation (5). Tightness around the line is indicative of a more predictive model.

2.4. Model Interpretation

We now intepret the estimated Expanded Nested-Solute-Concentration LFER model of Equation (5).

There are 15 rows in Equation (5), each representing the regression function for one combination of MWF/MWF concentration. For example, row one is for MWF mineral oil at concentration 0.05 percent, while row 15 is for MWF semi-synthetic oil at concentration five percent. Each row has a set of partial slopes that vary among the different combinations of MWF/MWF concentration. The estimates and associated standard errors of all partial slopes are shown in Table A2 in Appendix A.

To show how the partial slopes vary, in Figure 5 we plot 95 percent confidence intervals for each partial slope corresponding to E, S, A, B, V and log solute concentration across all 15 combinations of MWF/MWF concentration. The 95 percent confidence intevals are shown as vertical lines with two bars at the ends. A horizontal reference line of zero is also shown. There are some interesting trends seen in Figure 5.

For example, in Figure 5a, the partial slope of E generally decreases as MWF concentration increases within each MWF. In mineral oil, the effect (sign of β_1) of E (solute excess molar refractivity) even changes as MWF concentration increases. To be specific, using mineral oil at concentration of 0.05 percent, if we increase solute excess molar refractivity and other predictors are held fixed, then the partition coefficient is expected to increase (the 95 percent confidence interval lays above the reference line). On the other hand, using mineral oil at the higher concentration of five percent, if we increase solute excess molar refractivity, then we expect the partition coefficient to decrease (the 95 percent confidence interval lays below the reference line).

In Figure 5b, the partial slope of S generally increases as MWF concentration increases within mineral oil, soluble oil, and semi-synthetic oil, but partial slopes show no significant change as MWF concentration increases within polyethylene glycol-200 and synthetic oil. In general, S (solute dipolarity/polarizability) has an inverse relationship with expected partition coefficient, meaning that as S increases we expected a decrease in partition coefficient.

Figure 5c suggests increased levels of hydrogen bond acidity A are associated with decreased partition coefficients. However, the pattern of decrease changes according to the concentration of MWF. For example, in both mineral oil and soluble oil, higher MWF concentrations result in smaller decrease in partition coefficients. Figure 5d indicates that increased levels of hydrogen bond basicity B generally leads to decreased partition coefficients.

Figure 5e says larger molecules tend to have larger partition coefficients. In soluble oil, synthetic oil and semi-synthetic oil, the effect of molecule size V gets smaller as MWF concentration increases, resulting in less dramatic effect of molecule size on partition coefficients.

Figure 5f suggests that higher concentrations of solute generally result in lower partition coefficients. In both mineral oil and soluble oil, higher MWF concentrations result in stronger inverse relationships.

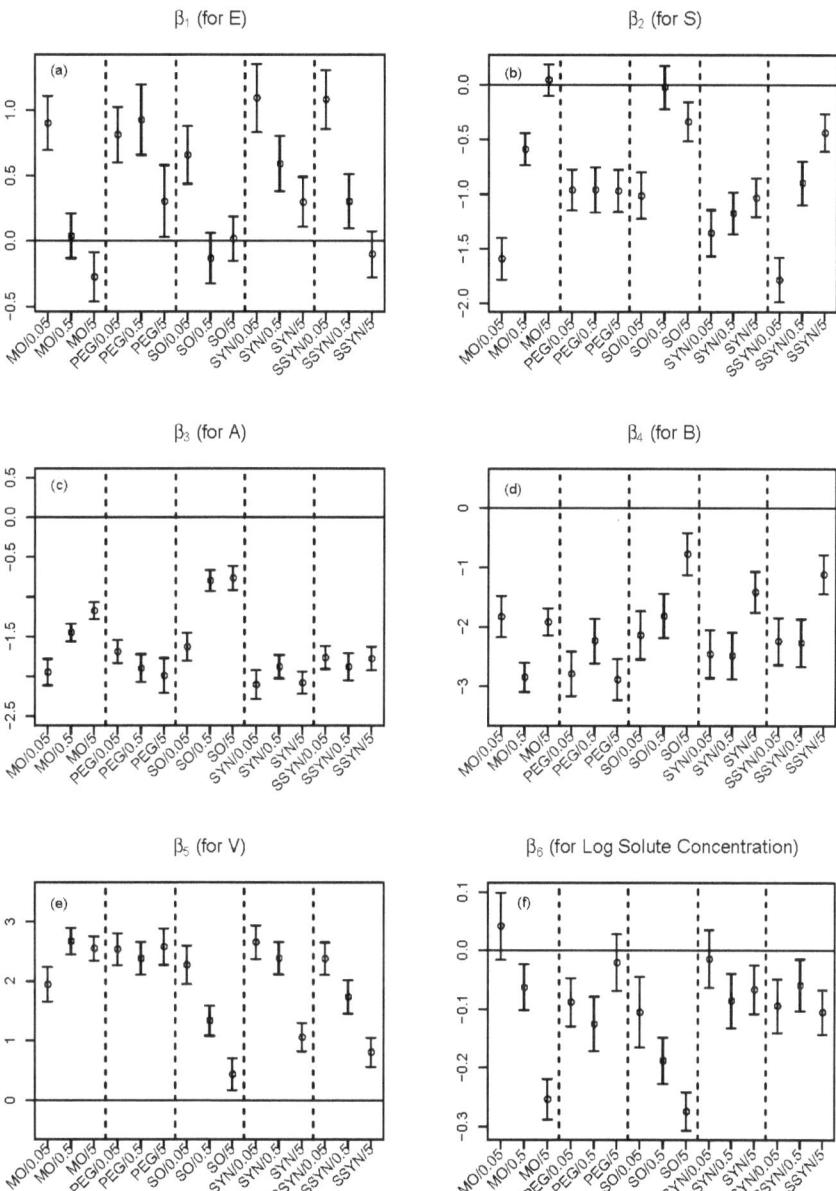

Figure 5. Estimated partial slopes (circles) corresponding to (**a**) E, (**b**) S, (**c**) A, (**d**) B, (**e**) V, and (**f**) log solute concentration from fitting the Expanded Nested-Solute-Concentration LFER model of Equation (5), for all 15 combinations of MWF/MWF concentration. Ninety-five percent (95%) confidence intervals are also shown, as vertical lines with two bars at the ends.

2.5. Validation of Partition Theory

2.5.1. Implication of Partition Theory

According to [14], it is assumed that the amount of solute extracted from the MCF, n^0, is proportional to the solute concentration, C_0, where the proportionality constant is not affected by C_0. Based on this assumption, we obtain $n^0 = pC_0$, where p is the proportionality constant and $0 \leq p \leq 1$. Applying this relationship to partition coefficients, we obtain:

$$K_{MCF/mix} = \frac{n^0 V_d}{V_m(C_0 V_d - n^0)} = \frac{pC_0 V_d}{V_m(C_0 V_d - pC_0)} = \frac{pV_d}{V_m(V_d - p)}. \tag{6}$$

Equation (6) suggests that $K_{MCF/mix}$ is independent of C_0, which suggests that irrespective of the solute concentration, the partition coefficient remains the same. This so-called "partition theory", if true, has practical meaning in the metalworking industry as it would indicate that increasing solute concentration has no impact on skin permeation ability of the solute. For example, higher concentrations of biocides might be preferred to extend preservation of fluids, while there is no detrimental effect of increasing the biocide's ability to permeate skin. As described in more detail in the methods section, the MCF consists of a PDMS coating that is 100 μm thick and 1 cm long on an inert silica fiber. Solute partitioning into this membrane is dependent on the many chemical-chemical interactions quantified by our Expanded LFER models. However, the membrane volume (V_m) suggests that this may be a limitation with increasing solute concentration. It was, therefore, interesting to see if this partition theory is supported by our data.

2.5.2. Violation from Experimental Data

Assume the Expanded Nested-Solute-Concentration LFER model of Equation (5). To test whether the partition theory holds, we simply tested whether the coefficients corresponding to any solute concentration terms are different from zero. If all coefficients corresponding to solute concentration terms equal zero in Equation (5), then log $K_{MCF/mix}$ will not change as solute concentration changes. More specifically, we test the following null hypothesis:

$$H_0: \beta_{6ij} = 0 \text{ for all } i = 1, 2, 3, 4, 5 \text{ and } j = 1, 2, 3.$$

The resulting p-value of less than 0.0001 allows us to strongly conclude that the solute concentration term for at least one combination of MWF/MWF concentration is significantly different from zero. In fact, the individual P-values for testing each $\beta_{6ij} = 0$ show that the solute concentration effect is significantly different from zero for 12 of the 15 combinations; nonsignificance is obtained only in MO/0.05, PEG/5 and SYN/0.05. These results are consistent with Figure 5f, where confidence intervals contain zero only for mineral oil at concentration 0.05, polyethylene glycol-200 at concentration 5 and synthetic oil at concentration 0.05.

Hoping to find that the partition theory holds true in either low or high solute concentrations, we considered subsets of data that contain only some of the solute concentrations. Detailed results are given in Table 5 of testing the null hypothesis that the partition theory holds for a number of different subsets of solute concentrations. For example, does the partition theory hold when considering only observations with solute concentrations less than or equal to 1 ppm? The answer is provided by row two of Table 5: with a p-value of less than 0.0001, the partition theory does not hold for solute concentrations less than or equal to 1 ppm, with violations happening in eight of the 15 combinations. In fact, the partition theory is violated in all subsets of solute concentrations.

Table 5. Testing the null hypothesis that the partition theory holds for a number of different subsets of solute concentrations.

Solute Concentrations	p-Value for H_0	Insignificant Conditions	Significant Conditions
All	< 0.0001	MO/0.05(265), PEG/5(246), SYN/0.05(256), SSYN/0.5(297)	MO/0.5(379), MO/5(397), PEG/0.05(313), PEG/0.5(261), SO/0.05(247), SO/0.5(341), SO/5(368), SYN/0.5(300), SYN/5(321), SYN/0.05(298), SSYN/5(357)
0.01, 0.05, 0.1, 0.5, 1	< 0.0001	MO/0.05(265), MO/0.5(345), PEG/0.05(279), PEG/5(218), SYN/0.05(240), SYN/5(287), SSYN/0.5(278)	MO/5(347), PEG/0.5(239), SO/0.05(242), SO/0.5(327), SO/5(292), SYN/0.5(283), SSYN/0.05(278), SSYN/5(303)
0.01, 0.05, 0.1, 0.5	< 0.0001	MO/0.05(226), MO/0.5(285), PEG/0.05(236), PEG/0.5(205), PEG/5(182), SO/0.05(212), SYN/0.05(199), SYN/0.5(243), SYN/5(234), SSYN/0.05(238), SSYN/0.5(246)	MO/5(261), SO/0.5(271), SO/5(219), SSYN/5(229)
0.01, 0.05, 0.1	< 0.0001	MO/0.05(183), PEG/0.05(185), PEG/0.5(165), PEG/5(135), SO/0.05(174), SYN/0.05(157), SYN/0.5(197), SYN/5(172), SSYN/0.05(191), SSYN/0.5(183)	MO/0.5(205), MO/5(176), SO/0.5(192), SO/5(145), SSYN/5(159)
0.05, 0.1, 0.5	< 0.0001	MO/0.05(173), MO/0.5(223), MO/5(208), PEG/0.05(184), PEG/0.5(155), PEG/5(147), SO/0.05(153), SYN/0.05(156), SYN/0.5(186), SYN/5(187), SSYN/0.05(180), SSYN/0.5(196), SSYN/5(187)	SO/0.5(141), SO/5(102)
0.01, 0.05	< 0.0001	MO/0.05(118), PEG/0.05(117), PEG/5(85), SO/0.05(127), SO/0.5(122), SYN/0.05(98), SYN/0.5(124), SYN/5(110), SSYN/0.05(123), SSYN/0.5(115), SSYN/5(99)	MO/0.5(129), MO/5(110), PEG/0.5(106), SO/5(91)
0.05, 0.1	< 0.0001	MO/0.05(130), MO/0.5(143), MO/5(123), PEG/0.5(115), PEG/5(100), SO/0.05(115), SO/5(102), SYN/0.05(114), SYN/0.5(140), SYN/5(125), SSYN/0.05(133), SSYN/0.5(133), SSYN/5(117)	PEG/0.05(133), SO/0.5(141)

The subset of solute concentrations is shown in the first column, with p-value given in the second column. MWF/MWF concentrations that support the partition theory (meaning their individual p-values are larger than 0.05/15, where division by 15 is to adjust for multiple testing) are shown in the third column (with sample sizes in parentheses). MWF/MWF concentrations that violate the partition theory are shown in the last column (with sample sizes in parentheses). The partition theory is violated in every subset, with the greatest support for the partition theory being achieved when limiting solute concentration to 0.05 or 0.1 or 0.5 ppm as the largest subset.

3. Materials and Methods

Our experiments were based on the MCF approach proposed in [14]. Only a single MCF was used, namely PDMS (polydimethylsiloxane). In the current study, solutes were dissolved into a particular formulation, then an MCF was placed in the vial to allow the solute to partition from the solute-spiked formulation into the MCF over a period of one to four hours; see Figure 6. Gas chromatography and mass spectrometry were then used to extract or desorb the solute from the MCF, and the amount extracted was recorded.

3.1. Solvent/Solute Preparation

Three industry generic metal working fluids (MWF) formulations; soluble oil, synthetic fluid, and semi-synthetic fluid were kindly supplied by from Cimcool Industrial Products LLC (Cincinnati, OH, USA). The precise composition for each of these three formulations is proprietary information. In general, soluble oil concentrates contained approximately 58% mineral oil along with various other performance additives such as sulfonates and ethanolamines, semi-synthetic fluid concentrates contained about 15% mineral oil along with other additives such as sulfonates and ethanolamines, and synthetic fluid concentrates contain no mineral oil but contained various carboxylic acid salts, ethanolamines, ethyleneglycols, and plant seed oils. This is typical of many commercial MWF formulations that fall into these three categories. In addition to these three MWFs, two laboratory prepared surrogate formulations, mineral oil and PEG-200 (Aldrich, St. Louis, MO, USA) were prepared volumetrically in 0.05%, 0.5%, and 5.0% formulations in ultrapure water (Pure Water Solutions, Hillsborough, NC, USA). Each of these formulations were then spiked to six concentrations in the range of 0.01–5.0 µg/mL ranges with a set of 37 solutes (Table 1). These solutes were chosen to represent a wide variety of physiochemical properties. All solutes were of the highest purity available for purchase (Sigma Aldrich, Milwaukee, WI, USA). The 37 solutes were also prepared in acetone in a 2000 µg/mL stock solution. Experimental solutions were prepared fresh and all samples were kept at ambient temperature prior to analysis by SPME/GC-MS. Liquid GC-MS injections of the same 37 solutes prepared in acetone (0.01–10.00 µg/mL) were run daily, as well as blank liquid (acetone) and SPME (prepared solvent without addition of 37 solute) injections.

3.2. SPME/GC-MS Analysis

SPME absorption and injection was performed by a CTC Analytics Comi-Pal auto injector (Varian Inc., Walnut Creek, CA, USA) outfitted with a 100 µm polydimethylsiloxane SPME unit (Supelco Analytical, Bellafonte, PA, USA). A 9 mL sample was first agitated in a 37 °C heating block for 5 min, the SPME MCF (Figure 6) was then inserted and exposed for 30 min at 37 °C with constant agitation. SPME and liquid (0.5 µL) injections were introduced into a Varian 1079 injector (Varian Inc., Walnut Creek, CA, USA) at 280 °C in a split less mode for five min, at 5.5 min the split was turned on to 100%. For the first 30 seconds a pressure pulse of 21.0 psi was applied. Column flow was maintained at a constant 1.0 mL/min using helium as the carrier gas (National Welders, Raleigh, NC, USA). The Varian CP-3800 GC oven (Varian Inc., Walnut Creek, CA, USA) was programmed to hold at 40 °C for the first minute, followed by a 20 °C/min ramp to 90 °C (3.5 min), at which time the ramp slowed to 2.5 °C/min until 127 C (18.30 min) was reached and the ramp was increased to 40 C/min until it reached 250 °C and held for 2.0 min (23.38 min), followed by another increased ramp of 40 C/min until 280 °C and held for 5.0 min (29.13 min). The Saturn 2200-MS (Varian Inc., Walnut Creek, CA, USA) was programmed to run in full scan mode (40–300 m/z) after the first 3.0 min. Individual solute peaks were identified/quantified by the Star v6.5 software (Varian Inc., Walnut Creek, CA, USA) using retention time and known quant ions as identified and confirmed in the initial method development. Our sensitivity was set at 0.01 µg/mL as we were working with solutes ranging in concentrations from 0.01–5.0 µg/mL. More importantly, no residues were detected in the second injection after each first test injection, which indicated that there was negligible carry over under the optimum desorption conditions.

Differential ability of the solute to dissolve into the MCF or remain in the formulation was measured using a partition ratio (coefficient) $K_{MCF/mix}$ between the equilibrium concentration of the solute in the MCF and the equilibrium concentration of the solute in the formulation. $K_{MCF/mix}$ was calculated, following [14], as:

$$K_{MCF/mix} = \frac{C_{pe}}{C_{me}} = \frac{n^0/V_m}{C_0 - n^0/V_d} = \frac{n^0 V_d}{V_m(C_0 V_d - n^0)} \qquad (7)$$

where n^0 is the amount (in µg) of solute extracted from the MCF, V_m is the volume (in mL) of the MCF, V_d is the volume (in mL) of formulation placed in the vial based on solute concentration C_0 (in µg/mL), $C_{pe} = n^0/V_m$ is the equilibrium concentration of solute in the MCF, and $C_{me} = C_0 - n^0/V_d$ is the equilibrium concentration of solute in the formulation.

ADME Boxes 4.95, commercial software from ACD/Labs [21], was used to identify the E, S, A, B, and V descriptors for all the 37 solutes used in the experiment.

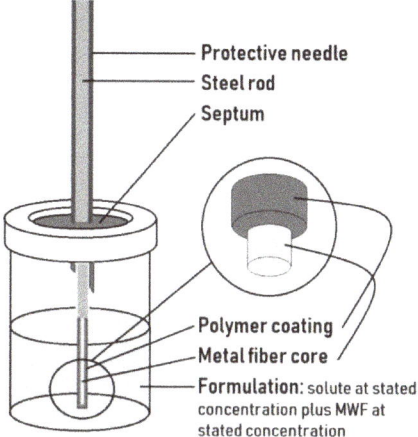

Figure 6. Membrane–coated fiber (MCF) and experimental setup. MWF = metal working fluid and the polymer coating is 100 μm thick polydimethylsiloxane (PDMS) that is part of the MCF.

4. Summary and Conclusions

The partition theory of [14] does not appear to hold for the current study, as evidenced by Figure 1, Figure 3, and Table 5. It is probable that there is a finite number of binding sites available in the coating of the fiber (i.e., in the MCF). As the solute concentration increases, the percentage of the solute that absorbs and/or adsorbs to the membrane coating decreases due to this finite number of binding sites.

Notwithstanding the complications that arise from violations of the partition theory, our Expanded LFER models are able to adequately capture the variability of partition coefficients as a function of solute properties and experimental conditions. The Expanded Crossed-Factors LFER model based on [19] is a vast improvement over the single LFER model, while the Expanded Nested-Solute-Concentration LFER model developed in this article is even more refined, more predictive, and offers simple interpretations. Table 3, Table 4, Figure 2, and Figure 4 provide strong evidence that the simple LFER model is not adequate in the presence of complicated experimental conditions.

Proper assessment of model prediction ability is demonstrated with Q^2_{LOSO} (previously $Q^2_{LOO-adj}$ in [19]), and this measure is contrasted with Q^2_{LOO} and the more familiar r^2 and Adj-r^2. The leave-one-solute-out strategy allows assessment to occur based on completely unseen solutes.

Author Contributions: J.M.H.-O. and G.X. conducted all predictive modeling and assessment; data was collected in REB's lab; and J.M.H.-O., G.X., and R.E.B. wrote, read, and approved of the paper.

Funding: This work was supported by National Institute of Occupational Safety and Health grant NIOSH R01-OH-03669.

Acknowledgments: We thank Cimcool Fluid Technology (Cincinnati, OH, USA) for contributing the generic MWFs (synthetic, semi-synthetic, and soluble oil) for this study. We also thank James Brooks for contributions to an earlier version of this manuscript and Beth Barlow for collecting the data.

Conflicts of Interest: The authors declare no conflicts of interest.

Appendix A

This appendix contains tables that describe experimental results.

Table A1. Ninety treatment combinations with observation counts: MO for mineral oil, PEG-200 for polyethyleneglycol-200; SO for soluble oil; SYN for synthetic oil, and SSYN for semi-synthetic oil. Ideally, each treatment combination would include 37 × 3 = 111 observed partition coefficients, but missing data issues result in N observations, with N specified in the table.

Treatment Combination	MWF	MWF Conc. (%)	Solute Conc. (ppm)	N
1	MO	0.05	0.01	53
2	MO	0.05	0.05	65
3	MO	0.05	0.1	65
4	MO	0.05	0.5	43
5	MO	0.05	1	39
6	MO	0.05	5	0
7	MO	0.5	0.01	62
8	MO	0.5	0.05	67
9	MO	0.5	0.1	76
10	MO	0.5	0.5	80
11	MO	0.5	1	60
12	MO	0.5	5	34
13	MO	5	0.01	53
14	MO	5	0.05	57
15	MO	5	0.1	66
16	MO	5	0.5	85
17	MO	5	1	86
18	MO	5	5	50
19	PEG	0.05	0.01	52
20	PEG	0.05	0.05	65
21	PEG	0.05	0.1	68
22	PEG	0.05	0.5	51
23	PEG	0.05	1	43
24	PEG	0.05	5	34
25	PEG	0.5	0.01	50
26	PEG	0.5	0.05	56
27	PEG	0.5	0.1	59
28	PEG	0.5	0.5	40
29	PEG	0.5	1	34
30	PEG	0.5	5	22
31	PEG	5	0.01	35
32	PEG	5	0.05	50
33	PEG	5	0.1	50
34	PEG	5	0.5	47
35	PEG	5	1	36
36	PEG	5	5	28
37	SO	0.05	0.01	59
38	SO	0.05	0.05	68
39	SO	0.05	0.1	47
40	SO	0.05	0.5	38
41	SO	0.05	1	30
42	SO	0.05	5	5
43	SO	0.5	0.01	51
44	SO	0.5	0.05	71
45	SO	0.5	0.1	70
46	SO	0.5	0.5	79
47	SO	0.5	1	56
48	SO	0.5	5	14
49	SO	5	0.01	43
50	SO	5	0.05	48
51	SO	5	0.1	54
52	SO	5	0.5	74
53	SO	5	1	73
54	SO	5	5	76
55	SYN	0.05	0.01	43

Table A1. *Cont.*

Treatment Combination	MWF	MWF Conc. (%)	Solute Conc. (ppm)	N
56	SYN	0.05	0.05	55
57	SYN	0.05	0.1	59
58	SYN	0.05	0.5	42
59	SYN	0.05	1	41
60	SYN	0.05	5	16
61	SYN	0.5	0.01	57
62	SYN	0.5	0.05	67
63	SYN	0.5	0.1	73
64	SYN	0.5	0.5	46
65	SYN	0.5	1	40
66	SYN	0.5	5	17
67	SYN	5	0.01	47
68	SYN	5	0.05	63
69	SYN	5	0.1	62
70	SYN	5	0.5	62
71	SYN	5	1	53
72	SYN	5	5	34
73	SSYN	0.05	0.01	58
74	SSYN	0.05	0.05	65
75	SSYN	0.05	0.1	68
76	SSYN	0.05	0.5	47
77	SSYN	0.05	1	40
78	SSYN	0.05	5	20
79	SSYN	0.5	0.01	50
80	SSYN	0.5	0.05	65
81	SSYN	0.5	0.1	68
82	SSYN	0.5	0.5	63
83	SSYN	0.5	1	32
84	SSYN	0.5	5	19
85	SSYN	5	0.01	42
86	SSYN	5	0.05	57
87	SSYN	5	0.1	60
88	SSYN	5	0.5	70
89	SSYN	5	1	74
90	SSYN	5	5	54

Table A2. Estimation details for the Expanded Nested-Solute-Concentration LFER model of Equation (5). Estimated coefficients (with standard errors in parentheses) are given that correspond to the 15 combinations of MWF/MWF concentrations of the nested model. The logarithm of solute concentration is denoted as t.

	β_{0ij} (Intercept)	β_{1ij} (for E)	β_{2ij} (for S)	β_{3ij} (for A)	β_{4ij} (for B)	β_{5ij} (for V)	β_{6ij} (for t)
MO/0.05	1.15(0.15)	0.9(0.11)	−1.59(0.10)	−1.95(0.08)	−1.83(0.18)	1.95(0.15)	0.04(0.03)
MO/0.5	0.10(0.11)	0.04(0.09)	−0.59(0.07)	−1.45(0.06)	−2.85(0.13)	2.67(0.11)	−0.06(0.02)
MO/5	−0.86(0.10)	−0.27(0.1)	0.04(0.07)	−1.17(0.05)	−1.91(0.12)	2.55(0.10)	−0.25(0.02)
PEG/0.05	−0.12(0.11)	0.81(0.11)	−0.96(0.10)	−1.69(0.07)	−2.79(0.19)	2.54(0.14)	−0.09(0.02)
PEG/0.5	−0.10(0.11)	0.93(0.14)	−0.96(0.11)	−1.90(0.09)	−2.23(0.19)	2.39(0.14)	−0.13(0.02)
PEG/5	0.15(0.15)	0.3(0.14)	−0.97(0.10)	−1.99(0.11)	−2.88(0.18)	2.58(0.16)	−0.02(0.02)
SO/0.05	0.30(0.13)	0.66(0.11)	−1.01(0.11)	−1.63(0.09)	−2.14(0.21)	2.27(0.17)	−0.11(0.03)
SO/0.5	0.69(0.11)	−0.13(0.10)	−0.02(0.10)	−0.79(0.07)	−1.82(0.19)	1.34(0.13)	−0.19(0.02)
SO/5	0.74(0.12)	0.02(0.09)	−0.34(0.09)	−0.76(0.08)	−0.77(0.18)	0.44(0.14)	−0.27(0.02)
SYN/0.05	0.03(0.12)	1.09(0.13)	−1.36(0.11)	−2.11(0.09)	−2.46(0.20)	2.66(0.14)	−0.01(0.03)
SYN/0.5	0.53(0.11)	0.59(0.11)	−1.17(0.10)	−1.88(0.07)	−2.48(0.20)	2.39(0.14)	−0.09(0.02)
SYN/5	1.55(0.11)	0.30(0.10)	−1.03(0.09)	−2.08(0.07)	−1.41(0.17)	1.06(0.12)	−0.07(0.02)
SSYN/0.05	0.53(0.11)	1.08(0.11)	−1.78(0.10)	−1.77(0.07)	−2.24(0.20)	2.38(0.14)	−0.09(0.02)
SSYN/0.5	0.90(0.12)	0.30(0.11)	−0.90(0.10)	−1.88(0.09)	−2.27(0.20)	1.74(0.14)	−0.06(0.02)
SSYN/5	1.03(0.11)	−0.10(0.09)	−0.44(0.09)	−1.78(0.08)	−1.11(0.17)	0.81(0.13)	−0.10(0.02)

References

1. McDougal, J.N.; Boeniger, M.F. Methods for assessing risks of dermal exposures in the workplace. *Crit. Rev. Toxicol.* **2002**, *32*, 291–327. [CrossRef] [PubMed]
2. Semple, S. Dermal Exposure to Chemicals in the Workplace: Just How Important Is Skin Absorption? *Occup. Environ. Med.* **2004**, *61*, 376–382. [CrossRef] [PubMed]
3. U.S. EPA Risk Assessment: Guidance for Superfund Volume I: Human Heatlh Evaluation Manual (Part E, Supplemental Guidance for Dermal Risk Assessment). Available online: https://www.google.com.tw/url?sa=t&rct=j&q=&esrc=s&source=web&cd=1&ved=2ahUKEwjm5Len6uneAhUIV7wKHXKLCX0QFjAAegQICRAC&url=https%3A%2F%2Fwww.epa.gov%2Fsites%2Fproduction%2Ffiles%2F2015-09%2Fdocuments%2Fpart_e_final_revision_10-03-07.pdf&usg=AOvVaw0icHqYHCaEqHYIwkv7WVCg (accessed on 23 November 2018).
4. Eisen, E.A.; Tolbert, P.E.; Monson, R.R.; Smith, T.J. Mortality Studies of Machining Fluid Exposure in the Automobile Industry I: A Standardized Mortality Ratio Analysis. *Am. J. Ind. Med.* **1992**, *22*, 809–824. [CrossRef] [PubMed]
5. Gordon, T. Metalworking Fluid-The Toxicity of a Complex Mixture. *J. Toxicol. Environ. Heal. Part A* **2004**, *67*, 209–219. [CrossRef] [PubMed]
6. Mehta, A.J.; Malloy, E.J.; Applebaum, K.M.; Schwartz, J.; Christiani, D.C.; Eisen, E.A. Reduced lung cancer mortality and exposure to synthetic fluids and biocide in the auto manufacturing industry. *Scand. J. Work. Environ. Health* **2010**, *36*, 499–508. [CrossRef] [PubMed]
7. Monteiro-Riviere, N.A.; Inman, A.O.; Barlow, B.M.; Baynes, R.E. Dermatotoxicity of Cutting Fluid Mixtures: In Vitro and In Vivo Studies. *Cutan. Ocul. Toxicol.* **2006**, *25*, 235–247. [CrossRef] [PubMed]
8. Feldstein, M.M.; Raigorodskii, I.M.; Iordanskii, A.L.; Hadgraft, J. Modeling of percutaneous drug transport in vitro using skin-imitating Carbosil membrane. *J. Control. Release* **1998**, *52*, 25–40. [CrossRef]
9. Moeckly, D.M.; Matheson, L.E. The development of a predictive method for the estimation of flux through polydimethylsiloxane membranes: I. Identification of critical variables for a series of substituted benzenes. *Int. J. Pharm.* **1991**, *77*, 151–162. [CrossRef]
10. Matheson, L.E.; Vayumhasuwan, P.; Moeckly, D.M. The development of a predictive method for the estimation of flux through polydimethylsiloxane membranes. II. Derivation of a diffusion parameter and its application to multisubstituted benzenes. *Int. J. Pharm.* **1991**, *77*, 163–168. [CrossRef]
11. Baynes, R.E.; Xia, X.-R.; Barlow, B.M.; Riviere, J.E. Partitioning Behavior of Aromatic Components in Jet Fuel into Diverse Membrane-coated Fibers. *J. Toxicol. Environ. Heal. Part A* **2007**, *70*, 1879–1887. [CrossRef] [PubMed]
12. Baynes, R.E.; Xia, X.R.; Imran, M.; Riviere, J.E. Quantification of Chemical Mixture Interactions Modulating Dermal Absorption Using a Multiple Membrane Fiber Array. *Chem. Res. Toxicol.* **2008**, *21*, 591–599. [CrossRef] [PubMed]
13. Xia, X.-R.; Baynes, R.E.; Monteiro-Riviere, N.A.; Riviere, J.E. An experimentally based approach for predicting skin permeability of chemicals and drugs using a membrane-coated fiber array. *Toxicol. Appl. Pharmacol.* **2007**, *221*, 320–328. [CrossRef] [PubMed]
14. Xia, X.; Baynes, R.E.; Monteiro-Riviere, N.A.; Leidy, R.B.; Shea, D.; Riviere, J.E. A Novel in-Vitro Technique for Studying Percutaneous Permeation with a Membrane-Coated Fiber and Gas Chromatography/Mass Spectrometry: Part I. Performances of the Technique and Determination of the Permeation Rates and Partition Coefficients of Chemical M. *Pharm. Res.* **2003**, *20*, 275–282. [CrossRef] [PubMed]
15. Potts, R.O.; Guy, R.H. Predicting Skin Permeability. *Pharm. Res.* **1992**, *9*, 663–669. [CrossRef] [PubMed]
16. Abraham, M.H. Scales of solute hydrogen-bonding: Their construction and application to physicochemical and biochemical processes. *Chem. Soc. Rev.* **1993**, *22*, 73–83. [CrossRef]
17. Vijay, V.; White, E.M.; Kaminski, M.D.; Riviere, J.E.; Baynes, R.E. Dermal Permeation of Biocides and Aromatic Chemicals in Three Generic Formulations of Metalworking Fluids. *J. Toxicol. Environ. Heal. Part A* **2009**, *72*, 832–841. [CrossRef] [PubMed]
18. Xu, G.; Hughes-Oliver, J.M.; Brooks, J.D.; Yeatts, J.L.; Baynes, R.E. Selection of appropriate training and validation set chemicals for modelling dermal permeability by U-optimal design. *SAR QSAR Environ. Res.* **2013**, *24*, 135–156. [CrossRef] [PubMed]

19. Xu, G.; Hughes-Oliver, J.M.; Brooks, J.D.; Baynes, R.E. Predicting skin permeability from complex chemical mixtures: Incorporation of an expanded QSAR model. *SAR QSAR Environ. Res.* **2013**, *24*, 711–731. [CrossRef] [PubMed]
20. Abraham, M.H.; Martins, F. Human Skin Permeation and Partition: General Linear Free-Energy Relationship Analyses. *J. Pharm. Sci.* **2004**, *93*, 1508–1523. [CrossRef] [PubMed]
21. Advanced Chemistry Development Inc. ACD/ADME BOXES, Version 4.95. Available online: www.acdlabs.com (accessed on 23 November 2018).

Sample Availability: Samples of the compounds are not available from the authors.

© 2018 by the authors. Licensee MDPI, Basel, Switzerland. This article is an open access article distributed under the terms and conditions of the Creative Commons Attribution (CC BY) license (http://creativecommons.org/licenses/by/4.0/).

Article

Volatile Terpenes and Terpenoids from Workers and Queens of *Monomorium chinense* (Hymenoptera: Formicidae)

Rui Zhao [1,2], Lihua Lu [2], Qingxing Shi [2], Jian Chen [3,*] and Yurong He [1,*]

1. Department of Entomology, College of Agriculture, South China Agricultural University, Tianhe District, Guangzhou 510642, China; 13570453805@163.com
2. Plant Protection Research Institute, Guangdong Academy of Agricultural Sciences, Tianhe District, Guangzhou 510640, China; lhlu@gdppri.com (L.L.); shiqingxing163@163.com (Q.S.)
3. National Biological Control Laboratory, Southeast Area, Agriculture Research Service, United States Department of Agriculture, 59 Lee Road, Stoneville, MS 38776, USA
* Correspondence: jian.chen@ars.usda.gov (J.C.); yrhe@scau.edu.cn (Y.H.)

Academic Editors: Constantinos K. Zacharis and Paraskevas D. Tzanavaras
Received: 8 October 2018; Accepted: 27 October 2018; Published: 1 November 2018

Abstract: Twenty-one volatile terpenes and terpenoids were found in *Monomorium chinense* Santschi (Hymenoptera: Formicidae), a native Chinese ant, by using headspace solid-phase microextraction (HS-SPME) coupled with gas-phase chromatography and mass spectrometry (GC-MS), which makes this ant one of the most prolific terpene producers in insect. A sesquiterpene with unknown structure (terpene 1) was the main terpene in workers and neocembrene in queens. Terpenes and terpenoids were detected in poison, Dufour's and mandibular glands of both workers and queens. Worker ants raised on a terpene-free diet showed the same terpene profile as ants collected in the field, indicating that *de novo* terpene and terpenoid synthesis occurs in *M. chinense*.

Keywords: terpenes; terpenoids; headspace solid phase microextraction; glandular source; *Monomorium chinense*

1. Introduction

Terpenes and terpenoids are the largest group of natural products, mostly produced by plants, but also identified in other eukaryotes such as fungi, insects, amoebae, marine organisms and even prokaryotes, such as, bacteria [1–3]. They have drawn great attention from academia and industry due to not only their economic importance in pharmacy, agriculture, food and perfumery industry, but also their ecological significance in mediating antagonistic and beneficial interactions among organisms [4–6].

Approximately 55,000 terpenes have been reported in nature [7]. According to our literature survey, a total of 220 terpenes and terpenoids were reported in 9 orders of insects (Blattodea, Coleoptera, Diptera, Heteroptera, Homoptera, Hymenoptera, Isoptera, Lepidoptera and Phasmatodea, Table S1). Among them, about forty-five terpenes or terpenoids originated from ants (Hymenoptera: Formicidae) (Table S1). Terpene and terpenoids play significant roles as pheromones and defense compounds. In the subfamily Formicinae, a wide variety of monoterpenes are utilized as alarm pheromones, such as citronellal, citronellol, α-pinene, β-pinene, limonene and camphene [8]. In genera of *Solenopsis* and *Monomorium* of the subfamily Myrmicinae, farnesenes are often used as trail pheromones [9,10]. In the subfamily Ectatomminae, isogeraniol, a monoterpene might function as a recruitment signal in ant *Rhytidoponera metallica* [11]. In addition to the pheromonal role, terpenes and terpenoids are used as defensive compounds, such as iridomyrmecin (cyclopentanoid monoterpenes) and iridodials in some ant species in the subfamily Dolichoderinae [12,13]. However, functions of many terpenes have not been elucidated, such as (E)-β-ocimene and geranylgeraniol in *Labidus praedator* of subfamily Ecitoninae and *Aenictus rotundatus* of subfamily Dorylinae, respectively [14,15].

Terpene and terpenoid biosynthesis have been well studied in plants and microorganisms due to their commercial applications [4]. Terpene biosynthesis and sequestration have also been studied in insects [16,17]. Most insects are herbivores, and previous research showed that terpene sequestration from host plants is common. For example, iridoid glycosides and grayanoid diterpenes were sequestered by certain lepidopteran insects in *Arichanna* and *Euphydryas* from their host plants *Chelone glabra* (Scrophulariaceae) and *Plantago lanceolata* (Plantaginaceae) [18] and the precursor for aggregation pheromone, (−)-*trans*-verbenol, by pine beetles, *Dendroctonus ponderosae*, from pine trees [19]. However, in some herbivorous insect species, terpenes and terpenoids can be produced *de novo*, such as bark beetles and flea beetles, but their biosynthesis pathway of terpenes diverges from that in plants [20,21]. So far, the biosynthesis of terpenes and terpenoids in omnivorous insects like ants has rarely been studied.

Monomorium chinense (Hymenoptera: Formicidae) is one of the most dominant ants in the ground ant community, distributed in Palaearctic and Oriental region, and China is the type locality [22]. Although its workers are tiny and look non-aggressive, the ant can succeed in the competition with the notorious invasive ant *Solenopsis invicta* [23]. It is assumed that the exocrine secretions may play a crucial role in the success of *M. chinense*; however, the chemistry of the exocrine glands of this ant species has not been studied. A preliminary study showed that the ant produces an extraordinary number of terpenes and terpenoids. The objective of this study was to identify these terpenes and terpenoids, determine their glandular origins and investigate the effect of diet on terpene composition in order to find out whether *de novo* terpene and terpenoid synthesis occurs in this species of ant.

2. Results

2.1. Identification of Terpenes and Terpenoids from Whole Bodies of Ants

Total ion chromatograms (TICs) of volatile compounds extracted by solid-phase microextraction (SPME) from *M. chinense* workers and queens are shown in Figure 1. In addition to alkaloids, terpenes and terpenoids were major volatile compounds which are listed in Table 1. For peaks 1, 5 to 9, 11 to 12, 14 to 16 and 18 to 19, compounds were identified as δ-elemene, β-acoradiene, α-neocallitropsene, β-chamigrene, γ-curcumene, aristolochene, β-himachalene, (Z)-α-bisabolene, β-curcumene, 7-epi-α-selinene, β-sesquiphellandrene, γ-cuprenene and 8-cedren-13-ol respectively by comparing their retention times (RTs), Arithmetic indexes (AIs), Kováts indexes (KIs) and mass spectra with compounds in the literatures (Figures S1, S5–S9, S11, S12, S14–S16, S18 and S19). For peak 2, 3 and 4, compounds were identified and confirmed as β-elemene, β-cedrene and (E)-β-farnesene respectively using authentic standards (Figures S2–S4). For peak 21, the compound was identified as neocembrene, a diterpene, since it had RT, AI, KI and mass spectrum matched with neocembrene gas chromatography and mass spectrometry (GC-MS) peak in *M. pharaonis* (Figure S21) [24].

The identities of peaks 10, 13, 17, and 20 could not be finalized because there was no match in RTs with available standards, no match in KIs and AIs with any terpenes and terpenoids in the literature. Therefore the mass spectra of those peaks are presented: peak 10 (terpene 1), [55(52), 79(88), 93(84), 105(92), 119(55), 133(87), 161(100), 175(51), 189(67), 204(78)]; peak 13 (terpene 2), [55(17), 79(22), 93(25), 105(70), 119(54), 133(35), 148(13), 161(97), 189(82), 204(100)]; peak 17 (terpene 3) [55(22), 79(28) 93(40), 105(93), 119(69), 133(44), 148(13), 161(100), 189(43), 204(77)]; and peak 20 (terpenoid 1) [55(49), 67(57), 81(69), 93(71), 107(100), 121(67), 147(34), 161(33), 175(40), 189(58), 217(38), 232(50)]. Peaks 10, 13, 17 are sequiterpenes with molecular ions at m/z 204, predicting their molecular formula $C_{15}H_{24}$. The mass spectrum of terpene 1 is similar to that of (5R^*,7R^*,10S^*)-selina-4(14), 11-diene found in *Nasutitermes* [25]. Peak 20 (terpenoids 1) was considered sesquiterpenoid based on its molecular ion at m/z 232, predicting its molecular formulas $C_{15}H_{20}O_2$. Mass spectra of other unknown terpenes from the ant were provided as well in the supporting material (Figures S10, S13, S17 and S20).

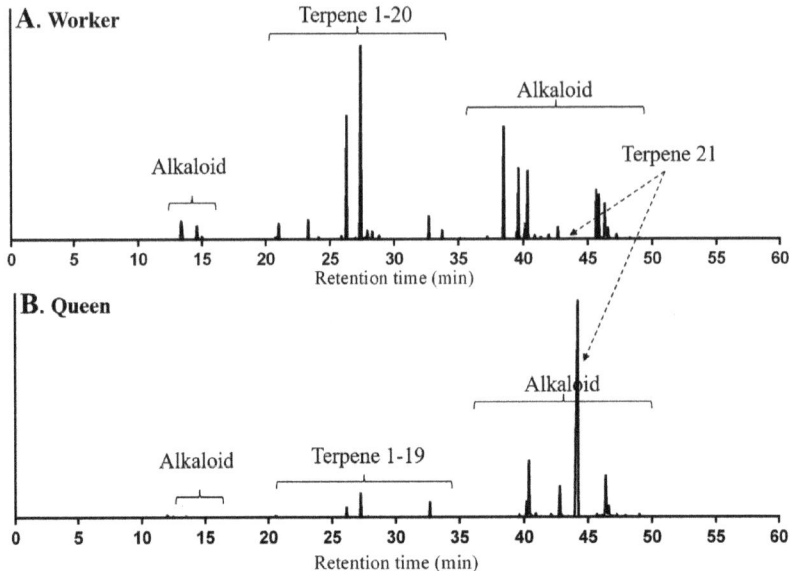

Figure 1. Total ion chromatogram of volatile compounds from *Monomorium chinense* workers (**A**) and queens (**B**) using solid-phase microextraction-gas chromatography-mass spectrometry (SPME-GC-MS) analysis with a DB-5 capillary column.

The squared Mahalanobis distance of terpene composition between workers and queens was 53.22 ($F = 29.94$, $p < 0.05$), indicating the significant difference between two groups. The relative content of terpene 1 in workers was 48.15 ± 2.97% of all compounds as a dominant terpene, followed by β-acoradiene (24.13 ± 2.11%). The other 19 terpenes and terpenoids in a small quantity accounted for 27.72% in average. Terpenoid 1 was only found in workers, but not in queens. For queens, neocembrene accounted for 89.00 ± 1.46% of all compounds as a dominant terpene, but the other 19 terpenes and terpenoids in a small amount accounted for 11.00% in average.

Table 1. Volatile terpenes and terpenoids from workers and queens of *Monomorium chinense*.

Peak No.	Compound	RT (min)	AI	KI	Identification Proposal *	Glandular Source **		Relative Content (Mean ± SE) (%)	
						Worker	Queen	Worker	Queen
1	δ-elemene	21.024	1338	1341	B	PG	Abd	2.7 ± 0.44	0.29 ± 0.14
2	β-elemene	23.292	1392	1393	A	PG	Abd	3.11 ± 0.77	0.3 ± 0.12
3	β-cedrene	24.436	1420	1421	A	MG	Head	0.71 ± 0.19	0.07 ± 0.05
4	(E)-β-farnesene	25.917	1457	1459	A	PG	Abd	0.76 ± 0.28	0.03 ± 0.01
5	β-acoradiene	26.323	1467	1469	B	MG, PG	PG, Head	24.13 ± 2.11	2.01 ± 0.18
6	α-neocallitropsene	26.569	1473	1475	B	Abd	Abd	0.24 ± 0.09	0.02 ± 0.00
7	β-chamigrene	26.708	1477	1478	B	Abd	Abd	0.7 ± 0.28	0.04 ± 0.01
8	γ-curcumene	26.836	1480	1481	B	Abd	Abd	0.22 ± 0.14	0.01 ± 0.01
9	aristolochene	27.012	1484	1485	B	Abd	Abd	0.62 ± 0.49	0.31 ± 0.05
10	terpene 1	27.4	1494	1494	C	PG	PG	48.15 ± 2.97	4.75 ± 0.48
11	β-himachalene	27.617	1499	1499	B	MG	Head	0.84 ± 0.72	0.14 ± 0.11
12	(Z)-α-bisabolene	27.747	1503	1503	B	Abd	Abd	0.63 ± 0.47	0.23 ± 0.12
13	terpene 2	27.974	1509	1509	C	PG	Abd	1.99 ± 0.85	0.15 ± 0.11
14	β-curcumene	28.093	1512	1512	B	Abd	Abd	0.33 ± 0.14	0.07 ± 0.01
15	7-epi-α-Selinene	28.311	1517	1518	B	PG	Abd	1.51 ± 0.2	0.07 ± 0.05
16	β-sesquiphellandrene	28.577	1524	1525	B	Abd	Abd	0.59 ± 0.44	0.05 ± 0.04
17	terpene 3	28.743	1529	1530	C	Abd	Abd	1.19 ± 0.81	0.01 ± 0.00
18	γ-cuprenene	28.853	1531	1533	B	MG, PG	Abd, Head	1.68 ± 0.72	0.09 ± 0.04
19	8-cedren-13-ol	32.721	1633	1634	B	MG	MG	7.04 ± 5.08	2.34 ± 0.62
20	terpenoid 1	33.751	1661	1662	C	DG	-	2.51 ± 0.67	0
21	neocembrene	43.951	1959	1959	B	DG	DG	0.33 ± 0.27	89.00 ± 1.46

* The reliability of the identification proposal is indicated by the following: A, mass spectrum, arithmetic index (AI) and Kováts index (KI) agreed with the standards; B, mass spectrum, arithmetic index and Kováts index agreed with literature data; C, unidentified terpenes or terpenoids, indicating no match of mass spectrum with standards & literature data & mass spectral database. ** Poison gland (PG); Dufour's gland (DG); mandibular gland (MG); abdomen (Abd).

2.2. Origin of Terpenes and Terpenoids

2.2.1. Body Parts

Twenty terpenes and terpenoids were found from whole body samples of both workers and queens, including nineteen sesquiterpenes and sesquiterpenoids (peak 1–19) and one diterpene (peak 21). One sesquiterpenoid, terpenoid 1 (peak 20) appeared only in the workers (Figure 2). TICs of different body parts revealed that abdomen and head were the major sources of terpenes, however no terpene or terpenoid was found in the thorax. For workers and queens, β-acoradiene (Peak 5) and γ-cuprenene (peak 18) were detected in both head and abdomen; and δ-elemene (peak 1), β-elemene (peak 2), (E)-β-farnesene (peak 4), α-neocallitropsene (peak 6), β-chamigrene (peak 7), γ-curcumene (peak 8), aristolochene (peak 9), terpene 1 (peak 10), (Z)-α-bisabolene (peak 12), terpene 2 (peak 13), β-curcumene (peak 14), 7-epi-α-selinene (peak 15), β-sesquiphellandrene (peak 16), terpene 3 (peak 17) and neocembrene (peak 21) only in the abdomen; β-cedrene (peak 3), β-himachalene (peak 11) and 8-cedren-13-ol (peak 19) only in the head. Terpenoid 1 (peak 20) was only found in the abdomen of workers.

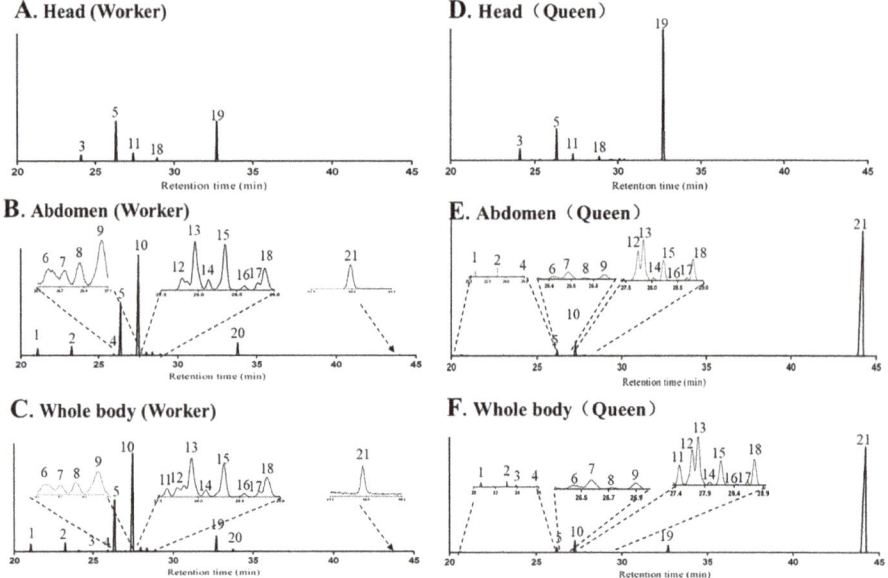

Figure 2. Total ion chromatograms (TICs) of volatile terpenes and terpenoids from the head (**A**), abdomen (**B**) and whole body (**C**) extracts in workers, head (**D**), abdomen (**E**) and whole body (**F**) extracts in queens of *Monomorium chinense* using SPME-GC-MS analysis with a DB-5 capillary column.

2.2.2. Exocrine Glands

In the workers of *M. chinense*, poison and Dufour's glands in the abdomen and mandibular gland in the head were dissected (Figure 3). Seven terpenes including δ-elemene (peak 1), β-elemene (peak 2), β-acoradiene (peak 5), terpene 1 (peak 10), terpene 2 (peak 13), 7-epi-α-selinene (peak 15) and γ-cuprenene (peak 18) were detected in the poison gland, terpenoid 1 (peak 20) and neocembrene (peak 21) in the Dufour's gland, and five terpenes and terpenoids including β-cedrene (peak 3), β-acoradiene (peak 5), β-himachalene (peak 11), γ-cuprenene (peak 18) and 8-cedren-13-ol (peak 19), in the mandibular gland (Figure 4A).

In queens, β-acoradiene (peak 5) and terpene 1 (peak 10) were detected in the poison gland, neocembrene (peak 21) in the Dufour's gland, and 8-cedren-13-ol (peak 19) in the mandibular gland (Figure 4B).

Figure 3. Lateral view of a worker (**A**), poison, Dufour's (**B**), and mandibular glands (**C**) of *Monomorium chinense* workers.

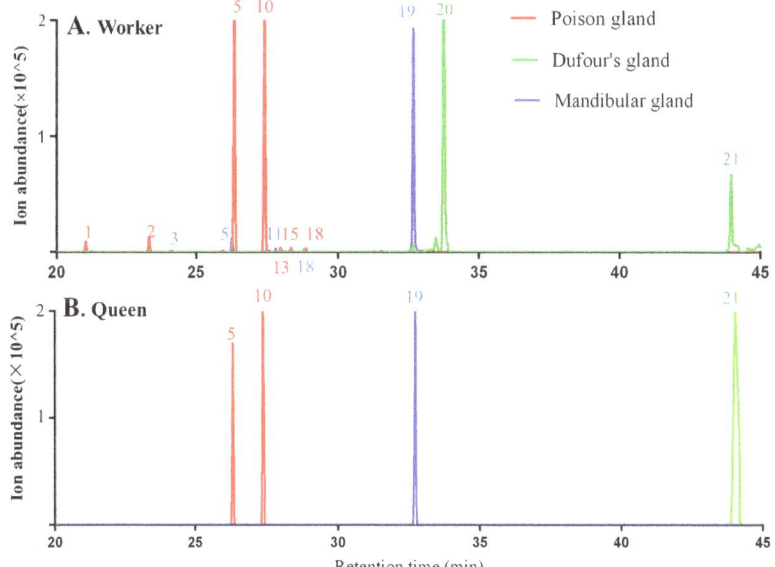

Figure 4. TICs of terpenes and terpenoids from the poison glands, Dufour's glands and mandibular glands of *Monomorium chinense* workers (**A**) and queens (**B**) using SPME-GC-MS analysis with a DB-5 capillary column.

Some terpenes and terpenoids found in ant body parts were not detected in gland samples. For example, nine terpenes in the abdomen of workers and queens were not detected in both poison and Dufour's glands, including α-neocallitropsene (peak 6), β-chamigrene (peak 7), γ-curcumene (peak 8), aristolochene (peak 9), (Z)-α-bisabolene (peak 12), terpene 2 (peak 13), β-curcumene (peak 14), β-sesquiphellandrene (peak 16) and terpene 3 (peak 17). Reduced abundance of all these compounds due to their evaporation during the dissection process may be the reason why they could not be detected by GC-MS in gland samples.

2.2.3. Influence of Diet on Terpene and Terpenoid Profile

TICs of whole-body samples of workers for a field colony (unknown diet), a laboratory colony (controlled diet: mealworm larvae and honey water) and an incipient colonies (terpene-free diet: sucrose water) are shown in Figure 5. Twenty-one terpenes and terpenoids were detected in both the field colonies and the laboratory colonies, in which natural diet were provided, all these compounds were observed also in the incipient colonies, which were fed with the sucrose solution. The squared Mahalanobis distance between field colonies and laboratory colonies, field colonies and incipient colonies, laboratory colonies and incipient colonies was 10.95 ($F = 3.65$, $p = 0.12$), 12.86 ($F = 4.29$, $p = 0.09$) and 7.88 ($F = 2.63$, $p = 0.18$), respectively and all p values were above 0.05, indicating that there was no significant difference of terpene contents among three treatments. Therefore, the terpenes and terpenoids, found in *M. chinense* workers in different treatments were not sequestered from their dietary sources.

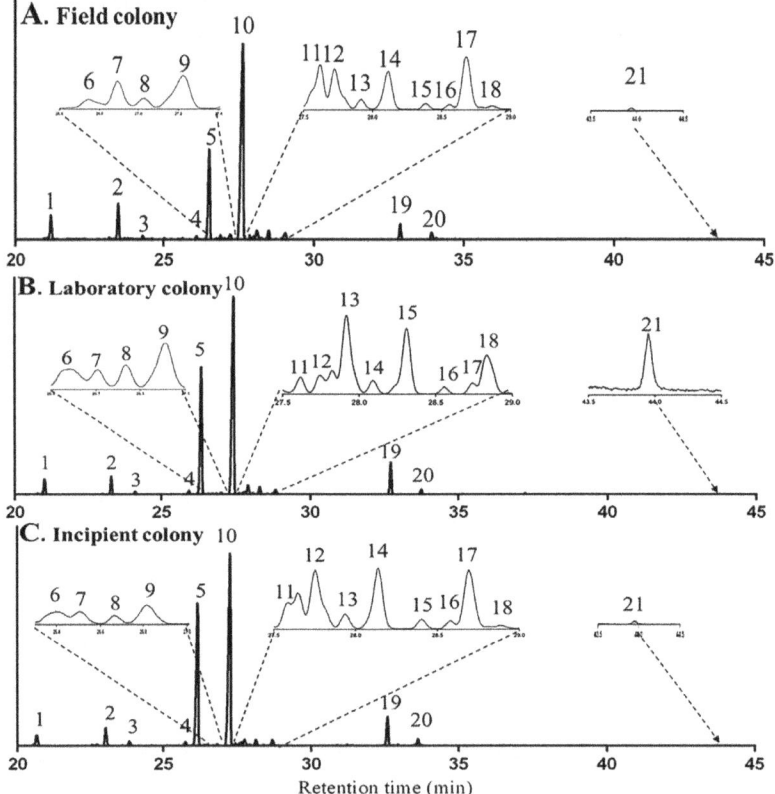

Figure 5. TICs of terpenes and terpenoids of *Monomorium chinense* workers from a field colony (**A**), a laboratory colony (**B**) and an incipient colony (**C**).

3. Discussion

Twenty-one terpenes and terpenoids were detected from workers and twenty from queens of *M. chinense*. Previous studies showed that *Pheidole sinaitica* and *Solenopsis geminata* are the top terpene producers in ants (Hymenoptera: Formicidae) in term of numbers of terpenes and terpenoids detected. For example, the minor workers of *P. sinaitica* contained a mixture of more than 11 sesquiterpenes (farnesene-type hydrocarbons) [26]. The *S. geminata* queens produced 11 sesquiterpenes in the venom secretion and among them β-elemene was only tentatively identified [27]. The results indicate that *M. chinense* is an exceptional terpene producing ant.

In insects, 60 terpenes and terpenoids have been discovered from papilionid larvae (Lepidoptera: Papilionidae) and 53 from termite soldiers (Isoptera: Rhinotermitidae) (Table S1). They seem to be the top terpene and terpenoid producers in insects. In addition to 21 terpenes and terpenoids identified in *M. chinense*, 10 terpenes and terpenoids have been reported in *M. minimum* and five in *M. pharaonic* [28]. These results suggest that *Monomorium* ants may be one of the most potent terpene producers in insects.

To the best of our knowledge, α-neocallitropsene, β-chamigrene and 8-cedren-13-ol have never been reported in insects. The following 10 sesquiterpenes were found for the first time in ants, including δ-elemene, β-cedrene, γ-curcumene, aristolochene, β-himachalene, (Z)-α-bisabolene, β-curcumene, β-sesquiphellandrene, γ-cuprenene, and 7-epi-α-selinene. Some terpenes found in this study have already been reported in other insects. For example, β-elemene was detected in termite *Reticulitermes speratus* (Isoptera: Rhinotermitidae), δ-elemene, γ-curcumene, β-cedrene, β-himachalene, and β-acoradiene in butterflies (Lepidoptera: Papilionidae), β-sesquiphellandrene in stinkbugs *Thyanta pallidovirens* and *Piezodorus guildinii* (Hemiptera: Pentatomidae), and β-curcumene in ciid beetles *Octotemnus glabriculus* and *Cis boleti* (Coleoptera: Ciidae). Terpenes play multiple functions in these insects, serving as inhibitory primer pheromones, queen-recognition and sex pheromones, defensive chemicals against natural enemies, and antimicrobials against pathogens [29–34]. Further research is needed to determine whether terpenes and terpenoids play similar roles in *M. chinense*.

Terpene 1, a sesquiterpene with unknown structure, was the main terpene in workers, followed by β-acoradiene and the remaining terpenes and terpenoids are all minor products. In addition to the major product, nearly half of all characterized monoterpene and sesquiterpene synthases in plants form significant number of minor products [35]. For example, the major sesquiterpene product of valencene synthase was identified as (+)-valencene (49.5% of total product), followed by (−)-7-epi-α-selinene (35.5%) along with five minor products [36]. It is possible that one or few terpene synthases in the workers of *M. chinense* may be responsible for such a diversity of terpenes and terpenoids.

Usually only one type of gland is involved in terpene and terpenoid production and/or storage in one species of insects, such as osmeteria glands in Papilionid (Lepidoptera: Papilionidae) larvae, frontal glands in termite soldiers (Isoptera: Rhinotermitidae, Serritermitidae, and Termitidae), and metasternal glands in longhorned beetles (Coleoptera: Cerambycidae) [37–39]. In contrast, in this study, the terpenes and terpenoids have been detected in three glands, including poison, Dufour's and mandibular glands in *M. chinense* workers and queens. A list of terpenes or terpenoids in ants with glandular source is summarized in Table S2. Monoterpenes and monoterpenoids have been discovered in rectum, mandibular, Dufour's, poison and pygidial glands in Formicinae, Myrmicinae, Dorylinae and Dolichoderinae. Although sesquiterpenes and diterpenes were mostly found from Dufour's gland, they were also detected in Mandibular, Dufour's or venom glands in Formicinae, Myrmicinae and Nothomyrmeciinae. The multiglandular origin of terpenes and terpenoids may make *M. chinense* a unique case in family Fomicidae, maybe even in the class Insecta.

Neocembrene was the major terpene produced in the Dufour's gland of *M. chinense* queens in contrast to its minor abundance in workers. This compound was found in the Dufour's gland in *M. pharaonis* queens, but not in the workers [24]. Whether neocembrene serves as queen pheromone in *M. pharaonis* remains questionable because it does not affect sexual brood rearing [40]. Besides two ant species mentioned above, neocembrene was detected in queens of other four species in the genus *Monomorium*, including *M. minimum* [41], *M. floricola*, *M. destructor* and *M. hiten* [42], indicating that neocembrene may be a genus- and queen-specific compound in genus *Monomorium*. Terpenes and terpenoids do not occur often in poison gland. When limonene, a monoterpene, was first found in poison glands of *Myrmicaria* species, it was considered an unusual case in Formicidae [43]. This study reports that sequiterpenes occur in the poison glands of worker ants. Typical poison gland chemistry of *Monomorium* species was dominated by alkaloids, which were believed to be the reason for them to successfully compete with the highly aggressive ant species [44,45]. This study reveals that not only alkaloids but also sesquiterpenes occur in the poison glands of *M. chinense* workers. Along with alkaloids, terpenes from the poison gland may act synergistically to provide higher toxicity or deterrence. However, the specific functions of these terpenes and terpenoids can only be clarified in the future research.

In plants, terpenoids function universally as primary metabolites, such as sterols, carotenoids, quinones, and hormones [46]. However, most of terpenes and terpenoids in plants are restricted to specific lineages and are involved in species-specific ecological interactions as secondary metabolites that may serve roles in plant defense and communication [47]. Terpenes identified in *M. chinense*, *M. pharaonis* and *M. minimum*, do not occur in genera outside *Monomorium*, indicating these terpenes and terpenoids may also lineage-specific (specialized) terpenoids. Thus, they are most likely also involved in the interaction with other organisms and environment, such as defense against enemies and diseases, or conspecific and heterospecific chemical communications.

All terpenes and terpenoids identified in this study have been found in plants. *M. chinense* is an omnivorous ant as other species in the genus *Monomorium* [48], so it was hypothesized that their diet is one potential source of these terpenes and terpenoids. However, ants raised on a terpene-free diet showed the same terpene profile as those of ants fed with natural diets, indicating that *de novo* terpene synthesis occurs in *M. chinense*. The terpene biosynthesis of bark beetles and flea beetles is well studied. Both beetles are oligophagous herbivores. Ivarsson et al. provided the first evidence that bark beetle *Ips duplicatus* can produce their main pheromone component, ipsdienol, a terpene alcohol [49]. Radiolabeling studies provided further evidence of the *de novo* biosynthesis of terpenes by bark beetles [50]. Geranyl diphosphate synthase of bark beetle *Ips pini* is the first animal prenyltransferase having terpene synthase activity [21]. No sesquiterpene synthases have been described in insects until the identification of an evolutionarily novel terpene synthase gene family in the striped flea beetle [51]. Terpene and terpenoid biosynthesis in ants have not really been studied by researchers. Considering the significance of terpenes and terpenoids in pharmacy, agriculture, food and perfumery industry, understanding and characterizing terpene synthases in ants may become important, since ant terpene synthase genes may provide us with new opportunities in bioengineering for production of high-valued terpenes and terpenoids. Due to its exceptional ability in terpene production, *M. chinense* may be a good model insect for study terpene biosynthesis in ants.

4. Materials and Methods

4.1. Ants

4.1.1. Maintenance of Field-Collected Ant Colonies

Nine colonies of *M. chinense* were collected in Guangzhou, Guangdong, China, and among them 3 colonies were collected from the campus of Guangdong Academy of Agriculture Science (GAAS) in July 2016, 3 from Baiyun district in April 2017 and 3 from Nansha district in April 2017. The colonies were reared in a 45 × 38 × 15 cm plastic container with the inner sides of the wall coated with Fluon F4-1 (Xingshengjie Sci and Tech Co., Ltd., Guangzhou, China) to prevent the escape of ants. Three glass test tubes (2.5 Φ × 19.5 cm) were placed in the container and used as artificial nests. Each tube was filled with 4–5 cm of water and a cotton plug was placed in the tube at the water level to retain the water. Tubes were covered with black paper to shield the light. Colonies were provided with minced mealworm, *Tenebrio molitor*, a cotton ball saturated with a 20% honey water solution, and a cotton ball with pure water in a Petri dish (7 × 1.5 cm). These colonies were maintained at 26 ± 2 °C and 12:12 (L:D) h photoperiod.

4.1.2. Establishment of Incipient Colonies

Incipient colonies were established by introducing newly dealate queens with 20 workers from the laboratory colonies into a container (45 × 38 × 15 cm). Once young workers emerged in the new colony, the old workers were removed. Colonies were provided with a cotton ball saturated with a 20% sucrose water solution, and a cotton ball with pure water in a Petri dish (7 × 1.5 m). They were maintained at 26 ± 2 °C and 12:12 (L:D) h photoperiod.

4.2. Chemical Analysis of Ant Volatile Terpenes and Terpenoids

4.2.1. Ant Sample Preparation and Extraction by HS-SPME

About 100 live ant workers or 10 queens were put into a 2 mL vial (Agilent Technologies, Santa Clara, CA, USA). In order to facilitate the release of volatiles from the sample into the head space, the vial was placed into a −80 °C refrigerator for 10 min [28]. Headspace solid-phase micro-extraction (HS-SPME) was then conducted on the sample at room temperature (25 ± 1 °C) for 12 h using an 85 µm Polyacrylate SPME fiber (Supelco Inc., Bellefonte, PA, USA). In order to add C_8 to C_{20} hydrocarbon standards to the sample, after the sample extraction, the same fiber was used to extract hydrocarbon standards for 1 min in another 2 mL vial. The hydrocarbon standards were prepared by adding 20 µL C_8 to C_{20} solution (Sigma-Aldrich, St. Louis, MO, USA, 40 mg/L) into the vial and letting solvent evaporate in a fume hood. In order to facilitate evaporation of the solvent, the capped vial was shaken for 5 s before it was opened in a fume hood. After 1 min of evaporation, the vial was capped and shaken again for 5 s before it was reopened in the fume hood for 9 min. Before each SPME sample extraction, a blank run was performed and the fiber was cleaned in the GC injector for 30 min. There were 5 replicates for each colony.

4.2.2. Determination of Glandular Sources of Terpenes and Terpenoids

Each worker or queen was cut into three major body parts (head, thorax and abdomen) by a razor blade. Each type of body parts was placed into 2 mL vial that was subjected to SPME extraction as described as above. Since terpenes and terpenoids were found in the head and abdomen, the chemistry of the poison gland and Dufour's gland in abdomen and mandibular gland in head were investigated. Because poison gland and Dufour's gland are connected, they were first removed from the body under a stereo microscope (SZ61, Olympus, Tokyo, Japan) by grasping the terminal abdominal segments or the stinger with fine forceps and pulling posteriorly. The poison and Dufour's glands were separated with a dissecting needle. The mandibular gland was removed by grasping the mandible away from head, then separating the gland using a dissecting needle. After separation, each gland was directly placed on the tip of the SPME fiber, which then was inserted into the inject port of the GC-MS system (Agilent Technologies, Santa Clara, CA, USA).

4.2.3. Gas Chromatography and Mass Spectrometry (GC-MS)

The samples were analyzed using GC-MS Agilent 7890A-gas chromatograph coupled with 5975B-mass spectrometer. The analytical conditions were used as follows, splitless injection at 250 °C, DB-5 column (30 m × 0.25 mm i.d., 0.25 µm film thickness), the temperature program was from 60 °C to 246 °C at 3 °C·min^{-1}. Injector temperature was 220 °C and transfer line temperature 240 °C. The mass spectrometer was operated at 70 eV in the electron impact mode.

4.3. Data Analysis

Arithmetic index (AI) and Kováts index (KI) of target compounds were calculated using the following formula [52]:

$$KI(x) = 100 P_Z + 100[(\log RT(x) - \log RT(P_Z))/(\log RT(P_{Z+1}) - \log RT(P_Z))]$$

$$AI(x) = 100 P_Z + 100[(RT(x) - RT(P_Z))/(RT(P_{Z+1}) - RT(P_Z))]$$

where: $RT(P_Z) \leq RT(x) \leq RT(P_{Z+1})$, and $P_8 \ldots P_{20}$ were *n*-paraffins. (up to N = 20 in the paper).

Terpenes and terpenoids were identified by comparing retention times (RT), AIs and KIs, and mass spectra of compounds with synthetic standards and compounds in literature [51] and libraries [NIST (National Institute of Standards and Technology, Gaithersburg, MD, USA) and Wiley (John Wiley & Sons, Inc., Hoboken, New Jersey, USA)]. Synthesized compounds of β-elemene, β-cedrene and (*E*)-β-farnesene were purchased from Sigma-Aldrich (St. Louis, MO, USA). Since neocembrene was originally identified in *Monomorium pharaonis* queens [24], the neocembrene extracted from *M. pharaonis* queens was used as a standard for identification of the compound in *M. chinense*. The pharaoh ant colonies were reared in the Laboratory of Biological Invasion, Plant Protection Research Institute, GAAS.

Relative peak area of each terpene or terpenoid was calculated in percentage over the total area of all peaks. To estimate the difference of terpene composition between worker and queens, and the difference among three groups (field colonies, laboratory colonies and incipient colonies), a total of 21 terpene peak relative contents were used as variables in a principal component analysis and the principal components extracted were used as independent variables in the subsequent discriminant analysis and the squared Mahalanobis distances (D2) between the clusters were calculated. There were 3 replicates for each colony. STATISTICA 10.0 (Palo Alto, CA, USA), was used in statistical analyses.

5. Conclusions

In summary, twenty-one volatile terpenes and terpenoids were found in the Chinese ant, *Monomorium chinense* using headspace solid-phase micro-extraction (HS-SPME) coupled with gas-phase chromatography and mass spectrometry (GC-MS). The discovery makes *M. chinense* the most prolific terpene producer in ants. A sesquiterpene with unknown structure terpene 1 and neocembrene are the main terpene in the workers and queens, respectively. Most terpenes and terpenoids were found in the poison, Dufour's and/or mandibular glands. *De novo* terpenes and terpenoids synthesis are demonstrated in in *M. chinense* its workers. These findings suggest *M. chinense* is a novel and promising organism for the study of terpene function and biosynthesis in ants.

Supplementary Materials: The following are available online. Mass spectra of terpenes and terpenoids showed in figures (Figures S1–S21); Terpenes and terpenoids in insects, terpenes and terpenoids in ants and their glandular source summarized in tables (Tables S1 and S2).

Author Contributions: Conceptualization, J.C. and Y.H.; methodology, J.C., R.Z. and L.L.; formal analysis, J.C., R.Z. and L.L.; investigation, R.Z.; data curation, R.Z. and Q.S.; writing—original draft preparation, R.Z.; writing—review and editing, R.Z., J.C., L.L., and Q.S.; supervision and project administration, L.L.

Funding: This work was financially supported by the National Science & Technology Pillar Program during the Twelfth Five-Year Plan Period (L.L., grant nr. 2015BAD08B02), the National Key R&D Program of China (2017YFC1200600) from the government of the Peoples Republic of China and Non-Funded Cooperative Agreement (J.C. & L.L., grant nr. 6066-22320-009-01-N) from the United Stated government.

Acknowledgments: We would like to thank Jian Yan, College of Natural Resources and Environment, South China Agricultural University, for comments and suggestions of the manuscript.

Conflicts of Interest: The authors declare no conflict of interest. The founders had no role in the design of the study; in the analyses, or interpretation of data; in the writing of the manuscript, or in the decision to publish the results.

References

1. Connolly, J.D.; Hill, R.A. *Dictionary of Terpenoids*, 1st ed.; Chapman & Hall: London, UK, 1991; ISBN 0-412-25770-X.
2. Chen, X.; Köllner, T.G.; Jia, Q.; Norris, A.; Santhanam, B.; Rabe, P.; Dickschat, J.S.; Shaulsky, G.; Gershenzon, J.; Chen, F. Terpene synthase genes in eukaryotes beyond plants and fungi: Occurrence in social amoebae. *Proc. Natl. Acad. Sci. USA* **2016**, *113*, 12132–12137. [CrossRef] [PubMed]
3. Yamada, Y.; Kuzuyama, T.; Komatsu, M.; Shinya, K.; Omura, S.; Cane, D.E.; Ikeda, H. Terpene synthases are widely distributed in bacteria. *Proc. Natl. Acad. Sci. USA* **2015**, *112*, 857–862. [CrossRef] [PubMed]
4. Breitmaier, E. *Terpenes: Flavors, Fragrances, Pharmaca, Pheromones*, 1st ed.; Wiley-VCH: Wallingford, UK, 2006; p. 214. ISBN 3-527-31786-4.
5. Gershenzon, J.; Dudareva, N. The function of terpene natural products in the natural world. *Nat. Chem. Biol.* **2007**, *3*, 408. [CrossRef] [PubMed]
6. Gnankiné, O.; Bassolé, I. Essential oils as an alternative to pyrethroids' resistance against *Anopheles* species complex giles (Diptera: Culicidae). *Molecules* **2017**, *22*, 1321. [CrossRef] [PubMed]
7. Leavell, M.D.; Mcphee, D.J.; Paddon, C.J. Developing fermentative terpenoid production for commercial usage. *Curr. Opin. Biotechnol.* **2016**, *37*, 114–119. [CrossRef] [PubMed]
8. Wilson, E.O.; Regnier, F.E. The evolution of the alarm-defense system in the Formicine ants. *Am. Nat.* **1971**, *105*, 279–289. [CrossRef]
9. Meer, R.K.V.; Alvarez, F.; Lofgren, C.S. Isolation of the trail recruitment pheromone of *Solenopsis invicta*. *J. Chem. Ecol.* **1988**, *14*, 825–838. [CrossRef] [PubMed]
10. Ritter, F.J.; Brüggemann-Rotgans, I.E.M.; Verwiel, P.E.J.; Persoons, C.J.; Talman, E. Trail pheromone of the Pharaoh's ant, *Monomorium pharaonis*: Isolation and identification of faranal, a terpenoid related to juvenile hormone II. *Tetrahedron Lett.* **1977**, *18*, 2617–2618. [CrossRef]
11. Meinwald, J.; Wiemer, D.F.; Hölldobler, B. Pygidial gland secretions of the ponerine ant *Rhytidoponera metallica*. *Naturwissenschaften* **1983**, *70*, 46–47. [CrossRef]
12. Cavill, G.W.K.; Robertson, P.L.; Brophy, J.J.; Clark, D.V.; Duke, R.; Orton, C.J.; Plant, W.D. Defensive and other secretions of the Australian cocktail ant, *Iridomyrmex nitidiceps*. *Tetrahedron* **1982**, *38*, 1931–1938. [CrossRef]
13. Welzel, K.F.; Lee, S.H.; Dossey, A.T.; Chauhan, K.R.; Choe, D.H. Verification of Argentine ant defensive compounds and their behavioral effects on heterospecific competitors and conspecific nestmates. *Sci. Rep.* **2018**, *8*, 1477. [CrossRef] [PubMed]
14. Oldham, N.J.; Morgan, E.D.; Gobin, B.; Schoeters, E.; Billen, J. Volatile secretions of old world army ant *Aenictus rotundatus* and chemotaxonomic implications of army ant Dufour's gland chemistry. *J. Chem. Ecol.* **1994**, *20*, 3297–3305. [CrossRef] [PubMed]
15. Keegans, S.J.; Billen, J.; Morgan, E.D.; Gökcen, O.A. Volatile glandular secretions of three species of new world army ants, *Eciton burchelli*, *Labidus coecus*, and *Labidus praedator*. *J. Chem. Ecol.* **1993**, *19*, 2705–2719. [CrossRef] [PubMed]
16. Hick, A.J.; Luszniak, M.C.; Pickett, J.A. Volatile isoprenoids that control insect behaviour and development. *Nat. Prod. Rep.* **1999**, *16*, 39–54. [CrossRef]
17. Kunert, M.; Søe, A.; Bartram, S.; Discher, S.; Tolzin-Banasch, K.; Nie, L.; David, A.; Pasteels, J.; Boland, W. De novo biosynthesis versus sequestration: A network of transport systems supports in iridoid producing leaf beetle larvae both modes of defense. *Insect Biochem. Mol.* **2008**, *38*, 895–904. [CrossRef] [PubMed]
18. Bowers, M.D.; Puttick, G.M. Fate of ingested iridoid glycosides in lepidopteran herbivores. *J. Chem. Ecol.* **1986**, *12*, 169–178. [CrossRef] [PubMed]
19. Taft, S.; Najar, A.; Erbilgin, N. Pheromone production by an invasive bark beetle varies with monoterpene composition of its naïve host. *J. Chem. Ecol.* **2015**, *41*, 540–549. [CrossRef] [PubMed]
20. Burse, A.; Boland, W. Deciphering the route to cyclic monoterpenes in Chrysomelina leaf beetles: Source of new biocatalysts for industrial application? *Z. Naturforsch. C* **2017**, *72*, 417–427. [CrossRef] [PubMed]
21. Gilg, A.B.; Tittiger, C.; Blomquist, G.J. Unique animal prenyltransferase with monoterpene synthase activity. *Naturwissenschaften* **2009**, *96*, 731–735. [CrossRef] [PubMed]
22. Bolton, B. A review of the *Solenopsis* genus-group and revision of Afrotropical *Monomorium* Mayr (Hymenoptera: Formicidae). *Bull. Br. Mus. Entomol.* **1987**, *39*, 263–452. [CrossRef]

23. Chen, Y.C.; Kafle, L.; Shih, C.J. Interspecific competition between *Solenopsis invicta* and two native ant species, *Pheidole fervens* and *Monomorium chinense*. *J. Econ. Entomol.* **2011**, *104*, 614–621. [CrossRef] [PubMed]
24. Edwards, J.P.; Chambers, J. Identification and source of a queen-specific chemical in the pharaoh's ant, *Monomorium pharaonis* (L.). *J. Chem. Ecol.* **1984**, *10*, 1731–1747. [CrossRef] [PubMed]
25. Everaerts, C.; Roisin, Y.; Quéré, J.L.L.; Bonnard, O.; Pasteels, J.M. Sesquiterpenes in the frontal gland secretions of nasute soldier termites from New Guinea. *J. Chem. Ecol.* **1993**, *19*, 2865–2879. [CrossRef] [PubMed]
26. Ali, M.F.; Jackson, B.D.; Morgan, E.D. Contents of the poison apparatus of some species of *Pheidole* ants. *Biochem. Syst. Ecol.* **2007**, *35*, 641–651. [CrossRef]
27. Cruz-López, L.; Rojas, J.C.; De, L.C.R.; Morgan, E.D. Behavioral and chemical analysis of venom gland secretion of queens of the ant *Solenopsis geminata*. *J. Chem. Ecol.* **2001**, *27*, 2437–2445. [CrossRef] [PubMed]
28. Chen, J. Freeze-thaw sample preparation method improves detection of volatile compounds in insects using headspace solid-phase microextraction. *Anal. Chem.* **2017**, *89*, 8366–8371. [CrossRef] [PubMed]
29. Borges, M.; Millar, J.G.; Laumann, R.A.; Moraes, M.C. A male-produced sex pheromone from the neotropical redbanded stink bug, *Piezodorus guildinii* (W.). *J. Chem. Ecol.* **2007**, *33*, 1235–1248. [CrossRef] [PubMed]
30. Frankfater, C.; Tellez, M.R.; Slattery, M. The scent of alarm: Ontogenetic and genetic variation in the osmeterial gland chemistry of *Papilio glaucus* (Papilionidae) caterpillars. *Chemoecology* **2009**, *19*, 81–96. [CrossRef]
31. Guevara, R.; Hutcheson, K.A.; Mee, A.C.; Rayner, A.D.M.; Reynolds, S.E. Resource partitioning of the host fungus *Coriolus versicolor* by two ciid beetles: The role of odour compounds and host ageing. *Oikos* **2000**, *91*, 184–194. [CrossRef]
32. Mcbrien, H.L.; Millar, J.G.; Rice, R.E.; Mcelfresh, J.S.; Cullen, E.; Zalom, F.G. Sex attractant pheromone of the red-shouldered stink bug *Thyanta pallidovirens*: A pheromone blend with multiple redundant components. *J. Chem. Ecol.* **2002**, *28*, 1797–1818. [CrossRef] [PubMed]
33. Mitaka, Y.; Mori, N.; Matsuura, K. Multi-functional roles of a soldier-specific volatile as a worker arrestant, primer pheromone and an antimicrobial agent in a termite. *Proc. Biol. Sci.* **2017**, *284*, 20171134. [CrossRef] [PubMed]
34. Hisashi, Ô.; Noguchi, T.; Nehira, T. New oxygenated himachalenes in male-specific odor of the Chinese windmill butterfly, *Byasa alcinousalcinous*. *Nat. Prod. Res.* **2016**, *30*, 406–411. [CrossRef]
35. Degenhardt, J.; Köllner, T.G.; Gershenzon, J. Monoterpene and sesquiterpene synthases and the origin of terpene skeletal diversity in plants. *Phytochemistry* **2009**, *70*, 1621–1637. [CrossRef] [PubMed]
36. Lücker, J.; Bowen, P.; Bohlmann, J. *Vitis vinifera* terpenoid cyclases: Functional identification of two sesquiterpene synthase cDNAs encoding (+)-valencene synthase and (−)-germacrene D synthase and expression of mono- and sesquiterpene synthases in grapevine flowers and berries. *Phytochemistry* **2004**, *65*, 2649–2659. [CrossRef] [PubMed]
37. Honda, K.; Hayashi, N. Chemical nature of larval osmeterial secretions of papilionid butterflies in the genera *Parnassius*, *Sericinus* and *Pachliopta*. *J. Chem. Ecol.* **1995**, *21*, 859–867. [CrossRef] [PubMed]
38. Krasulová, J.; Hanus, R.; Kutalová, K.; Jan, Š.; Sillam-Dussès, D.; Tichý, M.; Valterová, I. Chemistry and anatomy of the frontal gland in soldiers of the sand termite *Psammotermes hybostoma*. *J. Chem. Ecol.* **2012**, *38*, 557–565. [CrossRef] [PubMed]
39. Ohmura, W.; Hishiyama, S.; Nakashima, T.; Kato, A.; Makihara, H.; Ohira, T.; Irei, H. Chemical composition of the defensive secretion of the longhorned beetle, *Chloridolum loochooanum*. *J. Chem. Ecol.* **2009**, *35*, 250–255. [CrossRef] [PubMed]
40. Boonen, S.; Billen, J. Caste regulation in the ant *Monomorium pharaonis* (L.) with emphasis on the role of queens. *Insects Soc.* **2017**, *64*, 113–121. [CrossRef]
41. Chen, J.; Cantrell, C.L.; Oi, D.; Grodowitz, M.J. Update on the defensive chemicals of the little black ant, *Monomorium minimum* (Hymenoptera: Formicidae). *Toxicon* **2016**, *122*, 127–132. [CrossRef] [PubMed]
42. Zhao, R.; Lu, L.; Shi, Q.; Chen, J.; He, Y. Contents of Dufour's Gland of the Queens of Monomorium Species (Hymenoptera: Formicidae). **2018**, unpublished work.
43. Brand, J.M.; Blum, M.S.; Lloyd, H.A.; Fletcher, D.J.C. Monoterpene hydrocarbons in the poison gland secretion of the ant *Myrmicaria natalensis* (Hymenoptera: Formicidae). *Ann. Entomol. Soc. Am.* **1974**, *67*, 525–526. [CrossRef]

44. Jones, T.H.; Blum, M.S.; Howard, R.W.; Mcdaniel, C.A.; Fales, H.M.; Dubois, M.B.; Torres, J. Venom chemistry of ants in the genus *Monomorium*. *J. Chem. Ecol.* **1982**, *8*, 285–300. [CrossRef] [PubMed]
45. Andersen, A.N.; Blum, M.S.; Jones, T.H. Venom alkaloids in *Monomorium* "*rothsteini*" Forel repel other ants: Is this the secret to success by *Monomorium* in Australian ant communities? *Oecologia* **1991**, *88*, 157–160. [CrossRef] [PubMed]
46. Kirby, J.; Keasling, J.D. Biosynthesis of plant isoprenoids: Perspectives for microbial engineering. *Annu. Rev. Plant Biol.* **2009**, *60*, 335–355. [CrossRef] [PubMed]
47. Pichersky, E.; Gang, D.R. Genetics and biochemistry of secondary metabolites in plants: An evolutionary perspective. *Trends Plant Sci.* **2000**, *5*, 439–445. [CrossRef]
48. García-Martínez, M.A.; Martínez-Tlapa, D.L.; Pérez-Toledo, G.R.; Quiroz-Robledo, L.N.; Castaño-Meneses, G.; Laborde, J.; Valenzuela-González, J.E. Taxonomic, species and functional group diversity of ants in a tropical anthropogenic landscape. *Trop. Conserv. Sci.* **2015**, *8*, 1017–1032. [CrossRef]
49. Ivarsson, P.; Schlyter, F.; Birgersson, G. Demonstration of de novo pheromone biosynthesis in *Ips duplicatus* (Coleoptera: Scolytidae): Inhibition of ipsdienol and *E*-myrcenol production by compactin. *Insect Biochem. Mol.* **1993**, *23*, 655–662. [CrossRef]
50. Seybold, S.J.; Quilici, D.R.; Tillman, J.A.; Vanderwel, D.; Wood, D.L.; Blomquist, G.J. De novo biosynthesis of the aggregation pheromone components ipsenol and ipsdienol by the pine bark beetles *Ips paraconfusus* Lanier and *Ips pini* (Say) (Coleoptera: Scolytidae). *Proc. Natl. Acad. Sci. USA* **1995**, *92*, 8393–8397. [CrossRef] [PubMed]
51. Beran, F.; Rahfeld, P.; Luck, K.; Nagel, R.; Vogel, H.; Wielsch, N.; Irmisch, S.; Ramasamy, S.; Gershenzon, J.; Heckel, D.G. Novel family of terpene synthases evolved from trans-isoprenyl diphosphate synthases in a flea beetle. *Proc. Natl. Acad. Sci. USA* **2016**, *113*, 2922–2927. [CrossRef] [PubMed]
52. Adams, R.P. *Identification of Essential Oil Components by Gas Chromatography/Mass Spectrometry*, 4th ed.; Allured Business Media: Carol Stream, IL, USA, 2009; pp. 53–788. ISBN 0-931710-85-5.

Sample Availability: Samples of the compounds are not available from the authors.

© 2018 by the authors. Licensee MDPI, Basel, Switzerland. This article is an open access article distributed under the terms and conditions of the Creative Commons Attribution (CC BY) license (http://creativecommons.org/licenses/by/4.0/).

Communication

Quantification of VOC Emissions from Carbonized Refuse-Derived Fuel Using Solid-Phase Microextraction and Gas Chromatography-Mass Spectrometry

Andrzej Białowiec [1,*], Monika Micuda [1], Antoni Szumny [2], Jacek Łyczko [2] and Jacek A. Koziel [3]

1. Faculty of Life Sciences and Technology, Wrocław University of Environmental and Life Sciences, Wrocław 50-375, Poland; micuda.monika@gmail.com
2. Faculty of Biotechnology and Food Science, Wrocław University of Environmental and Life Sciences, Wrocław 50-375, Poland; Antoni.szumny@upwr.edu.pl (A.S.); jacek.lyczko@upwr.edu.pl (J.Ł.)
3. Department of Agricultural and Biosystems Engineering, Iowa State University, Ames IA 50011, USA; koziel@iastate.edu
* Correspondence: andrzej.bialowiec@upwr.edu.pl; Tel.: +48-71-320-5973

Academic Editors: Constantinos K. Zacharis and Paraskevas D. Tzanavaras
Received: 14 November 2018; Accepted: 4 December 2018; Published: 5 December 2018

Abstract: In this work, for the first time, the volatile organic compound (VOC) emissions from carbonized refuse-derived fuel (CRDF) were quantified on a laboratory scale. The analyzed CRDF was generated from the torrefaction of municipal waste. Headspace solid-phase microextraction (SPME) and gas chromatography-mass spectrometry (GC-MS) was used to identify 84 VOCs, including many that are toxic, e.g., derivatives of benzene or toluene. The highest emissions were measured for nonanal, octanal, and heptanal. The top 10 most emitted VOCs contributed to almost 65% of the total emissions. The VOC mixture emitted from torrefied CRDF differed from that emitted by other types of pyrolyzed biochars, produced from different types of feedstock, and under different pyrolysis conditions. SPME was a useful technology for surveying VOC emissions. Results provide an initial database of the types and relative quantities of VOCs emitted from CRDF. This data is needed for further development of CRDF technology and comprehensive assessment of environmental impact and practical storage, transport, and potential adoption of CRDF as means of energy and resource recovery from municipal waste.

Keywords: volatile organic compounds; torrefaction; waste to carbon; biochar; municipal solid waste; SPME

1. Introduction

Biochar is a fine-grained product characterized by a high content of organic carbon and low susceptibility to decomposition. It is obtained in the process of torrefaction, pyrolysis, or gasification of plant biomass, biodegradable waste, and sewage sludge [1]. The European Biochar Certificate [2] defines the carbon content above 50% of dry matter as the main requirement for biochar classification. Biochar has a wide range of applications with more than 50 already documented [3]. Biochars' intended use depends on the production process characteristics, primarily calorific value and the specific surface area [3]. The substrates used in the production of biochar include [4]: wood biomass, agricultural biomass (e.g., crop residues), energy crops (e.g., Miscanthus, energetic willow, Virginia mallow), organic waste including: organic fraction of municipal waste [5,6], waste from agro-food processing (e.g., oat fermentation, rice husks, nut shells, pomace), waste from poultry processing, animal manure, biomass from algae, digestate from biogas plants [7] and sewage sludge [8].

Municipal waste is used increasingly as a resource to recover energy and materials via thermal processes. The direction being actively pursued in the field of torrefaction and pyrolysis of municipal waste is the conversion of the fraction of combustible fraction of waste (a.k.a. refuse-derived fuel; RDF) into high-calorific solid fuel (CRDF) as a 'Waste-To-Carbon' waste management strategy [5,9].

One of the challenges related to the development of torrefaction and pyrolysis technology for municipal waste is the expected potential environmental impact of biochar through emissions of volatile organic compounds (VOCs). The VOCs are defined as any organic compound with an initial boiling point less than or equal to 250 °C measured at a standard pressure of 101.3 kPa [10], i.e., capable of off-gassing potentially hazardous compounds during production, storage, transportation, and use. This working hypothesis is derived by analogy to previous studies on the qualitative and quantitative analysis of VOCs content in biochars from biomass. From the research on other types of (pyrolyzed) biochar conducted so far, the occurrence of up to 140 [11] VOCs was observed, of which 74 were identified. The most frequently observed compounds in biochar from pyrolysis were acetone, benzene, methyl ethyl ketone, toluene, methyl acetate, ethanol, phenol, and cresols. Buss et al. [12] reported elevated levels of aliphatic acids and naphthalene. The 'Char Team 2015' reported 26 VOCs [13].

The problem of VOCs emissions from biochar was also reported by Taherymoosavi et al. [14], who analyzed biochar from compost. Particular attention has been paid to the generation of VOCs from the BTEX group in biocarbon as compared to raw materials. The content of VOCs in biochar depends on substrates as well as the process in which the char is produced [15]. Spokas et al. [11] compared the processes of biocarbon formation in terms of VOC content. For this purpose, biochar originating from various substrates, e.g., coconut, hardwood, and pig manure, produced at various process temperatures from 200 °C to 800 °C were subjected to analyses on GC-MS. A relationship was observed that the higher the temperature of the biochar formation process, the smaller the amount of VOCs emitted [11]. The highest number of absorbed VOCs is observed in the biochar derived under hydrothermal carbonization and rapid pyrolysis. These were primarily furans and aldehydes. Similar conclusions came from Wang et al. [15] analyzing the content of PAHs in biocarbon. The lowest concentration of PAH was observed in slow pyrolysis and longer retention (inside reactor) time. Thus, it is reasonable to expect that a torrefied (i.e., low temperature) process used for RDF production will result in greater VOC emissions.

To date, published literature on VOC emissions focuses on biochars produced from biomass, mainly via pyrolysis. However, no VOC emissions have been evaluated from biochars produced from municipal waste, and in particular from torrefied RDF, a new potential future fuel source in a circular economy. For this reason, the main purpose of the work was to identify VOCs emitted from carbonized-RDF (CRDF) biochar produced via torrefaction from RDF and quantify their emissions. Information is needed about the types and quantities of VOCs emitted. This, in turn, can address many practical questions about the potential toxicity; storage, transport, and adoption of CRDF as a future energy source.

2. Results

In this work, for the first time, emissions of VOCs from CRDF was studied qualitatively and quantitatively. Qualitative analysis consisted of identifying compounds based on MS spectral database and available literature (Kovats Retention C7-40 Index). Table 1 shows the VOCs emitted from the analyzed CRDF with the GC column retention time and the coefficient both in the literature and with the GC software presented in the database (Kovats Retention C7-40 Index). Also included was the internal standard (2-undecanone) added during analyzes (compound #80).

The 84 VOCs (without internal standard) have been identified. These compounds belong to various groups such as alcohols (e.g., pentanol), aldehydes, (e.g., nonanal, octanal, heptanal, hexanal, furfural), ketones (e.g., heptanone), aromatic compounds (including toluene and benzene derivatives), polycyclic aromatic hydrocarbons (PAHs; including naphthalene derivatives considered toxic), acids (e.g., acetic, benzoic), alkenes (e.g., styrene), phenols and a large group of heterocyclic compounds (including pyridine and pyrazine with derivatives).

The largest (by number) group were derivatives of benzene and naphthalene (e.g., tetralin). The highest density of peak elution of VOCs from the chromatographic column occurred between 7 to 12 min (Table 1). Most of the identified compounds had boiling points between 100 and 240 °C; i.e., the typical range of VOCs [16]. One compound was classified as very volatile (VVOCs) and one as a semi-VOC (Table 2). Among the identified compounds, many have been known to have a negative impact on human health and the natural environment, including mutagenic and carcinogenic aromatic compounds, e.g., toluene, benzene, ethylbenzene or cumene, and PAH, e.g., naphthalene.

The total mass of VOCs emitted from CRDF was 16.4 mg/kg (Table 2) based on 7 days of accumulation in the headspace of a sealed storage vessel. The top 10 compounds with the highest emissions were as follows: nonanal, octanal, heptanal, butylbenzene, hexanal, 1-methyl-4-prop-1-en-2-ylcyclohexene, benzaldehyde, decanal, toluene, and hexylbenzene. Among the analyzed compounds, the highest emission (as a group) from the CRDF was determined for aldehydes: nonanal, followed by octanal, and heptanal (Table 2). The top 10 of the most emitted VOCs consisted almost 65% of total emissions.

Table 1. VOCs emitted from torrefied carbonized refuse-derived fuel (CRDF).

#	Retention Time (min)	Compound Name, IUPAC	Retention Coefficient, KI Experimental (MS Database)	Retention Coefficient (Kovats C7-40 Index)	CAS Number
1	1.87	acetic acid	-	593	123-72-8
2	2.45	propanoic acid	700	700	79-49-4
3	2.93	pyrimidine	740	736	289-95-2
4	3.10	pyridine	753	746	110-86-1
5	3.29	pentan-1-ol	768	765	71-41-0
6	3.36	toluene	774	769	108-88-3
7	3.45	2-methylpropanoic acid	781	775	79-31-2
8	3.78	hexanal	804	800	66-25-1
9	4.23	2-methylpyrazine	826	831	109-08-0
10	4.41	furan-2-carbaldehyde	835	833	98-01-1
11	5.01	1,3-xylene	864	866	108-38-3
12	5.06	2-oxopropyl acetate	866	870	592-20-1
13	5.18	1,4-xylene	872	866	106-42-3
14	5.35	pentanoic acid	881	902	109-52-4
15	5.49	unknown compound	887	-	
16	5.63	heptan-2-one	893	891	110-43-0
17	5.68	styrene	896	893	100-42-5
18	5.78	1,2-xylene	900	887	95-47-6
19	5.88	heptanal	904	902	111-71-7
20	6.03	hexa-2,4-diene, (E,E)-	909	911	592-46-1
21	6.15	1-(furan-2-yl)ethanone	914	912	1192-62-7
22	6.24	2-ethylpyrazine	917	921	13925-00-3
23	6.34	2,5-dimethylpyrazine	920	925	123-32-0
24	6.55	cumene	927	926	98-82-8
25	6.64	1,4-dimethylpyridine	931	930	108-47-4
26	6.81	4,6,6-trimethylbicyclo[3.1.1]hept-3-ene	936	937	80-56-8
27	6.97	3-methylbutanoic acid	942	947	503-74-2
28	7.02	4-ethylpyridine	944	956	536-75-4
29	7.34	n-propylbenzene	955	953	103-65-1
30	7.53	benzaldehyde	962	963	100-52-7
31	7.59	5-methylfuran-2-carbaldehyde	964	965	620-02-0
32	7.77	1,3,5-trimethylbenzene	970	972	108-67-8
33	8.14	phenol	980	983	108-95-2
34	8.47	4-methyl-1-propan-2-ylcyclohexene	993	988	500-00-5

Table 1. *Cont.*

#	Retention Time (min)	Compound Name, IUPAC	Retention Coefficient, KI Experimental (MS Database)	Retention Coefficient (Kovats C7-40 Index)	CAS Number
35	8.53	1,2,4-trimethylbenzene	996	993	95-63-6
36	8.79	octanal	1005	1003	124-13-0
37	8.87	dec-3-yn-1-ol	1007	1011	51721-39-2
38	9.06	an unknown isomer of ethyldimethyl benzene	1013	-	-
39	9.45	1,3-diethylbenzene	1025	1025	141-93-5
40	9.50	1-methyl-4-propan-2-ylbenzene	1027	1026	99-87-6
41	9.61	1-methyl-4-prop-1-en-2-ylcyclohexene	1030	1031	138-86-3
42	9.87	2,3-dihydro-1*H*-indene	1037	1030	496-11-7
43	10.32	1,2-diethylbenzene	1051	1045	135-01-3
44	10.42	1-methyl-2-propylbenzene	1055	1047	1074-17-5
45	10.53	butylbenzene	1058	1054	104-51-8
46	10.61	1-ethyl-3,5-dimethylbenzene	1060	1058	934-74-7
47	10.70	2-ethyl-1,4-dimethylbenzene	1063	1071	1758-88-9
48	10.87	1-phenylethanone	1068	1065	98-86-2
49	11.23	2-ethyl-1,3-dimethylbenzene	1079	1080	2870-04-4
50	11.29	4-ethyl-1,2-dimethylbenzene	1081	1083	499-75-2
51	11.36	1-ethenyl-2,4-dimethylbenzene	1083	1084	2234-20-0
52	11.54	2-ethyl-1,4-dimethylbenzene	1089	1090	1758-88-9
53	11.63	2-methoxyphenol	1091	1090	90-05-1
54	11.70	1-undecyne	1093	1095	2243-98-3
55	11.84	methyl benzoate	1098	1095	93-58-3
56	12.00	undecane	1102	1100	1120-21-4
57	12.05	nonanal	1104	1103	124-19-6
58	12.25	1,2,4,5-tetramethylbenzene	1110	1116	95-93-2
59	12.33	an unknown isomer of diethylmethylbenzene	1113	-	-
60	12.54	unknown compound	1118	-	-
61	12.68	1,2,3,5-tetramethylbenzene	1122	1117	527-53-7
62	12.94	1,3-dimethyl-2,3-dihydro-1*H*-indene	1130	1135	4175-53-5
63	13.33	5-methyl-2,3-dihydro-1*H*-indene	1142	1136	874-35-1
64	13.49	1,3-diethyl-5-methylbenzene	1145	1147	2050-24-0
65	13.70	4-methyl-2,3-dihydro-1*H*-indene	1152	1148	824-22-6
66	13.90	1-methyl-1*H*-indene	1158	1157	767-59-9
67	13.94	pentylbenzene	1160	1158	538-68-1

Table 1. Cont.

#	Retention Time (min)	Compound Name, IUPAC	Retention Coefficient, KI Experimental (MS Database)	Retention Coefficient (Kovats C7-40 Index)	CAS Number
68	14.08	1,2,3,4-tetrahydronaphthalene	1163	1157	119-64-2
69	14.14	1,4-diethyl-2-methylbenzene	1165	1164	13632-94-5
70	14.28	2,4-diethyl-1-methylbenzene	1168	1166	1758-85-6
71	14.83	azulene	1185	1182	275-51-4
72	14.99	1-methyl-4-propan-2-yl-2-[(E)-prop-1-enyl]benzene	1190	1191	97664-18-1
73	15.18	2-ethyl-2,3-dihydro-1H-indene	1196	n.d.	56147-63-8
74	15.52	decanal	1203	1206	112-31-2
75	15.70	unknown compound	1212	-	-
76	17.42	hexylbenzene	1253	1260	1077-16-3
77	17.57	6-methyl-1,2,3,4-tetrahydronaphthalene	1266	1263	1680-51-9
78	17.66	5-methyl-1,2,3,4-tetrahydronaphthalene	1269	1276	2809-64-5
79	18.17	4,7-dimethyl-2,3-dihydro-1H-indene	1284	1282	6682-71-9
80	18.57	undecan-2-one (internal standard)	1296	1298	112-12-9
81	18.77	2-methyl-5-propan-2-ylphenol	1302	1299	499-75-2
82	19.11	1-methylnaphtalene	1314	1307	112-44-7
83	19.43	3,3-dimethyl-2H-inden-1-one	1325	1330	26465-81-6
84	19.70	1,5-dimethyl-1,2,3,4-tetrahydronaphthalene	1334	1341	21564-91-0
85	20.82	5,6-dimethyl-1,2,3,4-tetrahydronaphthalene	1373	1381	21693-54-9

Table 2. VOCs emissions (accumulated in a headspace of sealed vessel over 7 days of storage) from (torrefied) carbonized refuse-derived fuel ordered from the highest (μg of VOC per kg of CRDF) to lowest; % of total emissions, boiling point, VOC classification, and a comparison with VOCs emitted from other types of (pyrolyzed) biochar (woody biomass, algal biochar, and municipal solid waste (compost), respectively) [11,12,14].

Compound Name (IUPAC)	Emissions (μg/kg)	% of Total Emissions	Boiling Point (°C)	Type of VOC [1]	Observed in Emissions from Biochar (+, −, =, Yes, No)		
					[11]	[12]	[14]
Nonanal *	2860.00	17.400	195	VOC	−	−	−
Octanal *	1480.00	9.010	171	VOC	+	−	−
Heptanal *	1180.00	7.150	153	VOC	+	−	−
butylbenzene	1030.00	6.290	183	VOC	−	−	−
Hexanal *	843.00	5.120	130	VOC	+	−	−
1-methyl-4-prop-1-en-2-ylcyclohexene	789.00	4.800	176.5	VOC	−	−	−
Benzaldehyde *	777.00	4.720	179	VOC	+	−	−

Table 2. Cont.

Compound Name (IUPAC)	Emissions (µg/kg)	% of Total Emissions	Boiling Point (°C)	Type of VOC [1]	Observed in Emissions from Biochar (+, −, =, Yes, No)		
					[11]	[12]	[14]
Decanal *	554.97	3.373	208	VOC	−	−	−
Toluene *	535.78	3.257	110.6	VOC	+	−	−
hexylbenzene	521.82	3.172	228	VOC	−	−	−
4,6,6-trimethyl-bicyclo[3.1.1]hept-3-ene *	408.38	2.482	155.5	VOC	−	−	−
1,3,5-trimethylbenzene	387.43	2.355	165	VOC	−	−	−
1-undecyne	373.47	2.270	195	VOC	−	−	−
2-ethyl-2,3-dihydro-1H-indene	342.06	2.079	-	-	−	−	−
1-ethyl-3,5-dimethyl-benzene	246.07	1.496	184	VOC	−	−	−
4,7-dimethyl-2,3-dihydro-1H-indene	235.60	1.432	225.9	VOC	−	−	−
1,4-xylene	225.13	1.368	138	VOC	−	−	−
1-methyl-1H-indene	204.19	1.241	199	VOC	−	−	−
acetic acid *	**197.21**	**1.199**	**118**	**VOC**	**+**	**+**	−
heptan-2-one *	160.56	0.976	149	VOC	+	−	−
2-methyl-5-propan-2-ylphenol	139.62	0.849	236.5	VOC	−	−	−
4-methyl-1-propan-2-ylcyclohexene *	136.13	0.827	166.8	VOC	−	−	−
Undecane *	136.13	0.827	196	VOC	−	−	−
1,3-dimethyl-2,3-dihydro-1H-indene	122.16	0.743	208.7	VOC	−	−	−
pyrimidine *	118.67	0.721	124	VOC	−	−	−
2-ethyl-1,4-dimethylbenzene	115.18	0.700	187	VOC	−	−	−
furan-2-carbaldehyde	113.44	0.690	162	VOC	−	−	−
1,2,3,4-tetrahydro-naphthalene	109.95	0.668	207	VOC	−	−	−
1-ethenyl-2,4-dimethylbenzene	108.20	0.658	-	-	−	−	−
1,2,3,5-tetramethyl-benzene	108.20	0.658	198	VOC	−	−	−
1,2,4,5-tetramethyl-benzene	104.71	0.636	196.5	VOC	−	−	−
1,3-xylene	99.48	0.605	139	VOC	−	−	−
pentylbenzene	97.73	0.594	205	VOC	−	−	−
2-oxopropyl acetate	95.99	0.583	175	VOC	−	−	−
Phenol *	**95.99**	**0.583**	**182**	**VOC**	**−**	**+**	**+**
1,2-diethylbenzene	92.50	0.562	183	VOC	−	−	−
2-ethyl-1,3-dimethyl-benzene	87.26	0.530	190	VOC	−	−	−
unknown isomer of ethyldimethyl benzene	85.51	0.520	-	-	−	−	−
Styrene *	75.04	0.456	145.5	VOC	−	−	−
methyl benzoate	66.32	0.403	198.5	VOC	−	−	−

Table 2. Cont.

Compound Name (IUPAC)	Emissions (µg/kg)	% of Total Emissions	Boiling Point (°C)	Type of VOC [1]	Observed in Emissions from Biochar (+, −, =, Yes, No)		
					[11]	[12]	[14]
6-methyl-1,2,3,4-tetrahydronaphthalene	62.83	0.382	226	VOC	−	−	−
2-ethyl-1,4-dimethylbenzene	61.08	0.371	187	VOC	−	−	−
unknown compound	59.34	0.361					
2,3-dihydro-1H-indene	54.10	0.329	176	VOC	−	−	−
n-propylbenzene	52.36	0.318	159	VOC	−	−	−
1-methyl-4-propan-2-ylbenzene	50.61	0.308	177	VOC	−	−	−
1-(furan-2-yl)ethanone	48.87	0.297	168	VOC	−	−	−
2-methylpyrazine	47.12	0.286	135	VOC	−	−	−
4-methyl-2,3-dihydro-1H-indene	45.38	0.276	204	VOC	−	−	−
1,3-diethyl-5-methylbenzene	43.63	0.265	200.7	VOC	−	−	−
5-methyl-2,3-dihydro-1H-indene	41.88	0.255	204.1	VOC	−	−	−
unknown compound	41.88	0.255	-	-			
dec-3-yn-1-ol	36.65	0.223	130.5	VOC	−	−	−
1,4-dimetylopirydyne	34.90	0.212	159	VOC	−	−	−
pentan-1-ol *	33.16	0.202	138	VOC	−	−	−
azulene	24.43	0.148	242	VOC	−	−	−
1-methyl-4-propan-2-yl-2-[(E)-prop-1-enyl]benzene	22.69	0.138	-	-	−	−	−
propanoic acid *	22.69	0.138	141.5	VOC	−	−	−
1,3-diethylbenzene	20.94	0.127	182	VOC	−	−	−
unknown isomer of diethyl methylbenzene	20.94	0.127	-	-			
2,4-diethyl-1-methylbenzene	19.20	0.117	205	VOC	−	−	−
4-ethylpyridine	15.71	0.095	168	VOC	−	−	−
unknown compound	15.71	0.095					
1,2,4-trimethylbenzene	13.96	0.085	168	VOC	−	−	−
1,5-dimethyl-1,2,3,4-tetrahydronaphthalene	13.96	0.085	247.5	SVOC	−	−	−
5,6-dimethyl-1,2,3,4-tetrahydronaphthalene	13.96	0.085	-	-			
2-methylpropanoic acid	10.47	0.064	155	VOC	−	−	−
3,3-dimethyl-2H-inden-1-one	8.73	0.053	122	VOC	−	−	−
1-methylnaphtalene	8.73	0.053	120	VOC	−	−	−
5-methylfuran-2-carbaldehyde	6.98	0.042	188	VOC	−	−	−
2-ethylpyrazine	6.98	0.042	152.5	VOC	−	−	−
pyridine	6.98	0.042	115	VOC	−	−	−
1-methyl-2-propylbenzene	5.24	0.032	185	VOC	−	−	−

Table 2. *Cont.*

Compound Name (IUPAC)	Emissions (µg/kg)	% of Total Emissions	Boiling Point (°C)	Type of VOC [1]	Observed in Emissions from Biochar (+, −, =, Yes, No)		
					[11]	[12]	[14]
1,2-xylene	5.24	0.032	144	VOC	−	−	−
hexa-2,4-diene, (E,E)-	5.24	0.032	82	VVOC	−	−	−
1-phenylethanone	1.75	0.011	202	VOC	−	−	−
2,5-dimethylpyrazine	1.75	0.011	155	VOC	−	−	−
4-ethyl-1,2-dimethylbenzene	1.75	0.011	236.5	VOC	−	−	−
cumene	1.75	0.011	153	VOC	+	−	−
pentanoic acid *	1.75	0.011	110.5	VOC	−	−	−
1,4-diethyl-2-methylbenzene	0.10	0.001	207	VOC	−	−	−
2-methoxyphenol *	0.10	0.001	205	VOC	−	−	−
3-methylbutanoic acid	0.10	0.001	176	VOC	−	−	−
5-methyl-1,2,3,4-tetrahydronaphthalene	0.10	0.001	234	VOC	−	−	−
Total	16,452.46	-	-	-			

[1] —according to [16], where VVOC—very volatile organic compounds (0–100 °C), VOC—volatile organic compounds (100–240 °C), SVOC—semi-volatile organic compounds (240–400 °C); bold font = common compounds found in at least two other studies; * Identified using analytical standards.

3. Discussion

The determined composition of the VOCs mixture emitted from CRDF stored in a sealed vessel (this research) is unique because it was likely driven by the type of municipal waste and the process parameters used for its production. However, for illustrative purposes, it is useful to compare with VOCs emitted from other types of biochar. Spokas et al. [11] reported 140 different compounds, 74 were identified in all studied biochars, generated from 77 different materials; but without municipal solid waste and without fuels derived from municipal waste. Spokas et al. [11] have not found clear feedstock dependencies to the adsorbed VOC composition, suggesting a stronger linkage with biochar production conditions coupled with post-production handling and processing. Lower pyrolytic temperatures (≤ 350 °C) produced biochars with adsorbed VOCs consisting of short carbon chain aldehydes, furans, and ketones; elevated temperature biochars (>350 °C) typically were dominated by adsorbed aromatic compounds and longer carbon chain hydrocarbons.

In the present work, only eight compounds were also reported by Spokas et al. [11] (Table 2). This relatively small number of common VOCs corroborates the unique influence of feedstock type —CRDF (in this research), and torrefaction process (a lower temperature process different to pyrolysis, and gasification) on VOCs formation during waste/biomass thermal treatment. Similarly, to present studies [11] aldehydes were identified in biochars (Table 2).

Buss et al. [12] analyzed VOCs emitted from three algal biochars, including two contaminated by re-condensates during pyrolysis. Buss et al. [12] identified numerous compounds from phenol groups mainly methylated and ethylene (25 compounds, but only phenol was common with present study) and acids such as acetic, formic or propionic. Taherymoosavi et al. [14] used municipal waste (compost) for the production of biochar and thus, was closest (as a source) to this work. Taherymoosavi et al., [14] analyzed biochar formed in the pyrolysis process at temperatures from 105 to 650 °C and reported the presence of alkylbenzenes, methoxy alkylphenols, organic compounds containing nitrogen, furans, and aromatic compounds. However, only phenol was a common compound identified in the present study (Table 2). Compared results show that only two compounds acetic acid and phenol were identified in the present study and [11,12], and [12,14] respectively.

There is little research in literature related to the subject of qualitative and quantitative identification of VOCs emitted from the surface of biochar, especially from biochar produced from municipal solid waste such as CRDF. This is a relatively new topic related to the trend of using torrefaction, and low-temperature pyrolysis of municipal solid waste in recent years. These new trends in municipal solids treatment are being sought as an alternative to both energy production and 'Waste to Carbon' utilization (e.g., CRDF). Thus the interest in identifying and mitigating VOC emissions from biochar will likely increase. As biochar VOCs are still not deeply explored, it is required to continue research on the effects of feedstock type and thermal treatment conditions on VOCs formation and emission, especially in the contest on potential harmful effect to workers during biochar storage and transportation and end users.

4. Materials and Methods

4.1. CRDF Used in the Experiment

CRDF was produced in the torrefaction process at 260 °C and a 50 min retention time in a batch reactor, according to the procedure described by [5]. The analyzed CRDF from the torrefaction of municipal waste at 260 °C and 50 min of retention time was characterized by physicochemical properties similar to those described in the literature. CRDF with a lower heating value (LHV) of 25.95 MJ/kg was similar to CRDF obtained in earlier studies [5] and to biochar from grass produced in a similar temperature range (250 to 350 °C) by Weber and Quicker [17], which had a calorific value of 25 to 30 MJ/kg. The higher heating value (HHV) of CRDF used in this experiment (27.315 MJ/kg) could define it as a 'hard coal' (HHV > 23.9 MJ/kg), according to the IEA's classification [18]. The moisture content of the analyzed material (1.54%) was in the 1 to 6% range [19]. The proximate and ultimate properties of the CRDF used were summarized by Białowiec et al. [9].

4.2. Qualitative and Quantitative Analyses of VOC Emitted from CRDF

Measurements of VOCs were made using headspace (HS) solid-phase microextraction (SPME) technology for gas extraction and gas chromatography coupled with mass spectrometry (GC-MS) (Palo Alto, CA, USA) for analyses. SPME technology combines sampling and sample preparation and is suited for exploratory qualitative and quantitative work on VOC emissions from a wide range of sources such as contaminated soils [20,21], decaying animal carcasses [22,23], fermentation by-products in beverages and aromas in wines [24,25], biological fluids and gases [26–30]. A comprehensive review of SPME applications to food and environmental analysis was published by Merkle et al. [31]. The apparatus and reagents were as follows:

1) the internal standard—a solution of 2-undecanone at a ratio of 20 µg compound per 20 mL of distilled water;
2) water bath with a temperature of 40 °C with glycol;
3) manual holder for SPME;
4) universal SPME fiber 3-component DVB/CAR/PDMS 50/30 µm coating (Supelco Inc., Bellefonte, PA, USA);
5) 10 µL syringe for internal standard addition;
6) a laboratory incubator (Thermo Fisher Scientific Inc., Waltham, MA, USA) with a constant temperature of 23 °C.

4.3. Preparation of CRDF Samples

To prepare the samples for VOCs emission analysis, the CRDF was pre-treated and ground in a 2SIEL90L2 grinding mill (Celia Indukta, Bielsko-Biała, Poland) to homogenize the sample to size <0.5 mm. Next, 10 g of bulk 3 subsamples were placed in a sealed 1000 mL glass vessels. An internal standard, 10 µg of 2-undecanone (Sigma-Aldrich, St. Louis, MO, USA), was added to the vessels to account for the variability in emissions and to aid VOC quantification. Each sealed sample was stored in a laboratory incubator at a constant temperature of 23 °C for 7 days, after which it was removed for sampling. The VOCs extraction was carried out from the headspace of sealed vessel, by the SPME.

4.4. Solid-Phase Microextraction

After placing the sealed vessel with the sample in a water bath with glycol preheated to 40 °C, a 3-component universal fiber coating (DVB/CAR/PDMS 50/30 µm) was introduced into the vessel headspace. The SPME exposure lasted 20 min, similarly to the types of coatings and extraction times used for VOC emissions from solid, porous matter. The DVB/CAR/PDMS 50/30 µm SPME coating is often recommended and used for exploratory work on VOC emissions from unknown sources [25,26,28]. The coating represents a mixture of polymers capable of extracting VOCs with a wide range of properties, i.e., suitable for the work with CRDF. No specific optimization was made on sampling time. However, it was chosen based on practical considerations and preliminary trials aiming at reliably extracting the greatest number of VOCs in a relatively short extraction.

4.5. Gas Chromatography with Mass Spectrometry

The separation, identification and quantification of VOCs adsorbed on the fiber was conducted using a GC coupled to a MS detector (Saturn 2000 MS Varian Chrompack, Palo Alto, CA, USA) with ZB-5 (Phenomenex, Torrance, CA, USA) column (30 m × 0.25 µm film × 0.25 mm i.d.). Chromatographic conditions were performed according to Calin-Sanchez et al. [32]. Scanning (1 scan/s) was performed in the range of 35–400 m/z using electron impact ionization at 70 eV [33]. The analyses were performed using helium as a carrier gas at a flow rate of 1.0 mL/min, in splitless mode in SPME, and with the following program for the oven temperature: 50 °C at the beginning; 4 °C/min to 130 °C; and 10 °C/min to 180 °C and 20 °C/min to 280 °C with a hold for 4 min. The injector was held at 220 °C.

4.6. Data Analysis

The VOCs emitted from CRDF samples were identified using three independent analytical methods: retention indices (RI), GC–MS retention times of authentic chemical standards, mass spectra of compounds [34] and comparison with authentic standards, if possible.

The retention index standards used in this study consisted of a mixture of aliphatic hydrocarbons ranging from C-7 through C-40 dissolved in hexane [34].

The use of internal standard enabled quantitative analysis of VOCs. It was carried out using the Mnova MS 12.0.1 software (Mestrelab Research, S.L., Santiago de Compostela, Spain) based on the retention time of individual compounds, through the integration of the peak area of the chromatogram. The percentage ratio of individual VOC was determined. VOC emissions (on per mass of CRDF basis) were estimated based on the recovered internal standard. All raw data were shown as Supplementary Materials.

5. Conclusions

In the analyzed CRDF (biochar) from municipal waste, 84 VOCs have been identified, including many that are toxic, e.g., derivatives of benzene or toluene. The highest emission was measured for nonanal, octanal, heptanal. The top 10 of the most emitted VOCs consisted almost 65% of total emissions. The mixture of emitted from CRDF VOCs differed from those emitted by other types of biochars, produced from different types of feedstock, and under different pyrolysis conditions. SPME provided a useful tool for characterizing VOC emissions from CRDF, a new potential fuel exemplifying the 'Waste to Carbon' concept in a circular, zero-waste economy.

Supplementary Materials: The following files have been submitted as supplementary materials in zipped folder "supplementary materials.zip": explanatory file "readme.docx", raw data in files "CRDF MS raw data.jdx; CRDF MS raw data.csv; CRDF peaks raw data.xlsx" and tables (Tables S1 and S2) in the file "Tables.xlsx".

Author Contributions: Conceptualization, A.B., A.S.; methodology, A.B., M.M., A.S., J.Ł.; formal analysis, A.B., M.M., J.Ł; validation, A.B., M.M., A.S., J.K.; investigation, M.M., J.Ł; resources, A.B. A.S.; data curation, M.M., J.Ł, A.S., A.B.; writing—original draft preparation, A.B., M.M.; writing—review and editing, A.B., J.K., A.S., J.Ł; visualization, A.B., M.M., J.K.; supervision, A.B., A.S., J.K.

Funding: Authors would like to thank the Fulbright Foundation for funding the project titled "Research on pollutants emission from Carbonized Refuse Derived Fuel into the environment", completed at the Iowa State University. In addition, this project was partially supported by the Iowa Agriculture and Home Economics Experiment Station, Ames, Iowa. Project no. IOW05400 (Animal Production Systems: Synthesis of Methods to Determine Triple Bottom Line Sustainability from Findings of Reductionist Research) is sponsored by Hatch Act and State of Iowa funds."

Conflicts of Interest: The authors declare no conflict of interest. The funders had no role in the design of the study; in the collection, analyses, or interpretation of data; in the writing of the manuscript, or in the decision to publish the results.

References

1. International Biochar Initiative IBI 2015. IBI Biochar Standards—Standardized Product Definition and Product Testing Guidelines for Biochar that is used in Soil. v.2.1. Available online: https://www.biochar-international.org/wp-content/uploads/2018/04/IBI_Biochar_Standards_V2.1_Final.pdf (accessed on 28 October 2018).
2. EBC (2012). European Biochar Certificate—Guidelines for a Sustainable Production of Biochar. European Biochar Foundation (EBC), Arbaz, Switzerland. Available online: http://www.european-biochar.org/biochar/media/doc/ebc-guidelines.pdf (accessed on 28 October 2018).
3. Schmidt, H.P. 55 Uses of Biochar. *Ithaka J.* **2012**, *1*, 286–289.

4. Bird, M.; Wurster, C.; de Paula Silva, P.; Bass, A.; de Nys, R. Algal biochar—Production and properties. *Bioresour. Technol.* **2011**, *102*, 1886–1891. [CrossRef] [PubMed]
5. Białowiec, A.; Pulka, J.; Stępień, P.; Manczarski, P.; Gołaszewski, J. The RDF/SRF torrefaction: An effect of temperature on characterization of the product—Carbonized Refuse Derived Fuel. *Waste Manag.* **2017**, *70*, 91–100. [CrossRef] [PubMed]
6. Stępień, P.; Białowiec, A. Kinetic parameters of torrefaction process of alternative fuel produced from municipal solid waste and characteristic of carbonized refuse derived fuel. *Detritus* **2018**, *3*, 75–83. [CrossRef]
7. Wiśniewski, D.; Gołaszewski, J.; Białowiec, A. The pyrolysis and gasification of digestate from agricultural biogas plant. *Arch. Environ. Prot.* **2015**, *41*, 70–75. [CrossRef]
8. Pulka, J.; Wiśniewski, D.; Gołaszewski, J.; Białowiec, A. Is the biochar produced from sewage sludge a good quality solid fuel? *Arch. Environ. Prot.* **2016**, *42*, 125–134. [CrossRef]
9. Białowiec, A.; Micuda, M.; Koziel, J.A. Waste to carbon: Densification of torrefied refuse-derived fuels. *Energies* **2018**, *11*, 3233. [CrossRef]
10. Directive 2004/42/CE of the European Parliament and of the Council of 21 April 2004 on the limitation of emissions of volatile organic compounds due to the use of organic solvents in certain paints and varnishes and vehicle refinishing products and amending Directive 1999/13/EC. Available online: https://eur-lex.europa.eu/legal-content/EN/TXT/?uri=celex%3A32004L0042 (accessed on 13 November 2018).
11. Spokas, K.A.; Novak, J.M.; Stewart, C.E.; Cantrell, K.B.; Uchiyama, M.; Dusaire, M.G.; Ro, K.S. Qualitative analysis of volatile organic compounds on biochar. *Chemosphere* **2011**, *85*, 869–882. [CrossRef]
12. Buss, W.; Masek, O.; Graham, M.; Wüst, D. Inherent organic compounds in biochar-Their content, composition and potential toxic effects. *J. Environ. Manag.* **2015**, *156*, 150–157. [CrossRef]
13. Allaire, S.E.; Lange, S.F.; Auclair, I.K.; Quinche, M.; Greffard, L. Analyses of biochar properties. CRMR-2015-SA-5. Presented at the Char Team (2015) Report, Centre de Recherche sur les Matériaux Renouvelables, Université Laval, Quebec, QC, Canada, 2015; p. 59.
14. Taherymoosavi, S.; Verheyen, V.; Munroe, P.; Joseph, S.; Reynolds, A. Characterization of organic compounds in biochars derived from municipal solid waste. *Waste Manag.* **2017**, *67*, 131–142. [CrossRef]
15. Wang, C.; Wang, Y.; Herath, H.M.S.K. Polycyclic aromatic hydrocarbons (PAHs) in biochar—Their formation, occurrence and analysis: A review. *Org. Geochem.* **2017**, *114*, 1–11. [CrossRef]
16. World Health Organization Report. Indoor air quality, organic pollutants: Report on a WHO meeting. *Eur. Rep. Stud.* **1989**, *111*, 1–70.
17. Weber, K.; Quicker, P. Properties of biochar. *Fuel* **2018**, *217*, 240–261. [CrossRef]
18. OECD/IEA/EUROSTAT, Energy Statistics Manual. Available online: http://ec.europa.eu/eurostat/ramon/statmanuals/files/Energy_statistics_manual_2004_EN.pdf (accessed on 21 September 2018).
19. Jakubiak, M.; Kordylewski, W. Biomass torrefaction. *Polski Inst. Spal.* **2010**, *10*, 11–25.
20. Kenessov, B.; Koziel, J.A.; Grotenhuis, T.; Carlsen, L. Screening of transformation products in soils contaminated with unsymmetrical dimethylhydrazine using headspace SPME and GC-MS. *Anal. Chim. Acta* **2010**, *674*, 32–39. [CrossRef]
21. Orazbayeva, D.; Kenessov, B.; Koziel, J.A.; Nassyrova, D.; Lyabukhova, N.V. Quantification of BTEX in soil by headspace SPME-GC-MS using combined standard addition and internal standard calibration. *Chromatographia* **2017**, *80*, 1249–1256. [CrossRef]
22. Akdeniz, N.; Koziel, J.A.; Ahn, H.K.; Glanville, T.D.; Crawford, B.; Raman, R.D. Laboratory scale evaluation of VOC emissions as indication of swine carcass degradation inside biosecure composting units. *Bioresour. Technol.* **2010**, *101*, 71–78. [CrossRef] [PubMed]
23. Koziel, J.A.; Nguyen, L.T.; Glanville, T.D.; Ahn, H.K.; Frana, T.S.; van Leeuwen, J.H. Method for sampling and analysis of volatile biomarkers in process gas from aerobic digestion of poultry carcass using time-weighted average SPME and GC-MS. *Food Chem.* **2017**, *232*, 799–807. [CrossRef] [PubMed]
24. Onuki, S.; Koziel, J.A.; Jenks, W.S.; Cai, L.; Rice, S.; van Leeuwen, J.H. Optimization of extraction parameters for quantification of fermentation volatile by-products in industrial ethanol with solid-phase microextraction and gas chromatography. *J. Inst. Brewing.* **2016**, *122*, 102–109. [CrossRef]
25. Rice, S.; Lutt, N.; Koziel, J.A.; Dharmadhikari, M.; Fennell, A. Determination of selected aromas in Marquette and Frontenac wine using headspace-SPME coupled with GC-MS and simultaneous olfactometry. *Separations* **2018**, *5*, 20. [CrossRef]

26. Soso, S.B.; Koziel, J.A. Characterizing the scent and chemical composition of *Panthera leo* marking fluid using solid-phase microextraction and multidimensional gas chromatography-mass spectrometry-olfactometry. *Sci. Rep. UK* **2017**, *7*, 5137. [CrossRef] [PubMed]
27. Soso, S.B.; Koziel, J.A. Analysis of odorants in marking fluid of Siberian tiger (*Panthera tigris altaica*) using simultaneous sensory and chemical analysis with headspace solid-phase microextraction and multidimensional gas chromatography-mass spectrometry-olfactometry. *Molecules* **2016**, *21*, 834. [CrossRef] [PubMed]
28. Maurer, D.L.; Koziel, J.A.; Engelken, T.J.; Cooper, V.L.; Funk, J.L. Detection of volatile compounds emitted from nasal secretions and serum: Towards non-invasive identification of diseased cattle biomarkers. *Separations* **2018**, *5*, 18. [CrossRef]
29. Cai, L.; Koziel, J.A.; Davis, J.; Lo, Y.C.; Xin, H. Characterization of VOCs and odors by in vivo sampling of beef cattle rumen gas using SPME and GC-MS-olfactometry. *Anal. Bioanal. Chem.* **2006**, *386*, 1791–1802. [CrossRef]
30. Cai, L.; Koziel, J.A.; O'Neal, M.E. Studying plant-insect interactions with solid phase microextraction: Screening for airborne volatile emissions of soybeans to the soybean aphid, Aphis glycines Matsumura (Hemiptera: Aphididae). *Chromatography* **2015**, *2*, 265–276. [CrossRef]
31. Merkle, S.; Kleeberg, K.K.; Fritsche, J. Recent developments and applications of solid phase microextraction (SPME) in food and environmental analysis—A review. *Chromatography* **2015**, *2*, 293–381. [CrossRef]
32. Calín-Sánchez, Á.; Figiel, A.; Lech, K.; Szumny, A.; Martínez-Tomé, J.; Carbonell-Barrachina, Á.A. Dying methods affect the aroma of *Origanum majorana* L. analyzed by GC–MS and descriptive sensory analysis. *Ind. Crop. Prod.* **2015**, *74*, 218–227. [CrossRef]
33. Nöfer, J.; Lech, K.; Figiel, A.; Szumny, A.; Carbonell-Barrachina, Á.A. The influence of drying method on volatile composition and sensory profile of *Boletus edulis*. *J. Food Qual.* **2018**, *2018*. [CrossRef]
34. Stein, S.E. "Mass Spectra" by NIST Mass Spec Data Center. Available online: https://webbook.nist.gov (accessed on 13 November 2018).

Sample Availability: Samples of the compounds are available from the authors.

© 2018 by the authors. Licensee MDPI, Basel, Switzerland. This article is an open access article distributed under the terms and conditions of the Creative Commons Attribution (CC BY) license (http://creativecommons.org/licenses/by/4.0/).

Article

Application of Direct Immersion Solid-Phase Microextraction (DI-SPME) for Understanding Biological Changes of Mediterranean Fruit Fly (*Ceratitis capitata*) During Mating Procedures

Hasan Al-Khshemawee [1,2,†], Xin Du [1,†], Manjree Agarwal [1,*], Jeong Oh Yang [3] and Yong Lin Ren [1,*]

1. School of Veterinary and Life Science, Murdoch University, 90 South St., Murdoch, WA 6150, Australia; hasan_hadi1984@yahoo.com (H.A.-K.); b.du@murdoch.edu.au (X.D.)
2. College of Agriculture, Wasit University, Wasit 120, Iraq
3. Plant Quarantine Technology Centre, Animal and Plant Quarantine Agency (APQA), Gimcheon 39660, Korea; joyang12@korea.kr
* Correspondence: M.agarwal@murdoch.edu.au (M.A.); y.ren@murdoch.edu.au (Y.L.R.); Tel.: +61-93601354 (M.A.); +61-893601397 (Y.L.R.)
† Both Hasan AL-Khshemawee and Xin Du contributed equally as first authors.

Academic Editor: Constantinos K. Zacharis
Received: 7 October 2018; Accepted: 9 November 2018; Published: 12 November 2018

Abstract: Samples from three different mating stages (before, during and after mating) of the Mediterranean fruit fly *Ceratitis capitata* were used in this experiment. Samples obtained from whole insects were subjected to extraction with the two mixtures of solvents (acetonitrile/water (A) and methanol/acetonitrile/water (B)) and a comparative study of the extractions using the different solvents was performed. Direct immersion-solid phase microextraction (DI-SPME) was employed, followed by gas chromatographic-mass spectrometry analyses (GC/MS) for the collection, separation and identification of compounds. The method was validated by testing its sensitivity, linearity and reproducibility. The main compounds identified in the three different mating stages were ethyl glycolate, α-farnesene, decanoic acid octyl ester, 2,6,10,15-tetramethylheptadecane, 11-tricosene, 9,12-(Z,Z)-octadecadienoic acid, methyl stearate, 9-(Z)-tricosene, 9,11-didehydro-lumisterol acetate; 1,54-dibromotetrapentacontane, 9-(Z)-hexadecenoic acid hexadecyl ester, 9-(E)-octadecenoic acid and 9-(Z)-hexadecenoic acid octadecyl ester. The novel findings indicated that compound compositions were not significantly different before and during mating. However, new chemical compounds were generated after mating, such as 1-iodododecane, 9-(Z)-tricosene and 11,13-dimethyl-12-tetradecen-1-acetate which were extracted with both (A) and (B) and dodecanoic acid, (Z)-oleic acid, octadecanoic acid and hentriacontane which were extracted with (A) and ethyl glycolate, 9-hexadecenoic acid hexadecyl ester, palmitoleic acid and 9-(E)-octadecenoic acid, which were extracted with solvent (B). This study has demonstrated that DI-SPME is useful in quantitative insect metabolomics by determining changes in the metabolic compounds in response to mating periods. DI-SPME chemical extraction technology might offer analysis of metabolites that could potentially enhance our understanding on the evolution of the medfly.

Keywords: DI-SPME; GC-MS; Mediterranean fruit fly; extraction solvent; metabolites

---

1. Introduction

The developed analytical methods for the analysis of volatile and non-volatile compounds are increasingly being used as tools for the study of plant chemistry and the evolution of insect–plant interactions [1]. The development of sample preparation and extraction methodologies is one of the main challenges for metabolism studies [2] and has an enormous impact on the quality of the data. Biological samples should be unbiased and nonselective [3]. Solid-phase microextraction (SPME) has been used for rapid sample preparation and provides an efficient method to detect chemicals in detection and separation systems [3,4]. The extraction of samples can be performed using two methods. In the first method, headspace SPME (HS-SPME), the polymeric film is exposed to the gas phase that adsorbs the volatiles in the headspace of the liquid, gas or gaseous samples. The second method is direct immersion (DI-SPME), in which the fiber is directly immersed in a small volume of the liquid-extracted sample [5–7]. After the sample matrix and SPME coating achieves equilibrium, the extracted SPME is inserted into a gas chromatograph-mass spectrometer (GC-MS) for thermal desorption or into a desorption solvent for coupling with liquid chromatography (LC-MS) [8]. In addition, SPME has been used on environmental samples for the extraction of volatile organic compounds and has been a focus of interest in analytical biology, as well as pharmaceutical and food studies [4]. Some of the compounds function as species-specific signals, i.e., pheromones that provide intraspecific information [9]. Several studies have investigated the volatile pheromonal emissions released by the medfly as a potential source for an effective virgin female attractant. This would be useful such as an attractant might also find use in female annihilation programs and in mating disruption studies. The mating behavior of the mature male medfly, is also associated with the release of pheromonal volatiles attractive to the female fly [10]. Studies further describe this calling process and suggested that several abdominal glands present in the males were involved in production and released of the pheromone mixture [11,12]. A more extensive list of the biological activity of medfly including (f)-2-hexenoic acid, linalool, geranyl acetate, 2,3- and 2,5-dimethylpyrazines was reported by Jang et al. [12] and identification by Heath et al. [13]. Also, Baker et al. [14] have monitored the release of three of the major male medfly emission components which are ethyl (f)-3-hexenoate, geranyl acetate, and α-farnesene. In fruit flies, long-chain hydrocarbons on the adult fly cuticle are perceived by other flies over a short distance. Several studies have investigated the role of these compounds in chemical communication in the fruit fly [15]. Recently, SPME has been combined with capillary electrophoresis and liquid chromatography, and used for various biological samples, e.g., plasma and urine [16].

The development of analytical technology with powerful quantitative and qualitative capabilities, as well as high specificity, is required for the study of metabolic samples. This study investigates the feasibility of using DI-SPME high-resolution metabolism for profiling of fruit fly tissues at different stages of adulthood. Headspace-SPME (HS-SMPE) has been reported to be selective for volatile analyses, and is highly sensitive to volatile chemicals [17–19]. This approach enables SPME to identify substances with poor chromatographic behavior, thermal instability, or high reactivity [20]. Here, to ensure a high degree of sensitivity and chemical specificity, SPME with a GC-MS was used to capture metabolites [21]. The potential uses of DI-SPME for extracted insect samples were tested using statistical analysis to detect changes in the extraction samples before, during and after mating. The growth of metabolic and chemical analyses involving these low-dimensional score plots necessitates the use of quantitative statistical measures to describe significant differences between experimental groups such as PCA/PLS-DA score plots [22]. The PLS-DA is the first application of multivariate statistical methods for classification by ambient ionization but these methods have been applied previously to other MS imaging methods [23]. Principal component analysis has been used successfully as a multivariate statistical process control tool for detecting differences in processes with highly correlated variables [24]. Finally, DI-SPME was used in response to the need for the acquisition of representative metabolism data and for a better understanding of the encountered effects of extract samples.

2. Results and Discussion

The precision of DI-SPME was tested using biological sources and analysis of variation, to determine the analytical variability of the data generated when adult flies were sampled at different stages. In this experiment, the DI-SPME samples of mature medfly adults at three different mating stages (before, during and after), were analyzed using a GC-MS. To examine the effectiveness of DI-SPME, two different solvent extractions were used to compare the DI-SPME, which indicated quantitative and qualitative differences between these solvents in the type and peak areas of compounds. For further testing of DI-SPME, a GC-MS was used to compare the composition of two extracts solvents after directly immersing the SPME fiber in the extract. Comparison of these compound profiles revealed that DI-SPME had higher levels of the lighter chemicals and lower levels of ponderous chemicals. Firstly, the choice of the sealing and desorption time was carried out by fixing the time (2, 4, 8 and 16 h of sealing times). The best results were obtained with the recently developed 50/30 μm Carboxen/DVB/PDMS and, thus, 16 h sealing time was selected for further method development.

Overall, DI-SPME detected 110 compounds using the acetonitrile/water solvent and 86 compounds using the methanol/acetonitrile/water solvent at three different stages. In the first solvent extraction, 47, 26 and 37 compounds were identified from samples taken before, during and after mating, respectively. In the second solvent, 33, 31 and 22 compounds were identified from the samples taken before, during and after mating, respectively. The method has developed a strategy for rapid comparison of non-processed MS data files. To explain the differences between the samples, the method includes the following: baseline correction; alignment; time window determinations; alternating regression; PLS-DA. The identification of the retention time, the retention index, and mass spectral, MS structurally ordered separation windows in the chromatograms. For understanding the trends in analytical variability of our data set generated when different sides of solvents were sampled, chemically and functionally distinct metabolites were tentatively identified with retention index, the aid of mass spectral similarity, injection of authentic standards (C_7-C_{30}), and structurally ordered separations. The results showed significant correlations between metabolite molecular weight, the retention index and metabolites. The main compounds identified were ethyl glycolate, α-farnesene; decanoic acid octyl ester; 2,6,10,15-tetramethylheptadecane, 11-tricosene, 9,12-(Z,Z)-octadecadienoic acid, methyl stearate; 9-(Z)-tricosene, 9,11-didehydrolumisterol acetate; 1,54-dibromotetrapentacontane, 9-(Z)-hexadecenoic acid hexadecyl ester, 9-(E)-octadecenoic acid, and 9-(Z)-hexadecenoic acid octadecyl ester, (Tables 1 and 2).

Table 1. Significant compounds peak area (one unit corresponds to a 10^4 area) detected at three mating stages of medfly by DI-SPME-GC-MS in acetonitrile/water solvent.

Compounds	RI [a]	RT [b]	Mating Stages			p Value	FDR [d]
			Before	During	After		
Dodecanoic acid	1572.6	17.342	N.D [c]	N.D [c]	104.884	0.003	0.015
1-Iodododecane	1716.2	19.656	N.D [c]	N.D [c]	108.690	0.002	0.014
Tetracosane	2078.5	25.429	110.994	N.D [c]	N.D [c]	0.001	0.014
trans-13-Octadecenoic acid	2122.7	26.132	361.845	980.758	N.D [c]	0.002	0.014
(Z)-Oleic acid	2130.2	26.249	N.D [c]	N.D [c]	618.801	6.670	0.014
Octadecanoic acid	2142.1	26.434	N.D [c]	N.D [c]	209.611	0.005	0.018
9-(Z)-Tricosene	2244.1	28.066	N.D [c]	N.D [c]	211.876	0.001	0.014
Hexacosane	2268.5	28.452	96.895	N.D [c]	N.D [c]	0.002	0.014
1-Eicosanol, TBDMS derivative	2327.8	30.144	44.947	N.D [c]	N.D [c]	0.003	0.015
Supraene	2748.8	36.122	434.511	N.D [c]	N.D [c]	0.007	0.024
2-Methyloctacosane	2785.6	36.698	N.D [c]	44.210	N.D [c]	0.003	0.015
Diethyldecyloxyborane	2831.5	37.430	66.238	N.D [c]	N.D [c]	0.001	0.014
3,5-Cyclo-6,814,22-ergostatriene	2873.7	38.086	64.498	N.D [c]	N.D [c]	7.440	0.014
Hentriacontane	2969.3	39.616	N.D [c]	N.D [c]	403.452	0.009	0.024
Octatriacontyl pentafluoropropionate	2991.1	39.964	N.D [c]	N.D [c]	70.866	0.004	0.015
1,54-Dibromotetrapentacontane	3017.3	40.379	N.D [c]	N.D [c]	72.014	0.002	0.014
9-(Z)-Hexadecenoic acid hexadecyl ester	3131.3	42.196	55.305	214.519	1583.587	9.960	0.014
11,13-Dimethyl-12-tetradecen-1-acetate	3137.0	42.888	N.D [c]	N.D [c]	139.731	0.003	0.015
9-(E)-Octadecenoic acid	3251.9	44.119	N.D [c]	76.668	600.066	0.002	0.014

[a] RI id retention index; [b] RT is retention times; [c] N.D is not detected; [d] FDR is false discovery rate of data. Each number represent the mean of three biological replicates.

Table 2. Significant compounds peak area (one unit corresponds to a 104 area) detected at three mating periods of medfly by DI-SPME-GC-MS in methanol/acetonitrile/water solvent.

Name	RI [a]	RT [b]	Mating Stages			p Value	FDR [d]
			Before	During	After		
N-methyleneethanamine	749.4	1.312	162.767	N.D [c]	N.D [c]	0.005	0.023
Ethyl glycolate	780.5	1.954	N.D [c]	N.D [c]	259.978	0.011	0.031
2,5-Dihydroxybenzaldehyde	1123.1	8.720	100.924	N.D [c]	N.D [c]	6.250	0.003
Acetic acid 2-propyltetrahydropyran-3-yl ester	1181.3	9.551	N.D [c]	283.245	N.D [c]	0.010	0.031
Diclofop-methyl	1266.7	11.602	N.D [c]	78.171	N.D [c]	0.008	0.027
1,2-Dihydro-2,2,4-trimethylquinoline	1452.6	15.297	43.9242	119.575	N.D [c]	0.018	0.041
α-Farnesene	1513.7	16.367	281.554	190.567	N.D [c]	0.001	0.009
Decanoic acid octyl ester	1650.5	18.601	116.138	N.D [c]	N.D [c]	0.021	0.043
Dodecane, 1-iodo-	1716.2	19.656	N.D [c]	N.D [c]	108.690	0.003	0.019
Tetradecanoic acid	1765.4	20.432	70.0986	N.D [c]	88.350	0.005	0.024
2,6,10,15-Tetramethylheptadecane	1892.7	22.466	52.3699	1066.241	176.519	0.008	0.027
Hexadecanoic acid methyl ester	1917.5	22.861	759.908	1283.292	N.D [c]	0.022	0.043
Hexadecanoic acid pyrrolidide	1937.7	23.182	N.D [c]	1168.109	N.D [c]	0.000	0.008
9-Hexadecenoic acid pyrrolidide	1944.1	23.182	382.040	N.D [c]	757.991	0.001	0.027
1-Piperidin-1-yl-hexadecan-1-one	1958.6	23.518	982.573	N.D [c]	1095.741	0.008	0.027
9,12-(Z,Z)-Octadecadienoic acid	2078.2	25.428	356.887	2684.126	N.D [c]	0.021	0.043
Methyl stearate	2105.0	25.849	226.924	294.663	N.D [c]	0.006	0.026
Heneicosyl acetate	2181.3	27.073	56.3354	N.D [c]	N.D [c]	0.001	0.009
9-(Z)-Tricosene	2244.1	28.066	N.D [c]	N.D [c]	209.611	0.001	0.012
Trimesitylborane	2672.6	34.89	316.653	N.D [c]	1398.338	0.025	0.049
1,4-Benzenedicarboxylic acid bis-2-ethylhexyl ester	2679.7	35.001	N.D [c]	1345.4	N.D [c]	0.017	0.041
9,11-Didehydrolumisterol acetate	2865.1	37.957	652.982	N.D [c]	495.747	0.001	0.012
Stigmasta-3,5-diene	2967.1	39.578	N.D [c]	225.929		0.009	0.029
β-Sitosterol acetate	2968.5	39.601	86.1762	N.D [c]	403.452	0.013	0.034
Octatriacontyl pentafluoropropionate	2991.1	39.964	N.D [c]	N.D [c]	70.866	0.003	0.018
α-Tocopheryl acetate	2995.1	40.029	152.892	N.D [c]	N.D [c]	0.011	0.031
3β,22(E)-Ergosta-5,8,22-trien-3-ol	3055.5	40.981	N.D [c]	217.940	N.D [c]	0.007	0.009
3-Stigmasta-5,22-dien-3-ol acetate	3094.1	41.611	259.867	231.980	N.D [c]	0.004	0.023
9-Hexadecenoic acid hexadecyl ester	3131.3	42.196	N.D [c]	N.D [c]	509.690	0.014	0.036
11,13-Dimethyl-12-tetradecen-1-ol acetate	3137.0	42.888	N.D [c]	N.D [c]	176.459	0.003	0.018
Palmitoleic acid	3189.3	43.124	N.D [c]	N.D [c]	139.731	0.019	0.041
9-(E)-Octadecenoic acid,	3251.9	44.119	N.D [c]	N.D [c]	600.066	0.001	0.012
9-Hexadecenoic acid octadecyl ester	3257.9	44.219	303.882	N.D [c]	1583.587	0.002	0.015

[a] RI is retention index; [b] RT is retention times; [c] N.D is not detected; [d] FDR is false discovery rate of data. Each number represent the mean of three biological replicates.

For some compounds, there were significant differences observed between samples collected at different stages. In the first solvent extraction, tetracosane; diethyldecyloxyborane, 9-(Z)-tricosene, hexacosane; 9-(E)-octadecenoic acid, 1,54-dibromotetrapentacontane, trans-13-octadecenoic acid, 2-methyloctacosane, 11,13-dimethyl-12-tetradecen-1-ol acetate; TBDMS-1-eicosanol; octatriacontyl pentafluoropropionate; 1-iodododecane, octadecanoic acid, supraene and hentriacontane were significantly different between the collection periods (Table 1). In the second extraction solvent, 9-hexadecenoic acid pyrrolidide; diclofop-methyl; 1-piperidin-1-yl-hexadecan-1-one; stigmasta-3,5-diene; ethyl glycolate; 1,2-dihydro-2,2,4-trimethylquinoline, palmitoleic acid, 9,12-(Z,Z)-octadecadienoic acid, methyl stearate and trimesitylborane were identified (Table 2). Principal component analysis (PCA), sparse partial least squares-discriminant analysis (sPLS-DA), and heat map and ANOVA analyses were used in these experiments. PCA visualizes both the covariance and correlation between the metabolites and the modeled class designation. Thereby the PCA-plot helps to identify statistically significant and potentially biochemically significant metabolites, based both on contributions to the model and their reliability. An extension of PCA, the sPLS-DA-plot, is applied to compare the outcome of multiple classification models compared to a common reference, e.g. control.

The example used is a GC-coupled MS-based metabolomics study in extracted samples where two mating time lines are compared between extract solvents. The two principal components were plotted: the first solvent extraction had 56% and 11.1%, and the second extraction had 39.3% and 28.3% (Figure 1). The heat map showed a clear difference between the samples, particularly during and after mating stage (Figures 2 and 3).

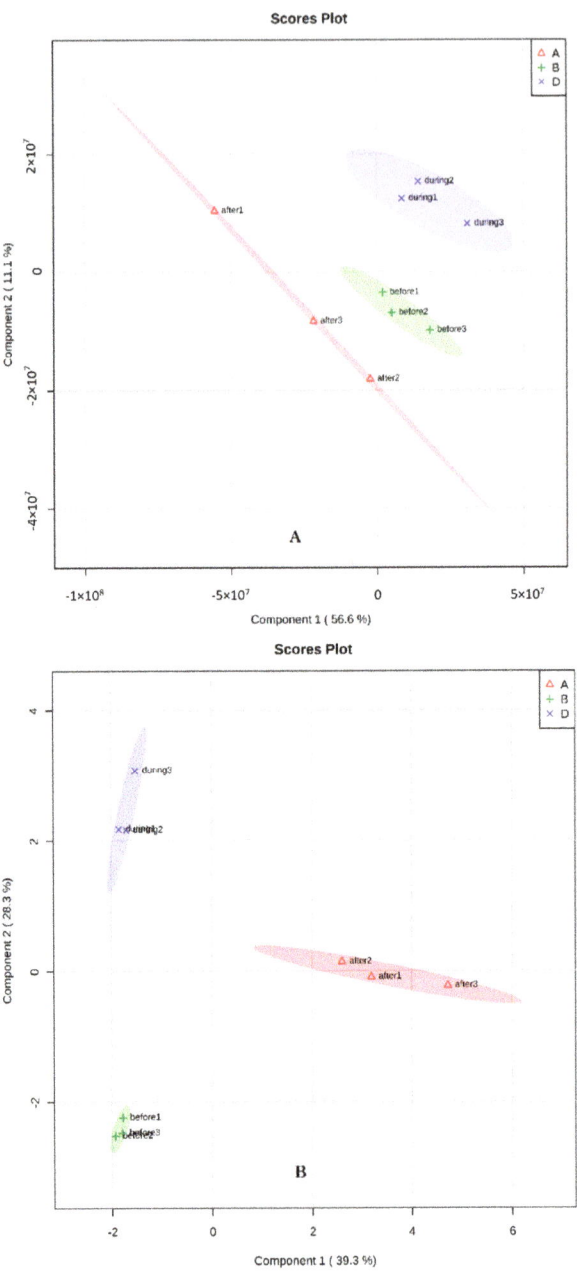

Figure 1. Score plots from Sparse Partial Least Squares-Discriminant Analysis (sPLS-DA) analyzed based on the total peak area obtained from GC-MS data of DI-SPME samples from three different mating stages of medfly: +, before mating; ×, during mating; Δ, after mating using two solvents (**A**) and (**B**). Three symbols in each group mean n = 3 biological replicates.

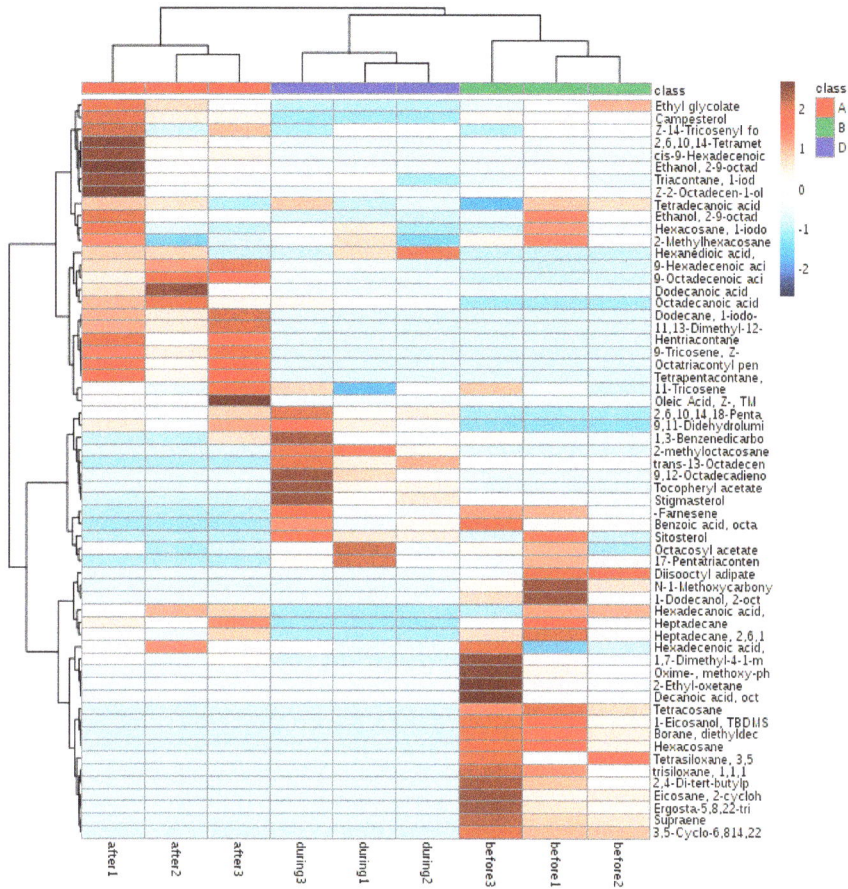

Figure 2. Heat map showing the changes of abundance values normalized to the compounds that are significantly influenced by extraction solvent and the time of insect sampling during mating stage. Three symbols in each group mean n=3 biological replicates.

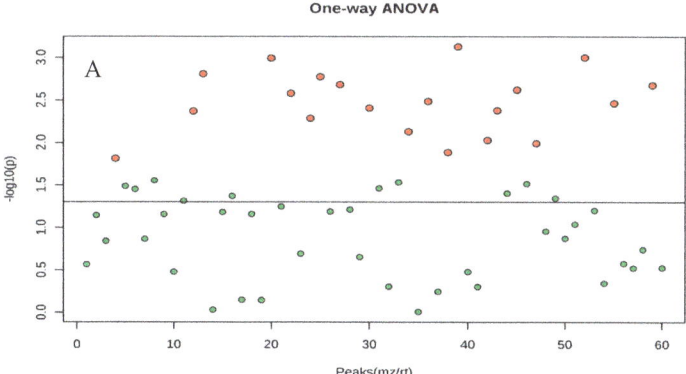

Figure 3. Cont.

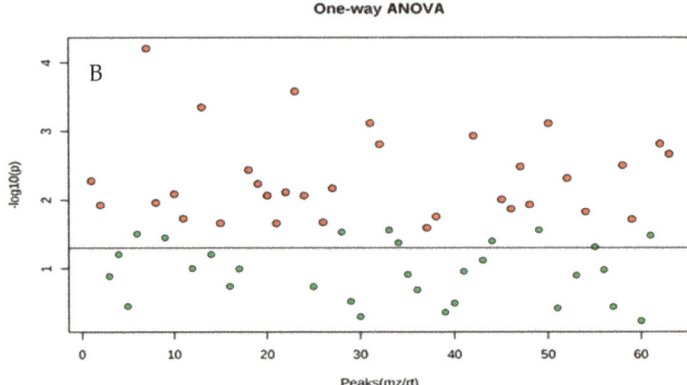

Figure 3. Red points represent significant compounds from the first solvent (**A**) and from the second solvent (**B**). Green points (**A**) and (**B**) are not significant. Each point represent three biological replicates.

Comparing the HS-SPME, compounds including: 2,3-hexanedione, *o*-dimethylbenzene, nonane, 2,3,4-trithiapentane, octanal, acetophenone, 2,6-dimethyl-(*E*,*Z*)-2,4,6-octatriene, 1*H*-pyrrole-2-carboxylic acid, 2,6,10-trimethyltridecane, dimethyl phthalate; farnesene, (*E*)-γ-bisabolene, 5-phenyl- undecane, carboric acid 2-ethylhexyl octyl ester, 2-ethylhexyl octyl ester; and 5-dodecyldihydro-2(3*H*)-furanone, were detected from medfly adults during mating stage (Table 3, Figure 4). Al-khshemawee et al. [7] reported that the compounds acetoin, 2,3-hexanedione, hexaldehyde, 4-hydroxybutanoic acid, 2,3,4- trithiapentane and octanal were identified from medfly adults using HS-SPME. Jacobson et al. [25] used HS-SPME to identify the pheromones from medfly adults. They found that methyl (*E*)-6-nonenoate and (*E*)-6-nonen-1-ol were the main compounds. Baker et al. [14] studied the volatile compounds emitted by sexually mature male Mediterranean fruit flies. They have been identified the key component involved in the sexual attraction of virgin female flies to males demonstrated to be the novel sex pheromone 3,4-dihydro-2*H*-pyrrole. Cossé et al. [26] reported that the male-produced volatiles eliciting responses from female were ethyl (*E*)-3-octenoate, geranyl acetate, (*E*,*E*)-α-farnesene, linalool, and indole, while Jang et al. [12] found and identified five major component groups that included ethyl hexenoates, hexanoates, methyl octenoates, monoterpenes and ketones.

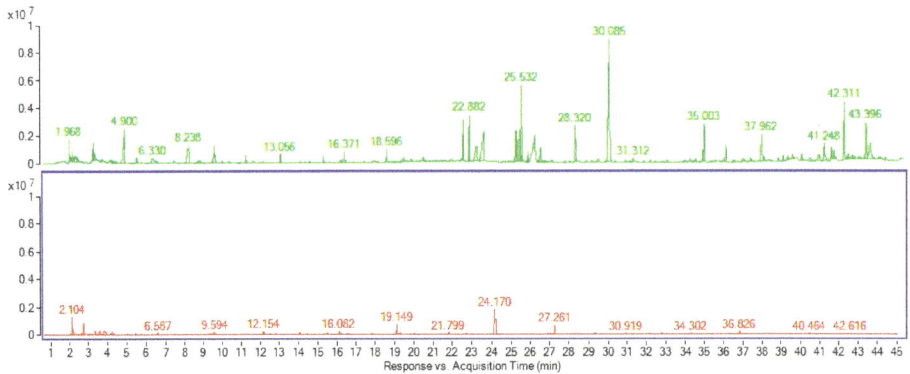

Figure 4. Chromatograms obtained after separation of compounds using DI-SPME and HS-SPME.

Table 3. Compounds identified from the adult stage of the medfly (one unit corresponds to a 10^5 area) determined by GC–MS using HS-SPME.

RT [a]	Compounds [b]	RI [c]	Peak Area
3.61	Acetoin	717	97.830
4.21	Toluene	755	20.493
5.54	Hexaldehyde	769	9.270
7.87	o-Dimethylbenzene	862	5.312
8.29	Nonane	900	4.095
9.67	4-Hydroxybutanoic acid	933	8.433
11.29	2,3,4-Trithiapentane	943	1.765
12.19	2,7-dimethyloctane	964	46.140
12.79	Octanal	982	2.035
13.75	4-Methyl-5-hexen-4-olide	996	3.624
14.57	Acetophenone	1049	0.851
15.52	3,3-Dimethylstyrene	1099	2.474
16.06	Cosmene	1134	5.422
16.52	2,6-Dimethyl-(E,Z)-2,4,6-octatriene	1292	4.970
19.08	2,6-Dimethylundecane	1214	1.554
19.26	1H-Pyrrole-2-carboxylic acid	1276	2.032
21.89	Tridecane	1300	2.965
22.66	2,6,10-Trimethyltridecane	1467	1.025
25.89	Dimethyl phthalate	1440	1.275
26.54	Cuparene	1496	1.677
27.01	Farnesene	1499	0.871
28.25	(E)-γ-Bisabolene	1523	1.849
30.27	5-Phenylundecane	1626	0.862
32.81	Tetradecanoic acid	1748	1.287
34.27	Carboric acid 2-ethylhexyl octyl ester	1857	0.422
36.82	n-Hexadecanoic acid	1968	2.129
38.56	5-Dodecyldihydro-2(3H)-furanone	2120	0.382
40.07	Octadecanoic acid	2187	1.743

[a] RT: retention time (min); [b] Compounds, name of compounds detected by GC-MS; [c] RI: retention indices. Each number of peak area represent three biological replicates.

Identifications are based on comparisons of both mass spectral data and GC retention indices with those of authentic reference compounds. Several components remain unidentified. Most of the unidentified run components were present at low concentrations, and were therefore thought to be contaminants. Some compounds were presented before mating, but they were missing during the mating stage. Some chemicals, were increased and some decreased within the mating stages (Tables 1 and 2). McDonald [27] reported that medfly males are stimulated to more frequent episodes of calling activity, when they are able to detect the presence of other medfly males. However, this interaction to visual and acoustic cues rather than to chemical communication. Jacobson et al. [25] and Ohinata et al. [28] studied which components are necessary to trigger an attractive response from female flies. This has been addressed to varying degrees except the present one, which is primarily a qualitative and semi-quantitative examination of the male emission complex. Ongoing laboratory evaluations of the major pheromone components identified indicate that many compounds contribute differentially, but synergistically to the pheromone's attractiveness for the female medflies. Other intermediate to low-concentration components may also be required to attain full parity with calling males. Flath et al. [29]; Al-khshemawee et al. [30] reported that three different medfly ages (5–6, 11–12, and 20–21 days old), and early-, mid-, and late-morning samples were used to collect volatiles. Thirty-two components were identified. However, propan-2-ol, hexanal, phenol, (Z,E)-α-farnesene, prop-2-yl-(E)-3-octenoate, ethyl (E)-2-octenoate, and propyl (E)-3-octenoate had been only partially identified in an earlier study. Quantitatively, ethyl acetate, 1-pyrroline, ethyl (E)-3-octenoate, geranyl acetate, and α-farnesene were the most abundant emission components from 5–6- and 11–12-day from old flies. The major compound for al fly ages was (2S)-2-hexenoic acid. Shelly [31] investigated the influence of α-copaene-containing plants on the mating system of *C. capitata* and the possibility of using attractants in prerelease exposure of males to increase the effectiveness of sterile insect release programs. Mature males were exposed to 20 µL of the attractant over a 6-h period and then were held for 2 d before testing. In field-cage trials, treated males (exposed to attractants) obtained significantly more matings than control males (no exposure) for all three substances. The potential exists for the development of an effective and useful female attractant, especially if essential components and their optimum release rates can be pinpointed and reproduced.

DI-SPME-GC-FID was first reported in an analysis of 13 commonly known benzodiazepines in urine [14]. The same group reported a modification of the method to analyze the hydrolysis of benzodiazepines from benzophenones extraction [32]. DI-SPME has been reported for quantitative analysis of biological samples including plant tissues [33], pesticides [4,34], milk [35], pharmaceuticals [36], wine [37] and water [38]. Myung et al. [39] optimized the DI extraction in blood samples for sorption of 1-octanol. Frérot et al. [40] used an organic solvent to soak or wash SPME in detected pheromones from the female abdominal tip of the Lepidopteran *Sesamia nonagrioides*. The pheromones of *Metamasius hemipterus* (Coleoptera) were sampled using SPME and compared to typical analytical methodologies. The SPME technique was shown to be cheaper, easier, faster and more reproducible [41]. SPME has been used to analyze cuticular hydrocarbons from ants [42]. DI-SPME has been used with pentane or hexane to analyze signaling chemicals and long-chain hydrocarbons from different parts of wasps' bodies [43]. The SPME technique has also been used to detect long-chain free fatty acids from insect exocrine glands, using a GC-MS [44]. Long chain fatty acids, such as oleic, palmitic, stearic, linoleic, and palmitoleic acids have been found in the exocrine secretions and cuticular extracts of many insects [45]. These compounds are important in intermediates and metabolites of biological pathways, and analytical techniques to study these compounds are of interest [46]. Filho et al. [47] showed that DI-SPME is more sensitive than HS-SPME, and it is thus the method of choice for the analysis of clean aqueous samples. The two extraction modes were evaluated and, despite being less sensitive than HS-SPME in the case of the more volatile compounds, DI-SPME mode successfully extracted 16 pesticides, while HS-SPME was able to extract only 12 compounds.

3. Materials and Methods

3.1. Insect Rearing

A medfly colony was obtained from the Department of Primary Industries and Regional Development (DPIRD), and flies were reared in the Post-harvest Biosecurity and Food Safety Laboratory at Murdoch University (Perth, Western Australia). All the flies were reared under the following conditions: temperature = 23 ± 2 °C, relative humidity = $75 \pm 5\%$, and light: dark cycle = 12:12-h [48]. Adults were placed in screen cages (40 cm cubes), each containing medfly food made from crystaline sugar (Bidvest, Sydney, Australia) and yeast hydrolysate (Australian Biosearch, Sydney, Australia) at a ratio of 4:1, and 50 mL water. Approximately 10–12 days after adult emergence from pupae and mating, eggs were collected each day. These were deposited on a mesh side of the cage and fell into a water tray kept adjacent to the cage.

3.2. DI-SPME Conditions

A GC-MS 7890B gas chromatograph equipped with a 5977B MSD mass spectrometer (Agilent Technologies, Santa Clara, CA, USA), with an Agilent HP-5MS column (30 m, 0.25 µm, 0.25 µm film thickness) was used in the experiments. The carrier gas used was helium at 99.999% (BOC, Sydney, Australia). The conditions for the GC-MS were as follows: injector port temperature of 270 °C; initial oven temperature of 60 °C, which increased to 320 °C (at 5 °C/min); MS Quad at 150 °C; MS source at 230 °C; pressure at 10.629 psi. The flow rate was 1.2 mL/min; the splitless was 30 mL/min at 1.0 min. The total run time was 45.40 min.

Standard *n*-alkane (C_7-C_{30}) reference material containing 1000 µg/mL of each component (decane, docosane, dodecane, eicosane, heneicosane, heptacosane, heptadecane, hexacosane, hexadecane, heptane, nonacosane, nonadecane, nonane, octacosane, octadecane, octane, pentacosane, pentadecane, tetracosane, tetradecane, triacontane, tricosane, tridecane and undecane) in hexane was purchased from Sigma-Aldrich (catalogue number 49451-U; Castle Hill, NSW, Australia), as was *n*-hexane (95%, catalogue number 270504-2L).

3.3. DI-SPME Procedure and Sampling Setup

SPME fiber 50/30 µm with Carboxen/DVB/PDMS (Sigma-Aldrich, Bellefonte, PA, USA) coating was inserted into extracted samples. SPME in the samples was conditioned at room temperature (25 ± 5 °C) for 16 h with a sampling depth of 3 cm. The DI-SPME extraction was carried out by immersing the fiber (length: 1.3 cm) into the extracted solution. After extraction for 16 h sealing time, the fiber was withdrawn into the needle, removed from the vial and immediately introduced into the GC injector port for thermal desorption. Samples in triplicate were used for extraction. For sample preparation, adult medflies (0.05 g) were taken before, during and after mating stages. Insects were grinded using tissuelyser at 270 rpm for 2 min. Two extraction solvents, acetonitrile/water (1:1) and methanol/acetonitrile/water (2:2:1) (CAS: 67-56-1, UN1230, Thermo Fisher Scientific, Perth, Australia), were used to extract the samples. Extraction solvent (1 mL) was added to the samples, and centrifuged at 2000 rpm for 5 min. The extraction samples were transferred to a 2 mL analytical vial. SPME was inserted directly into the vial for 16 h at room temperature. Then, the DI-SPME was analyzed using a GC-MS for 15 min desorption time. The samples were analyzed in biological triplicates.

3.4. Statistical Analysis

To observe the impact of observations, principal component analysis (PCA) with the correlation matrix method was used for statistical analysis using the online MetaboAnalyst 3.0 (2017) (Bellevue, Quebec, USA) tool, a comprehensive online tool for metabolomics analysis and interpretation. PCA was used to transfer the original data onto new axes where principal components corresponded to significant information represented by the original data. Three principal components are chosen from the result of PCA and sPLS-DA analysis based on Xia and Wishart [49]. The plots classifier was used to integrate the two components obtained from PCA and produce a segmented image. Since the heatmap centers were chosen randomly in the original means and the obtained results can be different for every run of the algorithm, the overall classification accuracies were averaged over different data.

4. Conclusions

In this study, two DI-SPME extraction solvents for were used at three different stages of the medfly adult life. The first extraction solvent was acetonitrile/water, and the second solvent was methanol/acetonitrile/water. Samples were collected before, during and after mating. This study compared these extraction solvents based on the metabolites extracted. The GC-MS analytical data showed a wide spectrum of compounds and DI-SPME sampling was developed to identify these compounds from medfly extracts. These results indicate that DI-SPME coupled with the GC-MS could be performed successfully on medfly extracts. Using DI-SPME with GC analysis of extracts, high sensitivity and good repeatability were obtained. This work is an example of the application of DI-SPME-GC in the analysis of complex samples and provides a way in which to prepare the samples of SPME coatings. Further development of DI-SPME is promising, and may provide an efficient extraction technique for biological samples.

Author Contributions: Conceptualization, H.A.-K. and X.D., Methodology and Writing-Original Draft Preparation, M.A.; Software, Data Curation and Supervision, Y.L.R.; Investigation, Writing-Review & Editing and Supervision, J.O.Y.

Funding: This research received no external funding.

Conflicts of Interest: The authors declare no conflict of interest.

References

1. Becerra, J.X. Insects on plants: Macroevolutionary chemical trends in host use. *Science* **1997**, *276*, 253–256. [CrossRef] [PubMed]
2. T'Kindt, R.; Morreel, K.; Deforce, D.; Boerjan, W.; Van Bocxlaer, J. Joint GC–MS and LC–MS platforms for comprehensive plant metabolomics: Repeatability and sample pre-treatment. *J. Chromatogr. B* **2009**, *877*, 3572–3580. [CrossRef] [PubMed]
3. Arthur, C.L.; Pawliszyn, J. Solid phase microextraction with thermal desorption using fused silica optical fibers. *Anal. Chem.* **1990**, *62*, 2145–2148. [CrossRef]
4. AL-Khshemawee, H.; Agarwal, M.; Ren, Y. Detection of Mediterranean Fruit Fly larvae *Ceratitis capitata* (Diptera: Tephritidae) in different types of fruit by HS-SPME GC-MS method. *J. Biosci. Med.* **2017**, *5*, 154–169. [CrossRef]
5. Aulakh, J.S.; Malik, A.K.; Kaur, V.; Schmitt-Kopplin, P. A review on solid phase micro extraction—High performance liquid chromatography (SPME-HPLC) analysis of pesticides. *Crit. Rev. Analy. Chem.* **2005**, *35*, 71–85. [CrossRef]
6. Bojko, B.; Reyes-Garcés, N.; Bessonneau, V.; Goryński, K.; Mousavi, F.; Silva, E.A.S.; Pawliszyn, J. Solid-phase microextraction in metabolomics. *TrAC Trends Anal. Chem.* **2014**, *61*, 168–180. [CrossRef]
7. AL-Khshemawee, H.; Agarwal, M.; Ren, Y. Evaluation of stable isotope $^{13}C_6$-glucose on volatile organic compounds in different stages of Mediterranean fruit fly (Medfly) *Ceratitis Capitata* (Diptera: Tephritidae). *Entomol. Ornith. Herpet. Curr. Res.* **2017**, *6*, 3–8.
8. Pawliszyn, J. *Solid Phase Microextraction: Theory and Practice*; John Wiley & Sons: Hoboken, NJ, USA, 1997.
9. Zhang, X.; Oakes, K.D.; Wang, S.; Servos, M.R.; Cui, S.; Pawliszyn, J.; Metcalfe, C.D. In vivo sampling of environmental organic contaminants in fish by solid-phase microextraction. *TrAC Trends Anal. Chem.* **2012**, *32*, 31–39. [CrossRef]
10. Feron, M. Chemical attraction of the *Ceratitis capitata* Wied. (Diptera: Tephritidae) male for the female. *C R. Acad. ScL Ser. D. (Paris)* **1959**, *248*, 2403–2404.
11. Lhoste, J.; Roche, A. Odoriferous organs of *Ceratitis capitata* males. (Diptera: Tephritidae). *Bull. Soc. Entomol. Fr.* **1960**, *65*, 206–209.
12. Jang, E.B.; Light, D.M.; Flath, R.A.; Nagata, J.T.; Mon, T.R. Electroantennogram responses of the Mediterranean fruit fly, *Ceratitis capitata* to identified volatile constituents from calling males. *Entomol. Exp. Appl.* **1989**, *50*, 7–19. [CrossRef]
13. Heath, R.R.; Landolt, P.J.; Tumlinson, J.H.; Chambers, D.L.; Murphy, R.E.; Doolittle, R.E.; Dueben, B.D.; Sivinski, J.; Calkins, C.O. Analysis, synthesis, formulation, and field testing of three major components of male Mediterranean fruit fly pheromone. *J. Chem. Ecol.* **1991**, *17*, 1925–1940. [CrossRef] [PubMed]
14. Baker, R.; Herbert, R.H.; Grant, G.G. Isolation and identification of the sex pheromone of the Mediterranean fruit fly, *Ceratitis capitata* (Wied). *J. Chem. Soc. Chem. Commun.* **1985**, *12*, 824–825. [CrossRef]
15. Ferveur, J.-F. Cuticular hydrocarbons: Their evolution and roles in drosophila pheromonal communication. *Behav. Genet.* **2005**, *35*, 279–295. [CrossRef] [PubMed]
16. Antony, C.; Jallon, J.-M. The chemical basis for sex recognition in drosophila melanogaster. *J. Insect Physiol.* **1982**, *28*, 873–880. [CrossRef]
17. Theodoridis, G.; Koster, E.D.; De Jong, G. Solid-phase microextraction for the analysis of biological samples. *J. Chromatogr. B Biomed. Sci. Appl.* **2000**, *745*, 49–82. [CrossRef]
18. Risticevic, S.; Lord, H.; Górecki, T.; Arthur, C.L.; Pawliszyn, J. Protocol for solid-phase microextraction method development. *Nat. Prot.* **2010**, *5*, 122–139. [CrossRef] [PubMed]
19. Risticevic, S.; DeEll, J.R.; Pawliszyn, J. Solid phase microextraction coupled with comprehensive two-dimensional gas chromatography–time-of-flight mass spectrometry for high-resolution metabolite profiling in apples: Implementation of structured separations for optimization of sample preparation procedure in complex samples. *J. Chromatogr. A* **2012**, *1251*, 208–218. [PubMed]
20. Risticevic, S.; Niri, V.H.; Vuckovic, D.; Pawliszyn, J. Recent developments in solid-phase microextraction. *Anal. Bioanal. Chem.* **2009**, *393*, 781–795. [CrossRef] [PubMed]

21. Seno, H.; Kumazawa, T.; Ishii, A.; Watanabe, K.; Hattori, H.; Suzuki, O. Detection of benzodiazepines in human urine by direct immersion solid phase micro extraction and gas chromatography. *Jpn. J. Forensic Toxicol.* **1997**, *13*, 207–210.
22. Worley, B.; Halouska, S.; Powers, R. Utilities for quantifying separation in PCA/PLS-DA scores plots. *Anal. Biochem.* **2013**, *433*, 102–104. [CrossRef] [PubMed]
23. Kano, M.; Hasebe, S.; Hashimoto, I.; Ohno, H. A new multivariate statistical process monitoring method using principal component analysis. *Com. Chem. Eng.* **2001**, *25*, 1103–1113. [CrossRef]
24. Dill, A.L.; Eberlin, L.S.; Costa, A.B.; Zheng, C.; Ifa, D.R.; Cheng, L.; Masterson, T.A.; Koch, M.O.; Vitek, O.; Cooks, R.G. Multivariate statistical identification of human bladder carcinomas using ambient ionization imaging mass spectrometry. *Chem. Eur. J.* **2011**, *17*, 2897–2902. [CrossRef] [PubMed]
25. Jacobson, M.; Ohinata, K.; Chambers, D.L.; Jones, W.A.; Fujimoto, M.S. Insect sex attractants. 13. Isolation, identification, and synthesis of sex pheromones of the male Mediterranean fruit fly. *J. Med. Chem.* **1973**, *16*, 248–251. [CrossRef] [PubMed]
26. Cossé, A.A.; Todd, J.L.; Millar, J.G.; Martínez, L.A.; Baker, T.C. Electroantennographic and coupled gas chromatographic-electroantennographic responses of the mediterranean fruit fly, *Ceratitis capitata*, to male-produced volatiles and mango odor. *J. Chem. Ecol.* **1995**, *21*, 1823–1836. [CrossRef] [PubMed]
27. McDonald, P.T. Intragroup stimulation of pheromone release by male Mediterranean fruit flies (Diptera: Tephritidae). *Ann. Entomol. Soc. Am.* **1987**, *80*, 17–20. [CrossRef]
28. Ohinata, K.; Jacobson, M.; Nakagawa, S.; Fujimoto, M.; Higa, H. Mediterranean fruit fly: Laboratory and field evaluations of synthetic sex pheromones. *J. Environ. Sci. Health* **1977**, *A12*, 67–78.
29. Flath, R.A.; Jang, E.B.; Light, D.M.; Mon, T.R.; Carvalho, L.; Binder, R.G.; John, J.O. Volatile pheromonal emissions from the male mediterranean fruit fly: Effects of fly age and time of day. *J. Agric. Food Chem.* **1993**, *41*, 830–837. [CrossRef]
30. AL-Kshemawee, H.; Agarwal, M.; Ren, Y. Optimization and validation for determination of volatile organic compounds from Mediterranean fruit fly (Medfly) *Ceratitis capitata* (Diptera: Tephritidae) by using HS-SPME-GC-FID/MS. *J. Biol. Sci.* **2017**, *17*, 347–352. [CrossRef]
31. Shelly, T.E. Exposure to α-Copaene and α-Copaene-containing oils enhances mating success of male Mediterranean fruit flies (Diptera: Tephritidae). *Ann. Entomol. Soc. Am.* **2001**, *94*, 497–502. [CrossRef]
32. Guan, F.; Ishii, A.; Seno, H.; Watanabe-Suzuki, K.; Kumazawa, T.; Suzuki, O. Use of an ion-pairing reagent for high-performance liquid chromatography–atmospheric pressure chemical ionization mass spectrometry determination of anionic anticoagulant rodenticides in body fluids. *J. Chromatogr. B Biomed. Sci. Appl.* **1999**, *731*, 155–165. [CrossRef]
33. Risticevic, S.; Souza-Silva, E.A.; DeEll, J.R.; Cochran, J.; Pawliszyn, J. Capturing plant metabolome with direct-immersion in vivo solid phase microextraction of plant tissues. *Anal. Chem.* **2015**, *88*, 1266–1274. [CrossRef] [PubMed]
34. Ai, Y.; Zhang, J.; Zhao, F.; Zeng, B. Hydrophobic coating of polyaniline-poly (propylene oxide) copolymer for direct immersion solid phase microextraction of carbamate pesticides. *J. Chromatogr. A* **2015**, *1407*, 52–57. [CrossRef] [PubMed]
35. González-Rodríguez, M.J.; Arrebola Liébanas, F.J.; Garrido Frenich, A.; Martínez Vidal, J.L.; Sánchez López, F.J. Determination of pesticides and some metabolites in different kinds of milk by solid-phase microextraction and low-pressure gas chromatography-tandem mass spectrometry. *Anal. Bioanal. Chem.* **2005**, *382*, 164–172. [CrossRef] [PubMed]
36. Snow, N.H. Solid-phase micro-extraction of drugs from biological matrices. *J. Chromatogr. A* **2000**, *885*, 445–455. [CrossRef]
37. Martínez-Uruñuela, A.; González-Sáiz, J.M.; Pizarro, C. Optimisation of a headspace solid-phase microextraction method for the direct determination of chloroanisoles related to cork taint in red wine. *J. Chromatogr. A* **2004**, *1056*, 49–56. [CrossRef] [PubMed]
38. López-Darias, J.; Pino, V.; Anderson, J.L.; Graham, C.M.; Afonso, A.M. Determination of water pollutants by direct-immersion solid-phase microextraction using polymeric ionic liquid coatings. *J. Chromatogr. A* **2010**, *1217*, 1236–1243. [CrossRef] [PubMed]
39. Myung, S.-W.; Min, H.-K.; Kim, S.; Kim, M.; Cho, J.-B.; Kim, T.-J. Determination of amphetamine, methamphetamine and dimethamphetamine in human urine by solid-phase microextraction (SPME)-gas chromatography/mass spectrometry. *J. Chromatogr. B Biomed. Sci. Appl.* **1998**, *716*, 359–365. [CrossRef]

40. Frérot, B.; Malosse, C.; Cain, A.H. Solid-phase microextraction (spme): A new tool in pheromone identification in lepidoptera. *J. High Resolut. Chromatogr.* **1997**, *20*, 340–342. [CrossRef]
41. Malosse, C.; Ramirez-Lucas, P.; Rochat, D.; Morin, J.P. Solid-phase microextraction, an alternative method for the study of airborne insect pheromones (metamasius hemipterus, coleoptera, curculionidae). *J. High Resolut. Chromatogr.* **1995**, *18*, 669–670. [CrossRef]
42. Monnin, T.; Malosse, C.; Peeters, C. Solid-phase microextraction and cuticular hydrocarbon differences related to reproductive activity in queenless ant dinoponera quadriceps. *J. Chem. Ecol.* **1998**, *24*, 473–490. [CrossRef]
43. Moneti, G.; Dani, F.R.; Pieraccini, G.; Turillazzi, S. Solid-phase microextraction of insect epicuticular hydrocarbons for gas chromatographic/mass spectrometric analysis. *Rapid Communi. Mass Spectro.* **1997**, *11*, 857–862. [CrossRef]
44. Maile, R.; Dani, F.R.; Jones, G.R.; Morgan, E.D.; Ortius, D. Sampling techniques for gas chromatographic–mass spectrometric analysis of long-chain free fatty acids from insect exocrine glands. *J. Chromatogr. A* **1998**, *816*, 169–175. [CrossRef]
45. Lockey, K.H. Lipids of the insect cuticle: Origin, composition and function. *Comp. Biochem. Physiol. B* **1988**, *89*, 595–645. [CrossRef]
46. Buckner, J.S. Cuticular polar lipids of insects. *Insect Lipids Chem. Biochem. Biol.* **1993**, 227–270.
47. Filho, A.M.; dos Santos, F.N.; Pereira, P.A.d.P. Development, validation and application of a method based on DI-SPME and GC–MS for determination of pesticides of different chemical groups in surface and groundwater samples. *Microchem. J.* **2010**, *96*, 139–145. [CrossRef]
48. Tanaka, N.; Steiner, L.; Ohinata, K.; Okamoto, R. Low-cost larval rearing medium for mass production of oriental and Mediterranean fruit flies. *J. Econ. Entomol.* **1969**, *62*, 967–968. [CrossRef]
49. Xia, J.; Wishart, D.S. Using MetaboAnalyst 3.0 for comprehensive metabolomics data analysis. *Curr. Protocol. Bioinform.* **2016**, *55*, 14.10.1–14.10.91. Available online: https://www.ncbi.nlm.nih.gov/pubmed/27603023 (accessed on 12 November 2018). [CrossRef] [PubMed]

Sample Availability: Samples of the compounds ethyl glycolate, α-farnesene, decanoic acid octyl ester, 2,6,10,15-tetramethylheptadecane, 11-tricosene, 9,12-(Z,Z)-octadecadienoic acid, methyl stearate, 9-(Z)-tricosene, 9,11-didehydro-lumisterol acetate; 1,54-dibromotetrapentacontane, 9-(Z)-hexadecenoic acid hexadecyl ester, 9-(E)-octadecenoic acid and 9-(Z)-hexadecenoic acid octadecyl ester., 1-iodododecane, 9-(Z)-tricosene and 11,13-dimethyl-12-tetradecen-1-acetate which were extracted with both (A) and (B) and dodecanoic acid, (Z)-oleic acid, octadecanoic acid and hentriacontane which were extracted with (A) and ethyl glycolate, 9-hexadecenoic acid hexadecyl ester, palmitoleic acid and 9-(E)-octadecenoic acid, which were extracted with solvent (B). All these compounds are available from the authors.

© 2018 by the authors. Licensee MDPI, Basel, Switzerland. This article is an open access article distributed under the terms and conditions of the Creative Commons Attribution (CC BY) license (http://creativecommons.org/licenses/by/4.0/).

Article

Analysis of the Volatile Profile of Core Chinese Mango Germplasm by Headspace Solid-Phase Microextraction Coupled with Gas Chromatography-Mass Spectrometry

Xiao-Wei Ma, Mu-Qing Su, Hong-Xia Wu, Yi-Gang Zhou and Song-Biao Wang *

Ministry of Agriculture Key Laboratory of Tropical Fruit Biology, South Subtropical Crops Research Institute, Chinese Academy of Tropical Agricultural Sciences, Zhanjiang 524091, China; maxiaowei428@126.com (X.-W.M.); pillar1984@163.com (M.-Q.S.); whx1106@163.com (H.-X.W.); zhouyigang@21cn.com (Y.-G.Z.)
* Correspondence: songbiaowang501@163.com; Tel.: +86-759-2859-312; Fax: +86-759-2859-312

Received: 9 May 2018; Accepted: 13 June 2018; Published: 19 June 2018

Abstract: Despite abundant published research on the volatile characterization of mango germplasm, the aroma differentiation of Chinese cultivars remains unclear. Using headspace solid phase microextraction (HS-SPME) coupled with gas chromatography–mass spectrometry (GC-MS), the composition and relative content of volatiles in 37 cultivars representing the diversity of Chinese mango germplasm were investigated. Results indicated that there are distinct differences in the components and content of volatile compounds among and within cultivars. In total, 114 volatile compounds, including 23 monoterpenes, 16 sesquiterpenes, 29 non-terpene hydrocarbons, 25 esters, 11 aldehydes, five alcohols and five ketones, were identified. The total volatile content among cultivars ranged from 211 to 26,022 μg/kg fresh weight (FW), with 123-fold variation. Terpene compounds were the basic background volatiles, and 34 cultivars exhibited abundant monoterpenes. On the basis of hierarchical cluster analysis (HCA) and principal component analysis (PCA), terpinolene and α-pinene were important components constituting the aroma of Chinese mango cultivars. Most obviously, a number of mango cultivars with high content of various aroma components were observed, and they can serve as potential germplasms for both breeding and direct use.

Keywords: mango germplasm; volatile compound; HS-SPME-GC-MS; multivariate analysis

1. Introduction

Sweetness, sourness and aroma constitute the main components of fruit flavour, with aroma being the most important contributing factor [1]. With the increasing requirement for fruit table quality and quality of processed products, fruit aroma has gained increasing research attention in recent years. Aroma components in fruits mainly consist of aldehydes, alcohols, esters, lactones, ketones, quinones and terpenes [2,3]. Each of these volatile compounds has a distinct odour, and their combinations, concentrations and ratios confer unique aroma characteristics to different fruits through cumulative, synergistic and masking effects [4]. The concentrations and composition of volatile compounds in fruits, although influenced by climatic and cultivation conditions [5–7], are mainly determined by the genetic background of the plants [8,9]. Therefore, the evaluation of volatile aroma compounds in fruits at the germplasm level is essential.

Mature mango possesses a rich flavour, which is a key characteristic which attracts consumers. Studies have indicated that mango aroma is the result of a mixture of terpenoids, alcohols, aldehydes, carbonyl compounds, esters, nitrogen-containing compounds, and other volatiles, with the composition and content of these aroma compounds in different cultivars being significantly distinct. At present, studies on aroma compounds in mango fruit at the germplasm level have focused on cultivars from India, Australia, the United States, Brazil and Cuba [10–13]. However, there are limited reports on the volatile profile of mango fruits at the germplasm level from China, which is an important producer of mango producing 129 million tonnes in 2013 according to the Food and Agriculture Organization.

In China, over the past 30 years, breeding objectives for mango have been primarily associated with yield, resistance, and appearance, with less emphasis on the improvement of flavour-associated traits such as the fruit aroma. As a result, superior flavour traits originally present in germplasm resources have been gradually lost during breeding, thus leading to largely similar fruit aroma amongst current commercial cultivars. At present, China's mango germplasm collection comprises 200 cultivars, and aroma sensory evaluations have revealed a marked variation in fruit flavour among these different cultivars which this affords the possibility of selecting potential parents for hybrid breeding. Phenotypic diversity assessment of fruit quality traits, for instance aroma, was the first step for effective germplasm conservation and utilisation. Although some literatures are available in the field of mango aroma, they are generally limited to particular cultivars [14].

In this study, 37 representative Chinese mango cultivars encompassing different maturation periods (early, moderate and late maturation), colour types (green, yellow and red) and genetic parents were selected as study materials [15,16]. The characteristics of volatile compounds in the fruits of different mango cultivars were studied using headspace solid-phase microextraction (HS-SPME) in conjunction with gas chromatography–mass spectrometry (GC-MS) with the aim of understanding characteristics of "good" mango fruit for ultimate flavour improvement.

2. Results and Discussion

2.1. Identity and Concentration of Volatile Compounds in Mango Cultivars

A total of 114 volatile compounds in the pulp were identified and relatively quantified, some of which were found only in a few of the cultivars in this study (see Tables 1–7). These compounds included 23 monoterpenes, 16 sesquiterpenes, 29 non-terpene hydrocarbons, 25 esters, 11 aldehydes, five alcohols and five ketones.

Table 1. Volatiles detected in fruits of all 37 cultivars.

Monoterpene	Code	Non-Terpene Hydrocarbon	Code	Decanoic Acid Ethyl Ester	E11
α-pinene	M1	Styrene	H1	Butanoic acid ethyl ester	E12
β-Ocimene	M2	2,4-dimethyl-Heptane	H2	Butanoic acid butyl ester	E13
Limonene	M3	Decane	H3	3-Hexen-1-ol acetate	E14
Terpinene	M4	Nonadecane	H4	Oxalic acid, 6-ethyloct-3-yl isohexyl ester	E15
β-Myrcene	M5	Tetradecane	H5	Hexanoic acid ethyl ester	E16
α-Terpinene	M6	Hexadecane	H6	Octanoic acid methyl ester	E17
3-Carene	M7	Heptadecane	H7	Propanoic acid 2-methyl-3-methylbutyl ester	E18
(1S)-(+)-3-Carene	M8	Octacosane	H8	Butanoic acid propyl ester	E19
γ-Terpinene	M9	2,4,6-trimethyl-Octane	H9	Ethyl 2-hexenoate	E20
4-carene	M10	Heptacosane	H10	Butanoic acid octyl ester	E21
β-pinene	M11	2,6,10-trimethyl-Pentadecane	H11	Tetradecanoic acid ethyl ester	E22
Camphene	M12	Pentadecane	H12	Propyl octanoate	E23
2-Thujene	M13	1-Fluorononane	H13	3-Hydroxymandelic acid ethyl ester	E24
Sylvestrene	M14	Eicosane	H14	1,2-Benzenedicarboxylic acid mono(2-ethylhexyl) ester	E25
Ocimene	M15	Heneicosane	H15	Aldehyde	Code
α-Phellandrene	M16	1,5,9,9-tetramethyl-1,4,7-Cycloundecatriene	H16	Heptanal	A1
Neoalloocimene	M17	1,3,8-p-Menthatriene	H17	2,6-Nonadienal	A2
β-Pinene	M18	1,3,5,8-Undecatetraene	H18	13-Octadecenal	A3

Table 1. Cont.

Monoterpene	Code	Non-Terpene Hydrocarbon	Code	Decanoic Acid Ethyl Ester	E11
D2-Carene	M19	10-Methylnonadecane	H19	Tetradecanal	A4
β-Terpinene	M20	1,5,5-Trimethyl-6-methylene-cyclohexene	H20	3,6-Nonadienal	A5
α-Pyronene	M21	1,3,5,7-Cyclooctatetraene	H21	2-Nonenal	A6
Artemisia triene	M22	5-Octadecene	H22	Nonanal	A7
1,3,5,5-tetramethyl-1,3-Cyclohexadiene	M23	2,2-dimethyl-3-methylene-Bicyclo[2.2.1]heptan	H23	Furfural	A8
Sesquiterpene	Code	2,6,10,15-tetramethyl-Heptadecane	H24	5-methyl-2-Furancarboxaldehyde	A9
Caryophyllene	S1	2,3,5-trimethyl-Decane	H25	5-Acetoxymethyl-2-furaldehyde	A10
α-Caryophyllene	S2	3-methyl-Dodecane	H26	Isopentyl hexanoate	A11
Germacrene D	S3	Undecane	H27	Alcohol	Code
α-Selinene	S4	1-methyl-2-(1-methylethyl)-Benzene	H28	2-propyl-1-Heptanol	B1
g-Selinene	S5	Tetratriacontane	H29	2-butyl-1-Octanol	B2
α-Gurjunene	S6	Ester	Code	1-Nonanol	B3
α-bulnsene	S7	Oxalic acid, isobutyl nonyl ester	E1	3,6-dimethoxy-9-(2-phenylethynyl)-Fluoren-9-ol	B4
Alloaromadendrene	S8	2-Propenoic acid, 2-ethylhexyl ester	E2	4-Ethyl-1-hexyn-3-ol	B5
α-Cubebene	S9	Oxalic acid, isohexyl pentyl ester	E3	Ketone	Code
Copaene	S10	Oxalic acid, isobutyl pentyl ester	E4	4-(2,6,6-trimethyl-1-cyclohexen-1-yl)-3-Buten-2-one	K1
β-Elemene	S11	2-Propenoic acid, 6-methylheptyl ester	E5	4-methoxy-2,5-dimethyl-3(2H)-Furanone	K2
Cubebene	S12	Oxalic acid, allyl nonyl ester	E6	1-(1,4-dimethyl-3-cyclohexen-1-yl)-Ethanone	K3
Calarene	S13	Dodecanoic acid, ethyl ester	E7	12-methyl-Oxacyclododec-9-en-2-one	K4
Epi-bicyclosesquiphellandrene	S14	Butanoic acid, 3-hexenyl ester	E8	2,3-dihydro-3,5-dihydroxy-6-methyl-4H-Pyran-4-one	K5
Isoledene	S15	Butanoic acid, hexyl ester	E9		
Aromadendrene	S16	Octanoic acid, ethyl ester	E10		

Monoterpenes, non-terpene hydrocarbons and esters were detected in all 37 cultivars, whereas most cultivars had no aldehydes, alcohols and/or ketones. The total volatile content showed great variation in different mango cultivars, and it ranged from 211.01 µg/kg fresh weight (FW) in Shengshi to 26,021.91 µg/kg FW in Xiaofei (Figure 1). The total aroma contents in Xiaofei, Guire 10, Jinhuang, Guire 7, and Guixiang cultivars were significantly higher than in the other cultivars. The fruit aroma is closely related to the content and number of volatile compounds. Significant differences of the number of volatiles were also found amongst the 37 cultivars. Xiaofei fruits contained the greatest number of volatiles (49), followed by Boluoxiang (48) and Tainong 1 (43), whereas Zaoshu had the least, with only 17 volatile compounds being detected. In previous sensory perception tests, Xiaofei fruit has been evaluated to exhibit obvious characteristics of aromatic flavour, which is associated with rich compounds and relatively high content of volatiles.

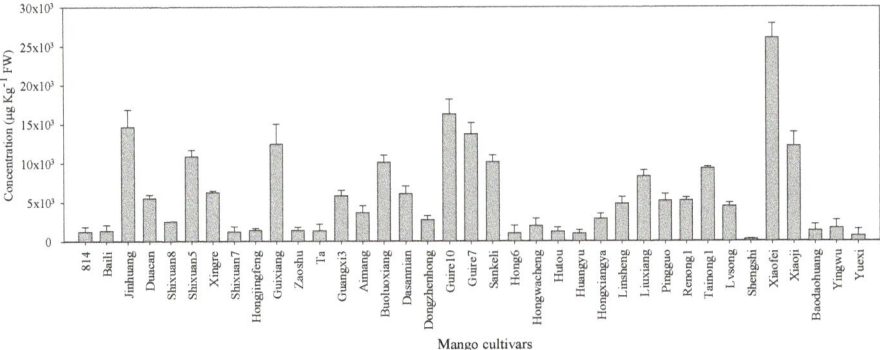

Figure 1. Concentrations (μg/kg fresh weight (FW) equivalent of nonyl acetate) of total volatile in 37 mango cultivars.

2.2. Relative Abundance of Different Classes of Volatile Compounds

Variation in the relative abundance of terpenes, non-terpene hydrocarbons, esters, aldehydes, alcohols and ketones amongst different mango cultivars was significant (Figure 2). Amongst the 37 mango cultivars examined, non-terpene hydrocarbons were the major volatiles in Shengshi, esters were the dominant volatiles in Boluoxiang fruit, while monoterpenes and sesquiterpenes were the dominant volatiles in other cultivars. Terpenoids are synthesised via two alternative pathways: the cytosolic mevalonate pathway and the plastidic methylerythritol-4-phosphate pathway [17]. Previous studies have shown that terpenoids, particularly monoterpenes, form the predominant volatile compounds in mango fruits [18,19]. Based on the content of terpenoids, Andrade [13] divided Brazilian mango varieties into three groups: group 1 with abundance of terpinolenes, group 2 with abundance of 3-carenein, and group 2 with abundance of myrcene. In the present study, Xiaofei fruit had the highest monoterpene content, with 23,726.61 μg/kg, whereas cultivar 814 had the lowest content with only 54.17 μg/kg FW. Monoterpenes accounted for 65.62–98.31% of the total concentrations of volatiles in all cultivars except for 814, Shengshi and Boluoxiang. In addition, α-pinene and terpinolene were identified in all cultivars and were considered to be important volatile components. Pingguo contained the highest level of α-pinene (1661.56 μg/kg FW; 31.62% of total volatiles), whereas Guire 10 had the highest terpinolene content (12,725.64 μg/kg FW; 78.04% of total volatiles) (see Table 2).

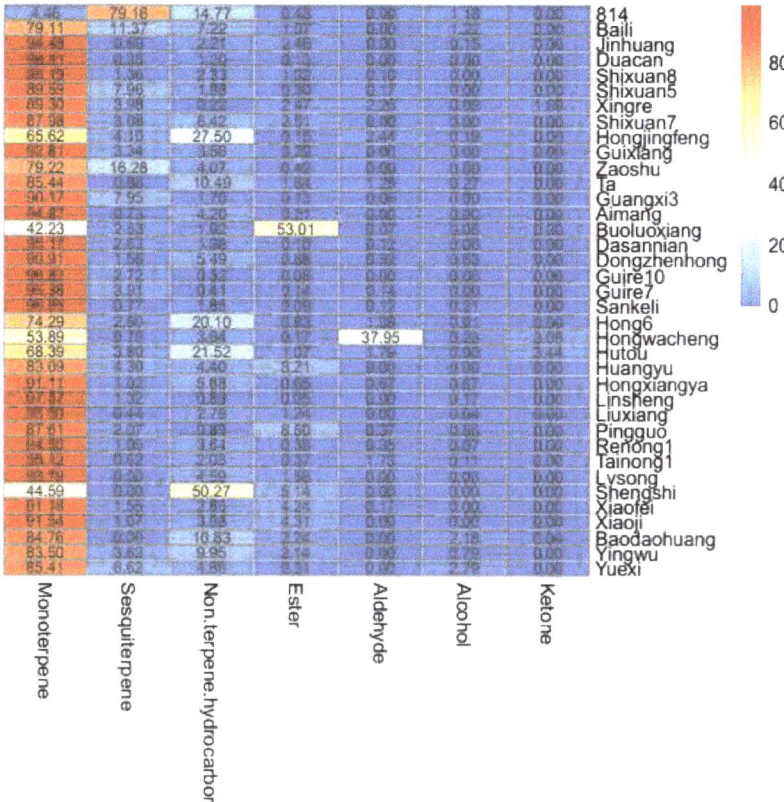

Figure 2. Relative abundance (%) and heatmap of different classes of volatile compounds in 37 mango cultivars.

No sesquiterpenes were detected in Lvsong, Shengshi and Baodaohuang cultivars. In 34 other cultivars, 814 mango had high total sesquiterpene content (961.96 µg/kg; 79.16% of total volatiles) mainly due to the high content of caryophyllene, α-caryophyllene and α-selinene. However, TA showed the lowest sesquiterpene content (11.66 µg/kg; 0.88% of total volatiles). Of the 26 sesquiterpenes, caryophyllene and α-caryophyllene were detected in 32 and 25 cultivars, respectively (see Table 3).

Significant differences of terpene constituents were also found amongst the cultivars. For example, Guixiang had a very high content of limonene (10,565.52 µg/kg), but 3-carene and β-piene were not detected. In Boluoxiang, 3-carene and 3,7,7-trimethyl-bicyclo[4.1.0]hept-3-ene were the dominant volatiles, but limonene was not detected. Liu et al. [14] detected selinene, eremophilene and aromadendrene only in Jinhuang mango, cinene only in Irwin and caryophyllene in Keitt. Fruit odours can be classified as fruity, green, spicy, woody and aldehydic based on sensory responses to aroma components with different chemical structures [20,21]. Odour components with different chemical structures elicit different sensory perceptions. For example, ocimene has a grassy odour, 3-carene has a sweet odour and cinene has a soft lemon odour [18]. Odour differences amongst mango cultivars may be associated with differences in terpenoid-type odour components.

Significant differences were found in both the total content and number of non-terpene hydrocarbons amongst different mango cultivars (see Table 4). The sum content of non-terpene hydrocarbons ranged from 13.60 µg/kg (Xingre) to 744.13 µg/kg (Xiaofei). Amongst all cultivars,

Hongjinfeng, Guixiang and Hutou were rich in non-terpene hydrocarbons, with 20, 18 and 18, respectively. Non-terpene hydrocarbons were the only predominant volatile compounds in Shenshi and represented 50.27% of the total volatiles. Meanwhile, for the remaining cultivars, the content of non-terpene hydrocarbons was low and accounted for less than 28% of the total volatiles.

Previous studies have shown that aldehydes are present at low concentrations, account for a small percentage of total volatiles in mango fruits and are important to mango flavour and aroma [22–24]. In this regard, Pino et al. [18] reported that the green, grassy odour of Cuban mangoes is derived primarily from aldehydes. In our study, aldehydes were present only in 19 cultivars, wherein their concentrations ranged from 2.42 µg/kg FW to 747.40 µg/kg FW (0.06–37.95% of total volatiles) in mango fruits. Amongst all cultivars studied, Hongwacheng with a sweet, fatty and green odour had the highest aldehyde content, particularly furfural (168.90 µg/kg FW) and 5-methyl-2-furancarboxaldehyde (566.11 µg/kg FW) (see Table 5).

Aldehydes are usually converted into alcohols or carboxylic acids, which are further converted into esters [25]. The present study could not determine which alcohols or esters were derived from aldehydes due to experimental limitations. Although certain alcohols are odour molecules, they do not substantially contribute to odour perception due to their higher odour threshold values compared with homologous aldehydes. A variety of esters are formed upon association of different alcohols with acetyl coenzyme A [26–28]. Aliphatic esters, which are mainly synthesised in actively growing tissues, are responsible for the odours of nearly all fruits. Lactones have classic fruity odours and are the second most abundant volatile aroma compounds in mangoes [29,30]. In this regard, Wilson et al. [31] asserted that esters in fruits could be detected at very low levels, and the contribution of lactones to mango aroma was second only to that of terpenoids. Even though esters are important odour components in mangoes, there is a large variation in the composition of esters amongst fruits of different origins and cultivars. In this regard, Pino detected 90 aliphatic, 16 aromatic and eight terpene esters from 20 Cuban mango cultivars. Amongst them, ethyl acetate and ethyl butanoate were the dominant ones. Eight lactones were found in 22 Indian and five non-Indian cultivarsbut they represented less than 1% of the total volatiles, and γ-butyrolactone was the major component [4]. In the present study, 24 aliphatic esters were detected in the mango germplasm, and no lactones were found. In different cultivars, the total contents of ester compounds varied widely, ranging from 2.01 µg/kg FW to 5361.09 µg/kg FW. Interestingly, Boluoxiang, which exhibited a characteristic fruity smell, had the highest content of esters (5361.09 µg/kg FW) and accounted for 53.01% of the total volatiles. Butanoic acid ethyl ester and octanoic acid ethyl ester were the major compounds in Boluoxiang (2916.45 and 1699.42 µg/kg FW, respectively). However, other cultivars had relatively low concentrations of ester compounds, i.e., 0.05–8.21% of the total volatiles (see Table 6).

The components and contents of alcohols and ketones in mango fruits are low [23,24]. Consistent with this trend, only five alcohols were found in 22 cultivars in this study. Alcohol concentrations ranged from 2.66 µg/kg FW to 29.22 µg/kg FW and accounted for 0.04–2.79% of the total volatiles. Ketone compounds were only found in Xingre, Boluoxiang, Hongwacheng and Hutou, with concentrations of 105.51, 19.92, 60.27 and 42.00 µg/kg FW, respectively (see Table 7).

Table 2. Concentrations (μg/kg fresh weight (FW) equivalent of nonyl acetate) of volatile monoterpenes in 37 mango cultivars.

	1	2	3	4	5	6	7	8	9	10	11	12	13	14	15	16	17	18	19
M1	3.48	7.95	56.77	84.52	263.53	581.23	39.28	3.14	5.02	551.59	7.34	5.06	416.76	9.5	25.54	1023.02	15.65	112.35	962.28
M2	24.23	-	14.7	3.92	-	6.86	50.78	-	1.76	23.08	3.43	-	2.91	-	-	439.03	1.57	172.44	345.75
M3	6.73	-	-	-	78.24	-	18.1	-	-	10,565.52	-	-	202	-	75.88	58.55	-	-	49.54
M4	19.73	756.93	8457.23	2339.88	107.19	506.2	23.55	17.58	23.57	33.15	923.16	19.02	260.25	1464.44	127	21.33	1664.68	12,725.64	21.04
M5	-	13.81	93.85	-	41.39	244.45	-	-	-	120.03	-	20.73	108.05	-	-	3502.42	23.17	184.04	5557.77
M6	-	81.25	297.25	154.56	1843.72	42.33	-	24.32	59.75	19.98	34.24	12.31	-	71.1	2326.81	27.06	65.61	577.17	-
M7	-	102.24	175.02	1297.64	-	7704.67	-	572.22	763.84	-	80.16	504.67	4086.38	147.86	1641.38	-	-	-	-
M8	-	54.19	717.56	72.99	10.54	34.47	-	444.71	-	-	-	566.63	20.99	0	-	26.04	594.49	1192.2	25.53
M9	-	38.92	217.64	1365.48	-	-	-	-	-	39.99	-	-	-	47.17	-	-	8.71	70.3	-
M10	-	-	3816.43	-	15.4	71.05	-	-	-	-	-	-	41.26	-	-	123.45	71.38	2.32	-
M11	-	-	-	65.29	-	13.53	5453.71	-	-	-	-	-	5.14	-	42.88	16.87	-	141.94	13.89
M12	-	-	-	-	-	335.18	-	16.89	0	-	29.3	-	-	-	-	-	-	-	5739.61
M13	-	-	-	-	-	12.01	-	-	30.63	40.36	-	-	-	14.63	-	-	-	-	14.05
M14	-	-	-	-	-	0	-	-	-	8.07	-	-	-	-	-	400.84	-	-	-
M15	-	-	-	-	37.47	176.39	-	-	-	-	-	-	104.18	39.65	30.91	152.19	29.61	-	280.91
M16	-	-	-	-	8.01	-	-	-	16.47	-	-	-	19.19	1610.71	-	12.18	-	601.89	104.79
M17	-	-	-	-	-	-	-	-	-	-	-	-	-	45.04	-	-	-	7.99	-

	Subtotal																		
Subtotal	54.17	1055.29	13,846.46	5447.44	-	9728.38	5585.43	1078.86	901.04	11,571.28	1077.65	1128.42	-	3450.10	4270.41	5802.99	2474.87	15,788.29	13,115.16

	20	21	22	23	24	25	26	27	28	29	30	31	32	33	34	35	36	37
M1	511.88	4.76	122.39	104.44	5.68	11.23	17.86	55.29	1661.56	25.5	2.43	12.71	17.43	66.83	76.25	34.94	167.23	15.31
M2	1065.36	-	98.58	160.64	-	-	-	-	12.68	-	-	-	75.43	44.57	306.66	21.03	1125.21	-
M3	47.54	-	6.19	7.14	-	-	-	-	-	-	-	-	-	-	-	-	-	-
M4	7.99	29.59	8.36	7.65	687.93	92.54	1000.18	6675.97	4.64	3192.13	7343.6	1554.95	1.22	7411.5	8703.66	933.59	6.28	496.25
M5	3257.4	10.74	234.33	-	27.38	28.79	36.64	80.65	-	0	89.25	19.33	-	96.28	100.72	-	18.29	-
M6	-	43.59	-	-	63.97	67.4	105.69	312.37	-	203.43	356.02	73.99	-	388.17	296.43	23.64	-	17.92
M7	-	663.45	-	-	-	1436.92	115.54	10.84	-	4.92	51.74	167.66	-	-	692.02	72.22	44.98	23.06
M8	-	-	-	-	-	932.55	267.87	-	-	1220.36	36.5	-	-	693.51	-	-	-	34.69
M9	28.27	-	-	-	-	-	91.91	581.44	-	18.78	651.32	141.31	-	767.38	477.88	33.57	-	-
M10	-	-	-	-	-	-	2940.17	-	9.92	64.63	-	-	-	13,945.6	-	-	-	-
M11	68.61	-	-	-	6.93	-	-	-	0	-	-	-	-	-	-	-	-	-
M12	10.69	-	-	8.79	-	-	-	-	1310.38	55.04	113.27	-	-	-	78.66	-	-	7.61
M13	3860.6	0	426.31	378.89	-	-	-	-	-	-	-	-	-	-	-	-	6.41	-
M14	-	-	-	-	-	-	-	-	-	-	-	-	-	-	-	4.12	-	-
M15	-	-	-	-	-	-	-	-	-	-	-	-	-	-	-	-	-	-
M16	-	-	-	155.33	-	-	-	-	-	-	-	-	-	-	24.35	-	20.68	-
M17	940.89	1.46	149.18	12.71	-	-	-	-	1557.46	-	-	19.63	-	-	98.67	6.42	6.39	-
M18	70.7	9.57	16.18	-	-	-	-	-	-	-	-	-	-	-	-	-	-	-

Table 2. Cont.

	20	21	22	23	24	25	26	27	28	29	30	31	32	33	34	35	36	37
M19	-	-	-	-	-	6.29	-	-	-	-	-	2195.17	-	-	-	-	-	-
M20	-	-	-	-	24.4	32.89	121	221.61	-	180.05	257.21	44.26	0	298.06	300.1	-	-	-
M21	-	-	-	-	-	-	-	3.29	-	-	-	-	-	10.74	-	-	4.98	-
M22	5.73	-	-	-	-	-	-	-	-	-	-	-	-	3.97	-	-	-	-
M23	-	-	-	-	-	-	-	2.96	-	0.92	4.01	-	-	-	-	-	-	-
Subtotal	9875.68	763.16	1061.52	835.59	816.29	2608.61	4696.86	7944.42	4556.64	4965.76	8905.35	4229.01	94.08	23,726.61	11,155.40	1129.53	1400.45	594.84

Numbers 1–37 represent the cultivars responding to the accession number in Table 2; The letter plus the number represent the compound corresponding to the code in Table 1; -—Indicated that the compound was not detected; The same as below.

Table 3. Concentrations (μg/kg fresh weight (FW) equivalent of nonyl acetate) of volatile sesquiterpenes in 37 mango cultivars.

	1	2	3	4	5	6	7	8	9	10	11	12	13	14	15	16	17	18	19
S1	445.68	20.93	84.46	15.04	-	293.95	61.65	22.28	33.38	125.14	41.45	11.66	167.24	20.44	106.89	52.46	11.75	21.49	319.07
S2	255.8	9.55	-	4.31	-	-	40.07	9.99	22.93	49.9	-	-	88.71	6.09	54.22	24.49	4.04	3.89	160.6
S3	4.59	-	-	-	-	177.13	-	-	-	36.44	5.87	-	76.74	-	63.46	17.7	5.86	-	-
S4	229.68	104.43	-	-	1.48	94.89	4.68	2.57	-	49.75	161.95	-	22.85	-	8.6	15.86	5.38	250.69	-
S5	26.22	-	-	-	-	-	-	-	-	29.05	-	-	-	-	-	-	-	9.15	-
S6	-	7.64	16.75	-	-	21.51	-	2.95	-	-	-	-	-	-	-	-	-	20.31	54.32
S7	-	9.19	-	-	12.2	102.39	-	-	-	74.4	12.24	-	38.16	-	8.16	24.93	10.63	-	4.02
S8	-	-	-	-	4.43	13.85	-	-	-	-	-	-	19.47	-	4	11.09	2.69	7.83	-
S9	-	-	-	-	9.26	12.72	-	-	-	-	-	-	0	-	-	-	2.09	73.4	-
S10	-	-	-	-	6.92	31.84	142.7	-	-	17.14	-	-	11.14	-	-	10.14	-	57.36	-
S11	-	-	-	-	-	6.65	-	-	-	-	-	-	0	-	-	-	-	-	-
S12	-	-	-	-	-	29.95	-	-	-	16.1	-	-	14.46	-	10.95	3.97	-	-	-
S13	-	-	-	-	-	16.83	-	-	-	6.94	-	-	11.37	-	-	-	-	-	-
S14	-	-	-	-	-	11.58	-	-	-	-	-	-	7.72	-	-	-	-	-	-
S15	-	-	-	-	-	39.38	-	-	-	11.19	-	-	0	-	-	-	-	-	-
S16	-	-	-	-	-	11.85	-	-	-	-	-	-	6.31	-	-	-	-	-	-
Subtotal	961.96	151.74	101.22	19.35	34.28	864.55	249.11	37.79	56.31	416.05	221.51	11.66	464.17	26.52	256.28	160.64	42.44	444.12	538.01

Table 3. *Cont.*

	20	21	22	23	24	25	26	27	28	29	30	31	32	33	34	35	36	37
S1	45.84	18.27	14.96	26.18	0	18.21	13.67	22.76	61.23	33.77	9.37	-	-	128.41	80.81	-	25.42	6.35
S2	28.86	10.46	0	17.85	0	11.1	0	13.46	32.15	21.33	4.85	-	-	53.67	44.76	-	-	2.29
S3	-	-	-	-	-	-	-	-	-	-	4.33	-	-	108.29	-	-	-	-
S4	-	-	-	-	35.15	-	45.89	-	-	-	-	-	-	35.49	-	-	-	34.65
S5	-	-	-	-	-	-	-	-	14.35	-	-	-	-	-	-	-	-	-
S6	4.18	-	-	2.35	3.74	-	4.04	-	-	-	-	-	-	-	5.08	-	-	2.84
S7	-	-	-	-	-	-	-	-	-	-	-	-	-	-	-	-	21.38	-
S8	-	-	-	-	-	-	-	-	-	-	-	-	-	-	-	-	10.96	-
S9	-	-	-	-	-	-	-	-	-	-	23.49	-	-	5.84	-	-	-	-
S10	-	-	-	-	-	-	-	-	-	-	10.31	-	-	6.36	-	-	-	-
S11	-	-	-	-	-	-	-	-	-	-	-	-	-	15.69	-	-	-	-
S12	-	-	-	-	-	-	-	-	-	-	5.29	-	-	23.32	-	-	2.89	-
S13	-	-	-	-	-	-	-	-	-	-	-	-	-	10.36	-	-	-	-
S14	-	-	-	-	-	-	-	-	-	-	-	-	-	16.07	-	-	-	-
S15	-	-	-	-	-	-	-	-	-	-	-	-	-	-	-	-	-	-
S16	-	-	-	-	-	-	-	-	-	-	-	-	-	-	-	-	-	-
Subtotal	78.88	28.73	14.96	46.38	42.25	29.31	63.60	36.22	107.73	55.10	57.64	-	-	403.50	130.65	-	60.65	46.13

Table 4. Concentrations (μg/kg fresh weight (FW) equivalent of nonyl acetate) of volatile non-terpene hydrocarbons in 37 mango cultivars.

	1	2	3	4	5	6	7	8	9	10	11	12	13	14	15	16	17	18	19
H1	5.66	-	-	-	-	-	-	-	-	-	-	-	-	-	7.42	14.05	5.48	7.97	-
H2	2.44	-	4.36	-	-	-	-	-	-	-	-	-	-	-	-	-	-	-	-
H3	6.05	-	-	-	-	-	-	3.36	-	-	-	-	-	-	-	4.06	-	-	-
H4	33.45	14.67	18.09	8.97	9.29	10.81	-	4.62	3.67	4.4	10.27	7.15	9.91	8.62	18.81	4.13	8.31	13.55	4.76
H5	10.5	12.52	8.61	5.17	2.16	-	-	7.88	7.8	1.21	2.31	32.22	6.23	22.41	13.3	-	10.96	12.91	-
H6	-	-	-	-	4.72	-	-	-	25.33	10.41	-	4.94	10	6.87	-	8.72	-	-	-
H7	8.3	6.01	-	-	-	-	-	-	11.63	6.21	-	-	-	22.71	-	-	-	-	-
H8	92.44	16.2	41.99	-	-	41.16	-	17.92	9.19	66.12	-	4.05	16.66	20.99	54.1	15.96	32.87	-	-
H9	-	-	-	5.79	-	-	-	0	64.66	12.99	-	28.64	-	-	-	-	5.01	-	-
H10	7.64	6.95	7.31	15.99	-	-	-	10.71	26.16	28.72	-	4.13	16.33	33.95	20.12	-	36.48	-	-
H11	10.94	-	-	-	-	-	-	0	13.28	10.86	-	-	-	-	10.28	-	-	-	-
H12	2.03	-	4.18	4.01	7.23	0	-	7.11	14.1	14.27	-	12.05	-	-	6.86	-	6.81	-	-
H13	-	3.49	7.52	-	15.55	2.55	-	0	2.62	14.03	-	-	-	-	-	-	-	-	-
H14	-	-	8.7	-	5.48	-	-	3.31	5.55	11.76	-	-	-	-	-	-	-	-	-
H15	-	-	22.08	-	-	-	-	0	53.86	10.59	-	30.07	-	22.65	-	-	-	-	-
H16	-	-	53.82	-	-	132.53	-	13.47	20.18	143.79	5.15	6.18	-	14.37	-	30.17	-	9.88	-
H17	-	-	28.72	5.72	-	-	-	-	-	-	-	-	-	-	-	17.36	-	-	-
H18	-	-	6.16	-	-	-	2.88	-	-	-	-	-	-	-	-	3.18	-	-	-
H19	-	33.26	-	11.61	-	-	-	-	23.98	-	-	-	12.37	-	12.89	-	20.86	-	17.13
H20	-	-	-	-	-	17.61	-	-	-	-	-	-	11.63	-	13.69	-	-	-	-
H21	-	3.26	3	-	-	-	10.72	3.35	3.95	-	-	-	-	-	3.6	-	5.43	7.68	12.34
H22	-	-	-	-	-	-	-	-	7.97	-	-	-	-	-	-	-	-	-	-
H23	-	-	-	-	-	10.83	-	-	-	-	31.89	-	7.52	-	-	22.94	-	-	-
H24	-	-	-	-	-	-	-	-	-	-	-	-	-	-	10.93	-	-	-	-
H25	-	-	-	4.75	-	-	-	-	16.2	9.89	-	-	-	-	-	-	6.38	-	10.68
H26	-	-	-	-	-	-	-	-	31.5	21.85	-	-	-	-	6.22	-	-	-	-
H27	-	-	5.03	4.75	14.52	-	-	7.04	22.7	14.01	-	9.06	2.96	-	15.6	-	10.96	-	12.11
H28	-	-	104.75	-	-	-	-	-	-	6.82	5.76	-	-	-	-	-	-	-	-
H29	-	-	-	-	-	-	-	-	13.28	56.11	-	-	5.82	-	-	-	-	-	-
Subtotal	179.45	96.37	324.34	66.75	58.97	215.49	13.60	78.77	377.61	444.04	55.38	138.50	99.43	152.57	193.81	120.57	149.55	51.98	57.03

Table 4. Cont.

	20	21	22	23	24	25	26	27	28	29	30	31	32	33	34	35	36	37
H1	6.72	2.56	16.52	4.66	6.15	7.57	8.88	8.42	6.8	6.14	9.01	7.88	5.18	7.4	3.52	4.72	17.62	6.09
H2	-	1.64	-	3.14	5.04	-	2.09	8.78	2.53	-	5.54	-	3.56	-	3.12	-	-	3.25
H3	4.54	4.24	-	-	4.84	-	3.44	6.81	-	-	-	3.43	6.9	5.66	4.51	11.74	4.83	-
H4	15.81	10.53	-	26.32	-	11.66	5.26	16.92	-	8.99	9.97	10.93	-	39.81	13.57	8.67	15.81	4.56
H5	9.28	8.71	11.36	15.82	8.74	8.99	-	13.76	5.74	9.09	9.26	11.34	9.61	12.83	9.2	10.6	-	-
H6	7.2	26.47	-	15	-	12.44	-	14.72	-	6.34	18.04	10.48	8.69	27.13	7.32	11.73	24.68	-
H7	10.22	24.71	-	7.4	-	-	-	12.73	-	14.03	14.39	-	-	-	12.28	14.78	-	-
H8	20.52	44.81	-	73.5	-	-	-	44.2	-	36.71	27.62	11.23	19.16	18.11	-	38.27	-	-
H9	-	-	2.46	-	-	22.81	-	-	-	-	6.36	-	6.24	0	-	-	-	-
H10	19.48	10.44	-	11.21	-	11.96	-	16.74	-	16.84	-	-	-	47.77	-	-	19.97	-
H11	-	-	-	4.22	-	29.87	-	-	-	8.55	8.59	8.14	-	4.92	-	3.98	-	-
H12	21.16	-	-	5.73	-	7.12	7.83	6.6	-	8.19	4.46	-	9.03	30.59	-	-	-	2.03
H13	6.07	-	-	-	-	4.9	-	-	-	-	-	-	4.84	-	3.7	-	-	11.67
H14	-	2.87	5.47	18.12	-	-	-	7.28	-	2.82	5.7	18.71	-	-	15.21	-	-	-
H15	15.23	9.31	17.18	13.43	-	14.99	-	26.15	-	14.57	24.3	54.47	13.73	11.58	44.1	-	12.77	-
H16	-	-	20.13	10.24	-	8.83	5.85	-	-	21.98	6.32	-	-	45.87	31.51	-	-	-
H17	-	-	-	-	4.14	-	-	22.39	-	-	8.18	-	-	-	4.26	-	-	-
H18	-	-	4.55	-	-	-	-	-	-	-	-	-	-	-	-	-	-	-
H19	17.1	17.35	-	-	-	-	-	13.79	19.2	5.86	10.42	36.55	-	11.52	77.77	-	15.94	-
H20	-	-	-	-	-	-	-	-	-	-	-	-	-	-	-	-	-	-
H21	-	-	-	-	6.29	-	-	2.59	5.56	-	-	8.84	-	14.52	7.02	-	-	-
H22	-	2.91	-	-	-	-	-	-	-	-	-	-	-	322.72	-	-	-	-
H23	10.05	-	-	-	-	-	-	-	-	-	-	-	-	-	-	-	-	-
H24	-	3.14	-	6.18	-	5.18	-	5.58	-	15.72	15.98	-	2.39	-	20.47	-	-	-
H25	12.78	-	-	8.3	-	5.95	-	-	-	-	-	-	-	13.33	-	-	-	-
H26	6.45	9.96	-	22.15	-	-	-	-	-	-	-	-	-	-	-	-	30.16	2.81
H27	5.8	14.28	-	4.41	8.07	10.28	-	4.62	-	15.33	7.79	-	9.51	118.18	117.53	11.5	3.48	3.43
H28	-	-	-	-	-	-	9.32	-	6.28	-	-	24.77	7.24	12.19	-	6.5	21.66	-
H29	-	12.52	-	13.1	-	-	-	-	-	-	-	-	-	-	-	21.84	-	-
Subtotal	188.42	206.45	77.67	262.93	43.27	162.55	42.67	232.08	46.11	191.16	191.93	206.77	106.08	744.13	375.09	144.33	166.92	33.84

Table 5. Concentrations (μg/kg fresh weight (FW) equivalent of nonyl acetate) of volatile aldehydes in 37 mango cultivars.

	1	2	3	4	5	6	7	8	9	10	11	12	13	14	15	16	17	18	19	20	21	22
A1	-	-	-	-	-	3.92	-	-	-	-	-	4.41	3.42	-	-	7.16	-	-	-	-	-	-
A2	-	-	-	-	2.42	14.29	-	-	18.57	-	-	-	-	-	-	-	14.03	-	19.82	11.75	-	-
A3	-	-	-	-	-	-	77.77	-	-	-	-	-	-	-	-	-	-	-	-	-	-	-
A4	-	-	-	-	-	-	63.52	-	-	-	-	-	-	-	-	-	-	-	-	-	-	-
A5	-	-	-	-	-	-	-	-	8.06	-	-	-	-	-	-	-	-	-	-	-	3.76	-
A6	-	-	-	-	-	-	-	-	-	-	-	12.5	-	-	-	-	-	-	-	-	-	-
A7	-	-	-	-	-	-	-	-	6.84	-	-	-	-	-	7.22	-	-	-	-	-	7.38	-
A8	-	-	-	-	-	-	-	-	-	-	-	-	-	-	-	-	-	-	-	-	-	168.9
A9	-	-	-	-	-	-	-	-	-	-	-	-	-	-	-	-	-	-	-	-	-	566.11
A10	-	-	-	-	-	-	-	-	-	-	-	-	-	-	-	-	-	-	-	-	-	12.39
A11	-	-	-	-	-	-	-	-	-	-	-	-	-	-	-	-	-	-	-	-	-	-
Subtotal	-	-	-	-	2.42	18.21	141.29	-	33.48	-	-	16.91	3.42	-	7.22	7.16	14.03	-	19.82	11.75	11.14	747.40

	23	24	25	26	27	28	29	30	31	32	33	34	35	36	37
A1	5.3	-	-	-	-	10.66	-	-	-	-	-	-	-	-	-
A2	4.91	-	14.02	-	-	-	-	108.12	-	-	-	-	-	-	-
A3	-	-	-	-	-	-	-	-	-	-	-	-	-	-	-
A4	-	-	-	-	-	-	-	-	-	-	-	-	-	-	-
A5	-	-	5.26	-	-	-	-	35.25	-	-	-	-	-	-	-
A6	-	-	-	-	-	-	-	-	-	-	-	-	-	-	-
A7	11.61	-	-	-	-	8.51	18.46	18.28	-	-	-	-	-	-	-
A8	-	-	-	-	-	-	-	-	-	-	-	-	-	-	-
A9	-	-	-	-	-	-	-	-	-	-	-	-	-	-	-
A10	-	-	-	-	-	-	-	-	-	-	-	-	-	-	-
A11	-	-	-	-	-	-	-	-	-	-	44.61	-	-	-	-
Subtotal	21.82	-	19.28	-	-	19.17	18.46	161.65	-	-	44.61	-	-	-	-

Table 6. Concentrations (μg/kg fresh weight (FW) equivalent of nonyl acetate) of volatile esters in 37 mango cultivars.

	1	2	3	4	5	6	7	8	9	10	11	12	13	14	15	16	17	18	19
E1	2.59	1.59	13.17	-	2.68	-	-	3.39	-	-	-	4	2.79	-	-	3.65	-	7.53	6.48
E2	2.69	-	-	2.07	2.35	-	-	1.92	-	1.62	-	2.77	-	-	2.66	2.49	2.39	-	3.1
E3	-	4.88	-	-	-	-	-	7.18	2.01	0.72	3.19	-	-	-	-	-	4.71	-	-
E4	-	-	-	5.35	1.91	1.25	2.23	-	-	-	2.52	2.18	2.34	2.81	-	-	1.42	2.48	-
E5	-	1.43	-	-	-	2.37	2.71	-	-	-	0	0	2.33	3.64	-	-	2.38	2.23	-
E6	-	3.36	-	-	3.79	-	-	-	-	-	-	-	-	-	-	-	-	-	-
E7	-	-	22.02	-	-	-	-	-	-	-	-	-	-	-	2.89	-	-	-	-
E8	-	-	5.02	-	15.04	3.29	43.13	7.15	-	34.21	-	-	-	-	56.69	-	-	-	-
E9	-	-	57.9	-	-	-	-	6.72	-	-	-	-	-	-	10.56	-	-	-	-
E10	-	-	7.8	-	-	3.63	9.16	-	-	-	-	-	-	-	140.4	-	-	-	10.35
E11	-	-	15.31	-	-	-	-	-	-	-	-	-	-	-	1699.42	-	-	-	-
E12	-	-	81.85	-	-	21.97	-	-	-	-	-	-	-	-	88.04	-	-	-	-
E13	-	-	157.96	-	-	-	-	-	-	-	-	-	-	-	2916.45	-	-	-	-
E14	-	-	-	-	-	-	97	-	-	-	-	-	-	-	-	-	-	-	-
E15	-	3.04	-	-	-	-	-	4.43	-	-	-	9.37	-	-	2.96	-	2.98	-	-
E16	-	-	-	-	-	-	-	-	-	-	-	-	-	1.11	273.04	-	-	-	-
E17	-	-	-	-	-	-	-	-	-	-	-	-	-	-	21.28	-	-	-	-
E18	-	-	-	-	-	-	-	-	-	-	-	-	-	-	29.01	-	-	-	-
E19	-	-	-	-	-	-	-	-	-	-	-	-	-	-	29.63	-	-	-	-
E20	-	-	-	-	-	-	-	-	-	-	-	-	-	-	18.57	-	-	-	-
E21	-	-	-	-	-	-	-	-	-	-	-	-	-	-	34.42	-	-	-	-
E22	-	-	-	-	-	-	-	-	-	-	-	-	-	-	20.2	-	-	-	-
E23	-	-	-	-	-	-	-	-	-	-	-	-	-	-	10.49	-	10.16	-	-
E24	-	-	-	-	-	-	-	-	-	-	-	-	-	-	4.4	-	-	-	-
E25	-	-	-	-	-	-	-	-	-	-	-	-	-	-	-	-	-	-	-
Subtotal	5.28	14.31	361.05	7.42	25.78	32.51	154.23	30.78	2.01	36.55	5.71	21.72	7.45	7.56	5361.09	6.14	24.04	12.24	19.93

Table 6. Cont.

	20	21	22	23	24	25	26	27	28	29	30	31	32	33	34	35	36	37
E1	-	-	-	-	-	-	-	-	-	-	-	-	2.41	-	-	-	-	-
E2	-	1.38	3.34	2.26	1.7	1.86	-	2.07	-	2.96	2.41	-	1.46	-	-	-	3.48	2.18
E3	5.42	-	-	-	-	-	-	-	-	6.65	2.79	6.09	-	-	-	3.04	-	-
E4	-	2.22	-	-	-	1.11	-	-	-	1	2.34	-	1.38	3.01	2.81	-	-	-
E5	1.96	-	-	-	-	-	2.44	-	1.84	2.41	-	-	1.61	-	-	-	3.73	-
E6	-	3.88	-	-	-	4.03	-	-	-	-	4.15	4.02	3.23	-	-	-	-	-
E7	-	-	-	-	-	-	-	2.95	-	-	-	-	-	-	37.02	-	-	-
E8	-	-	-	4.57	28.84	-	-	26.13	19.93	-	-	-	-	68.19	8.3	14.59	9.99	-
E9	-	-	-	-	-	-	-	0	43.61	-	-	15.36	-	116.78	167.06	-	-	-
E10	2.16	-	-	-	-	-	-	5.97	26.89	-	-	7.63	-	11.84	27.66	-	-	-
E11	-	-	-	-	12.96	-	-	0	5.12	-	-	3.71	-	42.69	27	-	-	-
E12	-	-	-	-	22.14	-	-	7.62	303.46	-	-	11.79	-	117.21	41.73	-	-	-
E13	-	-	-	-	-	-	-	19.17	-	-	-	21.9	-	277.79	214	-	10.85	-
E14	-	-	-	-	3.56	-	-	-	-	-	-	-	-	-	-	-	-	-
E15	-	-	-	-	-	-	-	-	-	-	-	-	-	-	-	-	-	-
E16	-	-	-	-	-	-	-	-	-	-	-	-	-	-	-	-	-	-
E17	-	-	-	-	-	-	-	-	-	-	-	-	-	-	-	-	-	-
E18	-	-	-	-	-	11.48	-	-	36.75	-	-	-	-	310.07	-	12.18	-	-
E19	-	-	-	-	-	-	-	-	-	-	-	-	-	-	-	-	-	-
E20	-	-	-	-	-	-	-	-	-	-	-	-	-	-	-	-	-	-
E21	-	-	-	-	-	-	-	-	-	-	-	-	-	113.3	-	-	-	-
E22	-	-	-	-	11.42	-	-	-	-	-	-	-	-	42.18	-	-	-	-
E23	-	-	-	-	-	-	-	-	-	-	-	-	-	-	-	-	-	-
E24	-	1.02	-	6.19	-	-	-	-	4.46	7.36	-	-	0.76	-	-	-	7.92	-
E25	-	-	-	-	-	-	-	39.05	-	-	23.32	-	-	-	-	-	-	-
Subtotal	9.54	8.50	3.34	13.02	80.62	18.48	2.44	102.96	442.06	20.38	35.01	70.50	10.85	1103.06	525.58	29.81	35.97	2.18

Table 7. Concentrations (μg/kg fresh weight (FW) equivalent of nonyl acetate) of volatiles alcohols and ketones in 37 mango cultivars.

	1	2	3	4	5	6	7	8	9	10	11	12	13	14	15	16	17	18	19	20	21	22
B1	14.36	-	-	-	-	-	-	-	-	-	-	-	-	-	-	0	17.25	-	-	-	-	-
B2	-	16.24	-	-	-	-	-	-	-	-	-	-	-	-	-	-	-	7.27	-	21.75	9.3	-
B3	-	-	22.27	-	-	-	5.74	-	2.66	-	-	3.58	-	-	4.72	-	-	3.29	-	-	-	4.47
B4	-	-	-	-	-	-	-	-	-	-	-	-	-	-	-	-	-	-	-	-	-	-
B5	-	-	-	-	-	-	-	-	-	-	-	-	-	-	-	-	-	-	-	-	-	-
Subtotal	14.36	16.24	22.27	-	-	-	5.74	-	2.66	-	-	3.58	-	-	4.72	-	17.25	10.56	-	21.75	9.30	4.47

	23	24	25	26	27	28	29	30	31	32	33	34	35	36	37							
B1	-	-	25.01	-	-	17.38	-	-	-	-	-	-	29.03	-	19.46							
B2	-	-	-	-	-	-	-	-	-	-	-	-	-	-	-							
B3	-	-	-	8.3	-	-	3.7	7.04	-	-	-	-	-	13.23	-							
B4	-	-	-	-	3.39	-	-	3.32	2.73	-	-	-	-	-	-							
B5	-	-	-	-	0	11.84	-	-	-	-	-	-	-	-	-							
Subtotal	-	-	25.01	8.30	3.39	29.22	3.70	10.36	2.73	-	-	-	29.03	13.23	19.46							

	1	2	3	4	5	6	7	8	9	10	11	12	13	14	15	16	17	18	19	20	21	22
K1	-	-	-	-	-	-	19.01	-	-	-	-	-	-	-	-	-	-	-	-	-	-	-
K2	-	-	-	-	-	-	82.42	-	-	-	-	-	-	-	-	-	-	-	-	-	-	-
K3	-	-	-	-	-	-	4.08	-	-	-	-	-	-	-	-	-	-	-	-	-	-	-
K4	-	-	-	-	-	-	0	-	-	-	-	-	-	-	19.92	-	-	-	-	-	-	60.27
K5	-	-	-	-	-	-	0	-	-	-	-	-	-	-	0	-	-	-	-	-	-	60.27
Subtotal	-	-	-	-	-	-	105.51	-	-	-	-	-	-	-	19.92	-	-	-	-	-	-	60.27

	23	24	25	26	27	28	29	30	31	32	33	34	35	36	37							
K1	-	-	-	-	-	-	-	-	-	-	-	-	-	-	-							
K2	-	-	-	-	-	-	-	-	-	-	-	-	-	-	-							
K3	-	-	-	-	-	-	-	-	-	-	-	-	-	-	-							
K4	-	-	-	-	-	-	-	-	-	-	-	-	-	-	-							
K5	42.00	-	-	-	-	-	-	-	-	-	-	-	-	-	-							
Subtotal	42.00	-	-	-	-	-	-	-	-	-	-	-	-	-	-							

2.3. Relationships Among Different Classes of Volatile Compounds

Correlation analysis was used to explore the relationships amongst different classes of volatile compounds (see Table 8). Sesquiterpene and non-terpene hydrocarbons were highly correlated with total monoterpene content (r = 0.374 and 0.569, respectively). Aldehydes were highly correlated with ketone content (r = 0.565, $p < 0.01$). Such correlation can facilitate the selection of cultivars with improved aroma quality because selection for one trait leads to the selection of genetically correlated traits. However, no significant relationships were found between aldehydes, esters and alcohols. The relationships amongst different classes of volatile compounds showed the complexity of fruit aroma metabolites.

Table 8. Linear correlation coefficients among different classes of volatile compounds.

	Monoterpene	Sesquiterpene	Non-Terpene Hydrocarbons	Ester	Aldehyde	Alcohol	Ketone
Monoterpene	1						
Sesquiterpene	0.374 *	1					
Non-Terpene Hydrocarbons	0.569 **	0.159	1				
Ester	0.122	0.085	0.185	1			
Aldehyde	−0.076	−0.105	−0.090	−0.045	1		
Alcohol	−0.102	−0.143	−0.149	−0.046	−0.052	1	
Ketone	−0.104	−0.025	−0.167	0.106	0.565 **	−0.107	1

* Significant correlation at $p < 0.05$. ** Significant correlation at $p < 0.01$.

2.4. Multivariate Data Analysis

Hierarchical cluster analysis (HCA) was used to analyse the data of 114 volatile compounds obtained from 37 mango cultivars (Figure 3). Thirty-seven mango cultivars could be divided into four groups. The first group included four cultivars: Xiaofei (33), Guire 10 (18), Jinhuang (3) and Xiaoji (34), which were characterized as having high concentrations of terpinolene (M4), δ-terpinolene (M9), α-terpinolene (M6), α-pinene (M1) and β-myrcene (M5). The second group included only one cultivar Boluoxiang (15) with high concentrations of butanoic acid ethyl ester (E12), 3-carene (M7), (1S)-3-carene (M8), octanoic acid ethyl ester (E10), limonene (M3) and α-pinene (M1). This cultivar was characterized by an extremely high esters content. The third group contained nine cultivars with high concentrations of α-pinene (M1), terpinolene (M4) and β-myrcene (M5). The other 23 cultivars were classified into group four. However, determining the dominant volatiles in this group was difficult. Although terpinolene and α-pinene were detectable in these cultivars, they were present only at very low levels except for 13 cultivars with high concentration of terpinolene. Thus, no characteristic volatile was observed for this group.

To further elucidate genetic clustering identified by HCA, we performed—principal component analysis (PCA) (Figure 4). However, the cumulative contribution of the first 14 components was only 80% (data not shown). This indicates the presence of relatively large variations in the composition and concentration of aromatic compounds in different cultivars, which results in scattered contribution rates of various aroma compounds and an insignificant cumulative contribution rate. Although two principal components (PC1 and PC2) represented only 23% of the variability, the mango germplasms can be divided into four groups based on the score scatter plot shown in Figure 3. In general, the PCA results were in accordance with the results of HCA.

Based on this study, terpinolene and α-pinene are the main volatile compounds responsible for Chinese mango aroma. The contributions of different volatile aroma compounds to fruit aroma are affected by the odour activity value, flavour dilution factors and aroma profile [32,33]. Given that the present study only analysed the content of aroma compounds, further studies are required to determine whether certain compounds act as characteristic aroma compounds in mangoes.

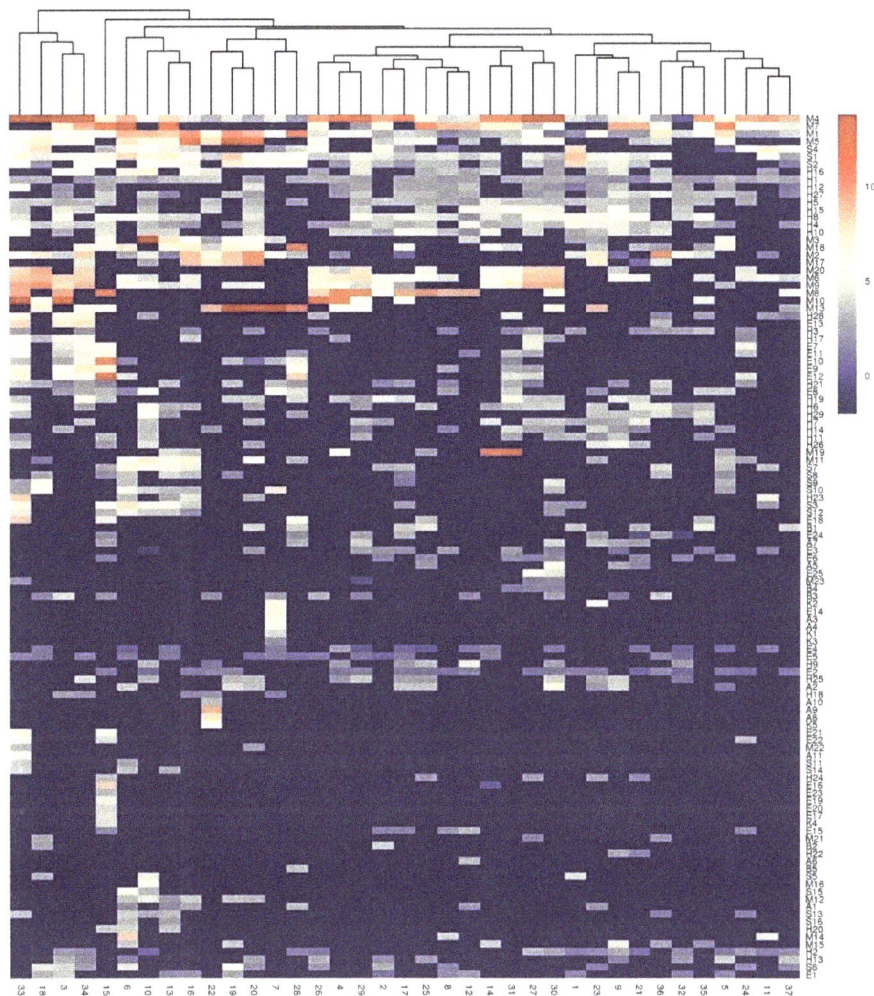

Figure 3. Hierarchical clustering (HCA) and heatmap of volatile compounds levels in 37 mango cultivars. values of all studied volatile compounds per cultivar are shown in the heatmap on a blue (negative) to red (positive) scale. The HCA and dendrogram of cultivars was according to Euclidean distance.

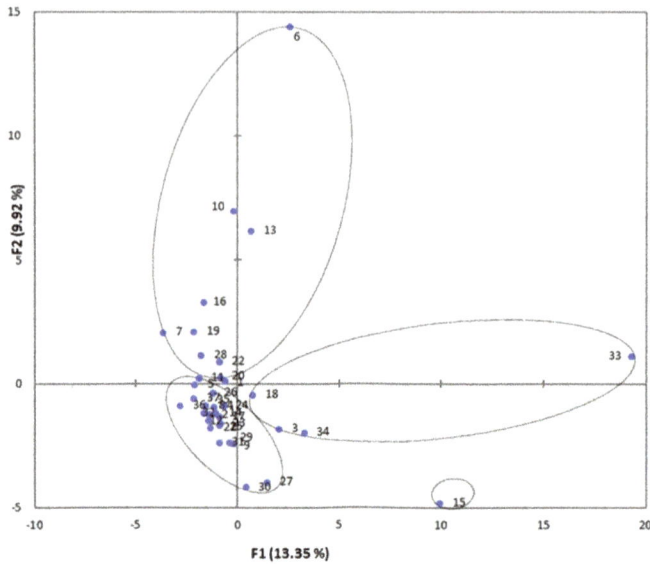

Figure 4. Positions of PC scores of all of the studied mango cultivars according to PC1 and PC2.

3. Materials and Methods

3.1. Materials

The 37 cultivars (see Table 9) considered in this study were cultivated in the orchards of the South Subtropical Crops Research Institute in Zhanjiang, China. All cultivars were grown under the same geographical conditions and with the same standard cultural practices. Based on production experience and days after pollination, the fruit of 37 cultivars were harvested at commercial maturity (with the flesh around the seeds starting to turn yellow) from June to July 2017. Three trees with moderate growth vigour were selected for each cultivar. Ten disease-free fruits of similar sizes were randomly picked from different locations on the crown of each tree and considered to be one replication, thus resulting in three replications per cultivar. All the fruits sampled were stored under a controlled atmosphere with day and night temperatures of 24 °C and 18 °C, respectively, for ripening, which was ascertained for each cultivar by conventional indices such as change in skin colour, smell, and softness to touch. Ripe fruits were peeled immediately, four slices were then taken by longitudinal cuts from different orientations of each fruit, and ground to powder in liquid nitrogen and stored at −70 °C for further studies.

Table 9. 37 cultivars used in this study. The number following the cultivars indicates the sampling order.

Cultivars	814	Baili	Jinhuang	Duacan	Shixuan8	Shixuan5	Xingre	Shixuan7
No.	1	2	3	4	5	6	7	8
Cultivars	Hongjinfeng	Guixiang	Zaoshu	TA	Guangxi3	Aimang	Boluoxiang	Dasannian
No.	9	10	11	12	13	14	15	16
Cultivars	Dongzhen Hong	Guire10	Guire7	Sankeli	Hong6	Hongwa Cheng	Hutou	Huangyu
No.	17	18	19	19	20	21	22	23
Cultivars	Hongxiangya	Linsheng	Liuxian	Pingguo	Renong1	Tainong1	Lvsong	Shengshi
No.	25	26	27	28	29	30	31	32
Cultivars	Xiaofei	Xiaoji	Baodaohuang	Yingwu	Yuexi			
No.	33	34	35	36	37			

3.2. Methods

3.2.1. Volatiles Extraction

For HS-SPME, the extraction of aroma volatiles was performed using an SPME fibre coated with polydimethylsiloxane-divinylbenzene (65 µm) (Supelco, Bellefonte, PA, USA). The fibre was preconditioned for 30 min per day at 250 °C according to the manufacturer's instructions. For each extration, 8 g pulp, 2 g NaCl and 30 µL internal standard (0.29 µg/mL nonyl acetate) were placed in a 20 mL capped SPME vial. The mixture was incubated in a water bath at 40 °C for 10 min with a magnetic stirrer. Next, the fibre was exposed for 40 min to the headspace. After extraction, the fibre was immediately inserted into the heated chromatograph injector port for desorption at 250 °C for 2 min in the splitless mode.

3.2.2. GC-MS Analysis

The volatile compounds were analysed by means of an Agilent 6890N gas chromatograph coupled with an Agilent 5973N mass selective detector (Agilent, Santa Clara, CA, USA) and equipped with a DB-5 MS (Supelco, Bellefonte, PA, USA) capillary column (30 m × 0.25 mm ID × 0.25 µm film thickness). The injector and detector temperatures were maintained at 220 °C and 250 °C. The oven temperature program were as follows: 50 °C for 1 min, increased at 5 °C/min to 140 °C, then increased at 10 °C/min to 250 °C and then kept for 10 min. Mass spectra conditions were as follows: electron impact mode at 70 eV, ion source temperature: 250 °C, mass scanning range: m/z 35–335 amu/s. The carrier gas was helium with a constant column flow of 1 mL/min [12,13]. The tentative identification of the volatile compounds was done by comparing the mass spectra with the data system library (NIST98) and linear retention index. Using a series of *n*-alkane standards (C8–C29), retention indices of each compound were determined. Semiquantitation was done by the internal standard method, where the relative content of each volatile compound was obtained as nonyl acetate equivalent by the GC peak area.

3.2.3. Data Analysis

Data for each cultivar were averages of three replication. Hierarchical cluster analysis (HCA) and principal component analysis (PCA) were carried out to detect clustering and establish relationships between cultivars and volatile compounds. HCA was performed using the Metabo Analyst 2.0 software package (www.metaboanalyst.ca). All volatile data were log10-transformed and used in the HCA analysis. PCA was processed using the XLSTAT (Addinsoft, Paris, France) software package.

4. Conclusions

There are quantitative and qualitative differences of volatile compounds among Chinese mango cultivars, and a 123-fold difference (max. and min. ratio) in the quantity of volatiles evolved from different cultivars. Among the 37 germplasm resources, with the exception of the Boluoxiang cultivar, which has a fruit aroma which is primarily dependent on esters, the fruit aromas of other cultivars are mainly dependent on monoterpenes and sesquiterpenes, followed by non-terpenoid hydrocarbons and esters, while lower diversity and concentrations of aldehydes, alcohols, and ketones are present in the fruits. Notably, the fruits of certain cultivars such as Xiaofei, Guire10, Jinhuang, Guire7 and Guixiang, possess a greater diversity and higher concentrations of volatile aromatic compounds. All the cultivars could be divided into four groups using HCA and PCA. In conclusion, this study provides a detailed database of volatile composition of Chinese mango germplasm, which can be used for breeding a more diversified set of mango flavourswhich will eventually satisfy our diet and industrial production.

Author Contributions: X.-W.M. and H.-X.W. conceived the experiments and X.-W.M. performed the experiments. M.-Q.S. and S.-B.W. provided help in the GC-MS analysis of volatiles. Y.-G.Z. prepared the samples for volatiles extraction.

Acknowledgments: This work was funded by the Natural Science Foundation of Guangdong Province, China (2016A030307005).

Conflicts of Interest: The authors declare no conflict of interest.

References

1. Monterocalderón, M.; Rojasgraü, M.A.; Martínbelloso, O. Aroma profile and volatiles odor activity along gold cultivar pineapple flesh. *J. Food Sci.* **2010**, *75*, S506–S512. [CrossRef] [PubMed]
2. Costa, F.; Cappellin, L.; Zini, E.; Patocchi, A.; Kellerhals, M.; Komjanc, M.; Gessler, C.; Biasioli, F. QTL validation and stability for volatile organic compounds (VOCs) in apple. *Plant Sci. Int. J. Exp. Plant Biol.* **2013**, *211*, 1–7. [CrossRef] [PubMed]
3. Negri, A.S.; Allegra, D.; Simoni, L.; Rusconi, F.; Tonelli, C.; Espen, L.; Galbiati, M. Comparative analysis of fruit aroma patterns in the domesticated wild strawberries "Profumata di Tortona" (*F. moschata*) and "Regina delle Valli" (*F. vesca*). *Front. Plant Sci.* **2015**, *6*, 56. [CrossRef] [PubMed]
4. Aragüez, I.; Valpuesta, V. Metabolic engineering of aroma components in fruits. *Biotechnol. J.* **2013**, *8*, 1144–1158. [CrossRef] [PubMed]
5. Kulkarni, R.S.; Chidley, H.G.; Pujari, K.H.; Giri, A.P.; Gupta, V.S. Geographic variation in the flavour volatiles of Alphonso mango. *Food Chem.* **2012**, *130*, 58–66. [CrossRef]
6. Liu, H.; Cao, X.; Liu, X.; Rui, X.; Wang, J.; Jie, G.; Wu, B.; Gao, L.; Xu, C.; Bo, Z. UV-B irradiation differentially regulates terpene synthases and terpene content of peach. *Plant Cell Environ.* **2017**, *40*, 2261–2275. [CrossRef] [PubMed]
7. Zhang, B.; Tieman, D.M.; Jiao, C.; Xu, Y.; Chen, K.; Fe, Z.; Giovannoni, J.J.; Klee, H.J. Chilling-induced tomato flavor loss is associated with altered volatile synthesis and transient changes in DNA methylation. *Proc. Natl. Acad. Sci. USA* **2016**, *113*, 12580–12585. [CrossRef] [PubMed]
8. Chai, Q.; Wu, B.; Liu, W.; Wang, L.; Yang, C.; Wang, Y.; Fang, J.; Liu, Y.; Li, S. Volatiles of plums evaluated by HS-SPME with GC–MS at the germplasm level. *Food Chem.* **2012**, *130*, 432–440. [CrossRef]
9. Wu, Y.; Duan, S.; Zhao, L.; Zhen, G.; Meng, L.; Song, S.; Xu, W.; Zhang, C.; Chao, M.; Wang, S. Aroma characterization based on aromatic series analysis in table grapes. *Sci. Rep.* **2016**, *6*, 31116. [CrossRef] [PubMed]
10. Andrade, E.H.A.; Maia, J.G.S.; Zoghbi, M.D.G.B. Aroma Volatile Constituents of Brazilian Varieties of Mango Fruit. *J. Food Compos. Anal.* **2000**, *13*, 27–33. [CrossRef]
11. Pino, J.A. Odour-active compounds in mango (*Mangifera indica* L. cv. Corazon). *Int. J. Food Sci. Technol.* **2012**, *47*, 1944–1950. [CrossRef]
12. Pino, J.A.; Mesa, J. Contribution of volatile compounds to mango (*Mangifera indica* L.) aroma. *Flavour Fragr. J.* **2006**, *21*, 207–213. [CrossRef]
13. Sagars, P.; Hemangig, C.; Rams, K.; Keshavh, P.; Ashokp, G.; Vidyas, G. Cultivar relationships in mango based on fruit volatile profiles. *Food Chem.* **2009**, *114*, 363–372.
14. Liu, F.X.; Fu, S.F.; Bi, X.F.; Chen, F.; Liao, X.J.; Hu, X.S.; Wu, J.H. Physico-chemical and antioxidant properties of four mango (*Mangifera indica* L.) cultivars in China. *Food Chem.* **2013**, *138*, 396–405. [CrossRef] [PubMed]
15. Shi, S.; Ma, X.; Xu, W.; Zhou, Y.; Wu, H.; Wang, S. Evaluation of 28 mango genotypes for physicochemical characters, antioxidant capacity, and mineral content. *J. Appl. Bot. Food Qual.* **2015**, *88*, 264–273.
16. Shi, S.Y.; Hong-Xia, W.U.; Wang, S.B.; Yao, Q.S.; Liu, L.Q.; Wang, Y.C.; Wei-Hong, M.A.; Zhan, R.-L. Fruit quality diversity of mango (*Mangifera indica* L.) germplasm. *Acta Hortic. Sin.* **2011**, *5*, 840–848.
17. Vranová, E.; Coman, D.; Gruissem, W. Network Analysis of the MVA and MEP Pathways for Isoprenoid Synthesis. *Ann. Rev. Plant Biol.* **2013**, *64*, 665–700. [CrossRef] [PubMed]
18. Munafo, J.P., Jr.; Didzbalis, J.; Schnell, R.J.; Schieberle, P.; Steinhaus, M. Characterization of the Major Aroma-Active Compounds in Mango (*Mangifera indica* L.) Cultivars Haden, White Alfonso, Praya Sowoy, Royal Special, and Malindi by Application of a Comparative Aroma Extract Dilution Analysis. *J. Agric. Food Chem.* **2014**, *62*, 4544–4551. [CrossRef] [PubMed]
19. Kulkarni, R.; Pandit, S.; Chidley, H.; Nagel, R.; Schmidt, A.; Gershenzon, J.; Pujari, K.; Giri, A.; Gupta, V. Characterization of three novel isoprenyl diphosphate synthases from the terpenoid rich mango fruit. *Plant Physiol. Biochem.* **2013**, *71*, 121–131. [CrossRef] [PubMed]

20. Lasekan, O.; See, N.S. Key volatile aroma compounds of three black velvet tamarind (Dialium) fruit species. *Food Chem.* **2015**, *168*, 561–565. [CrossRef] [PubMed]
21. Liu, C.; Cheng, Y.; Zhang, H.; Deng, X.; Chen, F.; Xu, J. Volatile Constituents of Wild Citrus Mangshanyegan (*Citrus nobilis* Lauriro) Peel Oil. *J. Agric. Food Chem.* **2012**, *60*, 2617–2628. [CrossRef] [PubMed]
22. Engel, K.H.; Tressl, R. Studies on the volatile components of two mango varieties. *J. Agric. Food Chem.* **1983**, *31*, 796–801. [CrossRef]
23. Matsui, K. Green leaf volatiles: Hydroperoxide lyase pathway of oxylipin metabolism. *Curr. Opin. Plant Biol.* **2006**, *9*, 274–280. [CrossRef] [PubMed]
24. Quijano, C.E.; Salamanca, G.; Pino, J.A. Aroma volatile constituents of Colombian varieties of mango (*Mangifera indica* L.). *Flavour Fragr. J.* **2007**, *22*, 401–406. [CrossRef]
25. Schauer, N.; Fernie, A.R. Plant metabolomics: Towards biological function and mechanism. *Trends Plant Sci.* **2006**, *11*, 508–516. [CrossRef] [PubMed]
26. Goepfert, S.; Poirier, Y. β-Oxidation in fatty acid degradation and beyond. *Curr. Opin. Plant Biol.* **2007**, *10*, 245–251. [CrossRef] [PubMed]
27. Lalel, H.J.D.; Singh, Z.; Tan, S.C.; Agustã, M. Maturity stage at harvest affects fruit ripening, quality and biosynthesis of aroma volatile compounds in 'Kensington Pride' mango. *J. Pomol. Hortic. Sci.* **2003**, *78*, 225–233. [CrossRef]
28. Zhang, B.; Shen, J.Y.; Wei, W.W. Expression of Genes Associated with Aroma Formation Derived from the Fatty Acid Pathway during Peach Fruit Ripening. *J. Agric. Food Chem.* **2010**, *58*, 6157–6165. [CrossRef] [PubMed]
29. Bonneau, A.; Boulanger, R.; Lebrun, M.; Maraval, I.; Gunata, Z. Aroma compounds in fresh and dried mango fruit (*Mangifera indica* L. cv. *Kent*): Impact of drying on volatile composition. *Int. J. Food Sci. Technol.* **2016**, *51*, 789–800. [CrossRef]
30. Pandit, S.S.; Kulkarni, R.S.; Chidley, H.G.; Giri, A.P.; Pujari, K.H.; Köllner, T.G.; Degenhardt, J.; Gershenzon, J.; Gupta, V.S. Changes in volatile composition during fruit development and ripening of 'Alphonso' mango. *J. Sci. Food Agric.* **2010**, *89*, 2071–2081. [CrossRef]
31. Cwiii, W.; Shaw, P.E.; Rjjr, K. Importance of some lactones and 2,5-dimethyl-4-hydroxy-3(2H)-furanone to Mango (*Mangifera indica* L.) aroma. *J. Agric. Food Chem.* **1990**, *38*, 1556–1559. [CrossRef]
32. Capone, S.; Tufariello, M.; Siciliano, P. Analytical characterisation of Negroamaro red wines by "Aroma Wheels". *Food Chem.* **2013**, *141*, 2906–2915. [CrossRef] [PubMed]
33. Wu, Y.W.; Pan, Q.H.; Qu, W.J.; Duan, C.Q. Comparison of volatile profiles of nine litchi (*Litchi chinensis* Sonn.) cultivars from Southern China. *J. Agric. Food Chem.* **2009**, *57*, 9676–9681. [CrossRef] [PubMed]

Sample Availability: not available.

© 2018 by the authors. Licensee MDPI, Basel, Switzerland. This article is an open access article distributed under the terms and conditions of the Creative Commons Attribution (CC BY) license (http://creativecommons.org/licenses/by/4.0/).

Article

Discrimination of Aroma Characteristics for Cubeb Berries by Sensomics Approach with Chemometrics

Huan Cheng [1,2], Jianle Chen [1,2], Peter J. Watkins [3], Shiguo Chen [1,2], Dan Wu [1,2], Donghong Liu [1,2] and Xingqian Ye [1,2,*]

1. College of Biosystems Engineering and Food Science, Zhejiang University, Hangzhou 310058, China; huancheng@zju.edu.cn (H.C.); 3090100118@zju.edu.cn (J.C.); chenshiguo210@163.com (S.C.); wudan2008@zju.edu.cn (D.W.); dhliu@zju.edu.cn (D.L.)
2. National-Local Joint Engineering Laboratory of Intelligent Food Technology and Equipment, Fuli Institute of Food Science, Zhejiang Key Laboratory for Agro-Food Processing, Zhejiang Engineering Laboratory of Food Technology and Equipment, Hangzhou 310058, China
3. CSIRO Agriculture and Food Nutrition Unit, 671 Sneydes Road, Werribee 3030, Australia; Peter.Watkins@csiro.au
* Correspondence: psu@zju.edu.cn; Tel.: +86-139-5805-5601

Academic Editors: Constantinos K. Zacharis and Paraskevas D. Tzanavaras
Received: 14 June 2018; Accepted: 2 July 2018; Published: 4 July 2018

Abstract: The dried cubeb berries are widely used as medicinal herb and spicy condiment with special flavor. However, there is a significant definition discrepancy for cubeb berries. In this study, an efficient analytical method to characterize and discriminate two popular cubeb fruits (*Litsea cubeba* and *Piper cubeba*) was established. The aroma profiles of cubeb berries were evaluated by different extraction methods including hydro-distillation, simultaneous distillation/extraction, and solid-phase micro-extraction followed by gas chromatography-mass spectrometry-olfactometry (GC-MS-O). In total, 90 volatile compounds were identified by HD, SDE, and SPME combined with GC-MS. Principal component analysis was further applied and discriminated ambiguous cubeb berries by their unique aromas: *Litsea cubeba* was characterized by higher level of D-limonene ("fruit, citrus"), citral ("fruit, lemon") and dodecanoic acid; α-cubebene ("herb") was identified as a marker compound for *Piper cubeba* with higher camphor ("camphoraceous"), and linalool ("flower"). Flavor fingerprint combined with PCA could be applied as a promising method for identification of cubeb fruits and quality control for food and medicinal industries.

Keywords: cubeb berry; principal component analysis (PCA); solid-phase microextraction (SPME); hydro-distillation (HD); simultaneous distillation/extraction (SDE); gas chromatography-mass spectrometry-olfactometry (GC-MS-O)

1. Introduction

Litsea cubeba (Lour.) Pers. (Lauraceae) gives off an aromatic odor and smells similar to an intensely lemonlike, spicy aroma. *Litsea cubeba* (*L. cubeba*) is a promising industrial crop as its fruit is rich in valuable essential oil. Recently, many reports have demonstrated the bioactivities of essential oil in *L. cubeba* [1-4]. *L. cubeba* has been widely employed in a flavoring or herbal medicinal industries and could be used as an ingredient in ionone flavors, botanical insecticides, food spices, and personal-care products.

The dried berry of *Piper cubeba* (Piperaceae), known as the 'cubeb pepper' or 'tailed pepper', have been widely used as a popular spice, with beneficial properties, including anti-inflammatory, analgesic, anti-proliferative, and leishmanicidal activities [5,6], and a flavoring agent for gins and cigarettes consumed throughout Europe as well as in many other Polynesian countries [7].

L. cubeba has been described as the cubeb berry in the Chinese Pharmacopoeia, whereas *P. cubeba* has been also listed as the origin of cubeb berries. Both of the two cubeb berries have provided special flavors for daily life. It is well known that the flavor of food is tightly related to the stimulation of the human chemical senses, odor and taste; meanwhile, the odor is mainly caused by different volatile compositions [8]. Therefore, it is an appropriate method to discriminate the two ambiguous cubeb berries by the identification of volatiles.

The volatile profiles of cubeb berries were studied previously using hydro-distillation (HD) for the extraction of essential oil. Li et al. [1] investigated the inhibitory activities of *L. cubeba* essential oil and found the main chemical composition including limonene oxide, D-limonene. Hydro-distillation (HD) and simultaneous distillation/extraction (SDE) have been the common methods for the volatile extraction of different materials [9]. SDE united the advantages of liquid–liquid and steam distillation-extraction and has been widely recognized as a relative convenient extraction method for essential oils. However, distillation at elevated temperatures may lead to the loss of some compounds and generate artefacts due to thermal changes. Headspace solid-phase microextraction (HS-SPME) based on the distribution coefficient of analytes among the sample matrix, the gas phase, and the fiber coating was a simple, rapid, and solvent-free technique [10]. SDE and SPME methods have been widely used to extract the volatiles of potato products [11], milk products [12], tea [13], and meat products [14]. However, to the best of our knowledge, headspace volatiles of cubeb fruit, directly contacting human olfactory receptors and closely associated with an overall special spicy aroma, were still not analyzed until now. Rather than rate techniques (HD, SDE, and SPME) as more superior to another in performance, HD, SDE, and SPME could be regarded as the techniques that provide complementary information for each other [14]. In present study, we evaluated the volatile profiles of different cubeb berries by the use of three extraction methods (HD, SDE, and SPME) coupled to GC-MS and combined with the principal component analysis, which would provide more comprehensive data for the discrimination of ambiguous cubeb fruit.

It is known that only a small portion of the large number of volatiles in a food matrix contribute to its overall perceived odor. GC-O is an appropriate analytical solution, as the eluted substances are perceived simultaneously by two detectors, one of them being the human olfactory system. Therefore, GC-O provides not only an instrumental, but also a sensorial analysis [8]. The GC-O technique has been widely used to the identification of aroma-active compounds from different fruits [15–18]. However, so far, GC-O technique has not been applied to the identification of aroma-active compounds of cubeb fruit.

In this essay, we report our latest study of characterization of cubeb fruit, which consisted of the following steps: (a) different pretreatment methods (HD, SDE, and SPME) were applied to obtain the volatiles and essential oils of different cubeb fruits; (b) GC-MS-O were adopted to characterize aroma-active compounds of the cubeb fruits; (c) specific aromas contributed to the discrimination of *L. cubeba* and *P. cubeba* were identified by chemometrics, which would provide helpful clues for the characteristics of aroma in different cubeb fruits and provide accurate information on the authenticity of the cubeb products.

2. Results and Discussion

2.1. Optimization of SPME

The volatile compounds in cubeb fruits were extracted using HS-SPME and the highest peak area response was selected in order to optimize the main parameters. These different desorption times, incubation times, extraction temperatures, and extraction times were optimized based on the total ion response in the GC-MS [19]. As shown in Figure 1a, study of desorption time including 1, 2, 3, and 4 min was tested. The peak areas of different volatiles were not significantly affected by the desorption time ($p < 0.05$). In order to clean the fiber sufficiently, 3 min was chosen as the desorption time in the present work.

Figure 1b shows the effect of incubation time (10, 15, 20, and 25 min) on the detection of total volatiles. The total volatile amounts significantly rose with increasing incubation time. However, there was no significant difference between 15, 20, and 25 min ($p < 0.05$), which indicated that a 15 min incubation time would allow distributions between the fiber, the vial headspace, and the analytes to reach an equilibrium.

Study of the extraction temperature including 40, 50, 60, and 70 °C was investigated as illustrated in Figure 1c. The peak areas of volatiles were significantly affected by the extraction temperature. In order to avoid aroma changes cause by higher temperature, 60 °C was chosen as the extraction temperature.

Similarly, Figure 1d showed that there was no significant difference between 30 min and 40 min ($p < 0.05$). In order to avoid the fiber desorption caused by the long time exposure in the vial, 30 min was chosen as the optimum extraction time.

Therefore, the optimal extraction conditions were as follows: desorption time, 1 min; incubation time, 15 min; extraction temperature, 60 °C; and extraction time, 30 min. These conditions were applied during the headspace extraction of volatile compounds from cubeb fruits.

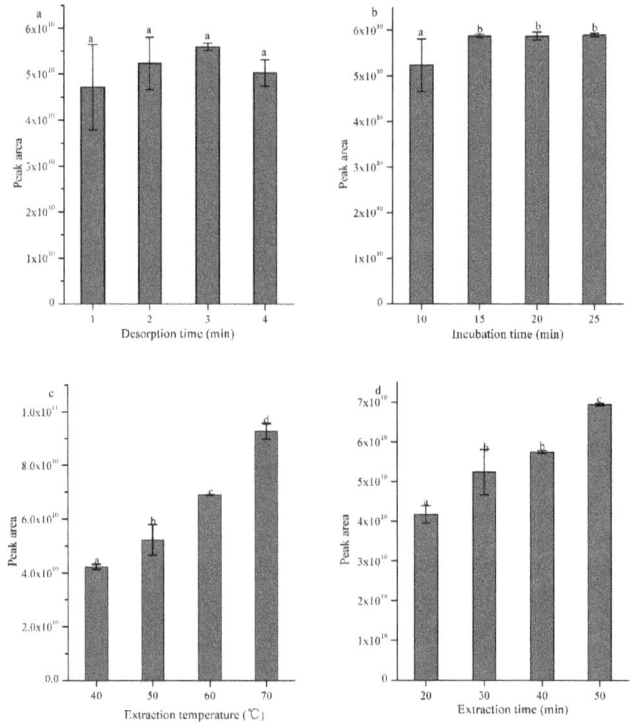

Figure 1. Effects of different factors—(**a**) desorption time; (**b**) incubation time; (**c**) extraction temperature; and (**d**) extraction time—on the peak areas of different volatile compounds of cubeb berries captured by DVB/CAR/PDMS. Different letters (a, b, c, and d) on the top of columns indicate significant differences ($p < 0.05$)

2.2. Identification of Aroma Compounds

To obtain a wider volatile profile and better discriminate the two cubeb berries, three extraction methods (HD, SDE, and SPME) were used (Table 1). In total, 90 volatile compounds were identified by HD, SDE, and SPME combined with GC-MS. Seventy-three volatile compounds belonged to different chemical families: terpenes (8.82–80.65% for *C. cubeba*; 18.69–52.27% for *P. cubeba*), ketones (1.49–3.24% for *C. cubeba*; 17.97–20.99% for *P. cubeba*), alcohols (4.16–10.56 % for *C. cubeba*; 16.68–28.97% for *P. cubeba*), aldehydes (2.02–23.79% for *C. cubeba*; 1.71–6.43% for *P. cubeba*), esters (0–1.06% for *C. cubeba*; 0.98–2.57% for *P. cubeba*), and acids (2.39–48.55% for *C. cubeba*; 0–18.69% for *P. cubeba*). In order to highlight the differences between the two cubeb fruit with different extraction techniques in a simple and immediate way, Figure 2 showed the comparison of the relative percentages of the main chemical families present in cubeb berries.

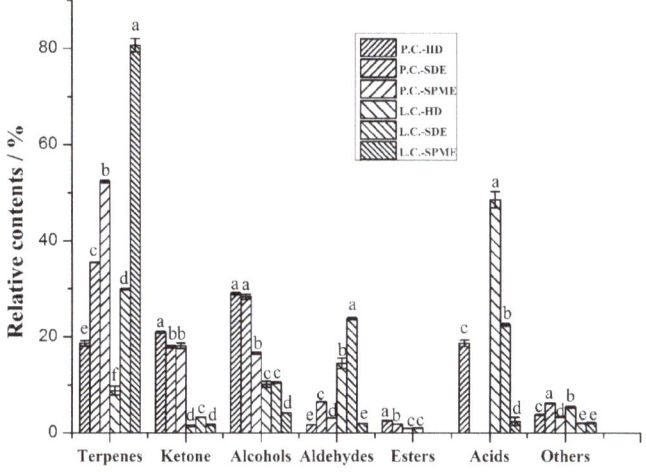

Figure 2. Chemical compositions of volatile compounds in different extracts (HD, SDE, and SPME) of *L. cubeba* and *P. cubeba*. Different letters (a, b, c, d, and e) indicate significant differences ($p < 0.05$).

One of the most abundant chemical families identified in cubeb fruits was terpenes. To our knowledge, the literature dealing with the comparison of different extraction methods of volatile compounds in cubeb berries is scare. Wang et al. [20] studied the chemical composition of the essential oil obtained only by HD of different parts (root, stem, leaf, flower, and fruit) of *L. cubeba* and showed that citral and limonene were the main constituents. In the present work, for *L. cubeba*, D-limonene was one of the main terpenes for the SDE (10.57%) and SPME extracts (38.89%), but it was not detected in the HD extract. Wang et al. [21] reported that some other terpenes, such as α-pinene, β-pinene, and β-caryophyllene, were also detected as the main volatile compounds in the bio-oils produced from *L. cubeba* seed by hydrothermal liquefaction. α-Pinene (5.74%), β-pinene (6.22%), D-limonene (9.84%), and caryophyllene (5.57%) were also the main terpenes for the SPME extract of *P. cubeba*. γ-Terpinene (A13, 0.12–0.39%), α-copaene (A19, 0.13–1.15%), caryophyllene (A23, 5.53–9.11%), and humulene (0.77–4.33%) were the common terpenes detected in all the three different extracts of two cubeb berries. Three terpenes could only be detected in all the extracts of *P. cubeb*, including α-cubebene (A18, 0.4–0.52%), bicyclosesquiphellandrene (A27, 2.41–5.28%), and α-farnesene (A30, 0.13–0.16%). Cubebene was also identified in the direct analysis in real time mass spectrometry (DART-MS) fingerprint of *P. cubeba* studied by Kim et al. [7]. The extracts obtained by SPME were rich in terpenes (80.65% for *L. cubeba* and 52.27% for *P. cubeba*) in comparison with those from HD (8.82% for *L. cubeba* and 18.69% for *P. cubeba*) and SDE (29.9% for *L. cubeba* and 35.4% for *P. cubeba*). However, SDE with solvents tends to extract a higher amount of the volatile monoterpenes than SPME in bay leaf [22], French beans [23], and wines [24]. These differences may be due to the matrix effect in releasing volatile compounds as each spice had a characteristic plant tissue structure. Meanwhile, strong oxidation and degradation of terpenes may occur for HD and SDE extracts because of higher temperature and longer time [25].

Table 1. Relative concentrations of volatile compounds identified in cubeb fruits by different preparation methods (HD, SDE, SPME).

Codes	RI [1]	Volatile Compounds	Chemical Formula	Concentration for Litsea cubeba (%) [2]			Concentration for Piper cubeba (%)		
				HD [3]	SDE [3]	SPME [3]	HD [3]	SDE [3]	SPME [3]
Terpenes									
A1	925	α-Pinene	$C_{10}H_{16}$	nd [4]	1.44 ± 0.01	10.81 ± 0.36	0.09 ± 0.01	1.06 ± 0.02	5.74 ± 0.11
A2	944	Camphene	$C_{10}H_{16}$	nd	0.52 ± 0	3.37 ± 0.16	0.07 ± 0	0.47 ± 0.01	2.09 ± 0.03
A3	971	β-Phellandrene	$C_{10}H_{16}$	nd	0.3 ± 0	nd	0.11 ± 0	0.93 ± 0.02	nd
A4	973	β-Pinene	$C_{10}H_{16}$	nd	1.22 ± 0.01	7.33 ± 0.13	0.21 ± 0.01	1.02 ± 0.01	6.22 ± 0.11
A5	991	β-Myrcene	$C_{10}H_{16}$	nd	1.45 ± 0.02	6.99 ± 0.42	0.16 ± 0.03	0.45 ± 0.01	1.97 ± 0.04
A6	1002	α-Phellandrene	$C_{10}H_{16}$	nd	nd	0.13 ± 0.07	0.46 ± 0.01	1.75 ± 0.02	4.55 ± 0.07
A7	1008	3-Carene	$C_{10}H_{16}$	nd	0.23 ± 0	nd	nd	0.08 ± 0	nd
A8	1015	2-Carene	$C_{10}H_{16}$	nd	0.11 ± 0	nd	0.07 ± 0	0.09 ± 0	nd
A9	1023	Cymene	$C_{10}H_{16}$	nd	0.14 ± 0	nd	0.24 ± 0.02	0.39 ± 0.01	0.7 ± 0.02
A10	1027	D-Limonene	$C_{10}H_{16}$	nd	10.57 ± 0.09	38.89 ± 1.67	nd	1.9 ± 0.03	9.84 ± 0.09
A11	1038	trans-beta.-Ocimene	$C_{10}H_{16}$	nd	nd	0.16 ± 0.02	0.16 ± 0	0.41 ± 0.01	0.64 ± 0.02
A12	1048	β-Ocimene	$C_{10}H_{16}$	nd	4.62 ± 0.04	0.1 ± 0.01	1 ± 0.03	2.5 ± 0.01	5.35 ± 0.04
A13	1057	γ-Terpinene	$C_{10}H_{16}$	0.12 ± 0.02	0.25 ± 0	0.26 ± 0.01	0.17 ± 0.01	0.29 ± 0.01	0.39 ± 0.01
A14	1070	4-Carene	$C_{10}H_{16}$	nd	nd	nd	0.15±0.01	nd	nd
A15	1087	Terpinolene	$C_{10}H_{16}$	nd	nd	nd	nd	0.27 ± 0.01	0.63 ± 0.01
A16	1130	2,4,6-Octatriene, 2,6-dimethyl-	$C_{10}H_{16}$	nd	nd	nd	nd	nd	0.5 ± 0.02
A17	1255	3-Carene	$C_{10}H_{16}$	nd	nd	0.14 ± 0.01	nd	nd	0.79 ± 0.03
A18	1349	α-Cubebene	$C_{15}H_{24}$	nd	nd	nd	0.4 ± 0.01	0.43 ± 0.02	0.52 ± 0.02
A19	1375	α-Copaene	$C_{15}H_{24}$	0.13 ± 0.01	1.07 ± 0.12	1.09 ± 0.65	0.75 ± 0.01	1.15 ± 0.02	0.95 ± 0.02
A20	1385	(+)-3-Carene	$C_{10}H_{16}$	nd	nd	nd	nd	nd	0.33 ± 0.02
A21	1389	Bicyclosesquiphellandrene	$C_{15}H_{24}$	nd	nd	nd	nd	0.4 ± 0.01	nd
A22	1392	β-Elemene	$C_{15}H_{24}$	nd	nd	1.16 ± 0.13	nd	nd	nd
A23	1419	β-Caryophyllene	$C_{15}H_{24}$	5.53 ± 0.23	6.41 ± 0.06	9.11 ± 0.37	5.84 ± 0.06	5.58 ± 0.02	5.57 ± 0.1
A24	1438	α-Guaiene	$C_{15}H_{24}$	nd	nd	nd	nd	nd	0.16 ± 0.01
A25	1453	Humulene	$C_{15}H_{24}$	3.04 ± 0.98	0.77 ± 0.01	0.88 ± 0.06	4.33 ± 0.68	1.76 ± 0.01	1.43 ± 0.03
A26	1481	β-Copaene	$C_{15}H_{24}$	nd	nd	nd	0.08 ± 0	2.74 ± 0.01	nd
A27	1497	Bicyclogermacrene	$C_{15}H_{24}$	nd	nd	nd	4.15 ± 0.04	5.28 ± 0.07	2.41 ± 0.07
A28	1500	α-Muurolene	$C_{15}H_{24}$	nd	nd	nd	nd	0.22 ± 0.01	0.13 ± 0.01
A29	1506	Cedrene	$C_{15}H_{24}$	nd	nd	nd	nd	nd	0.1 ± 0
A30	1509	α-Farnesene	$C_{15}H_{24}$	nd	nd	nd	0.13 ± 0	0.16 ± 0.01	0.16 ± 0.01
A31	1516	α-selinene	$C_{15}H_{24}$	nd	0.11 ± 0	0.11 ± 0.01	nd	nd	nd
A32	1575	γ-Muurolene	$C_{15}H_{24}$	nd	nd	nd	nd	1.09 ± 0.03	nd
A33	1582	Alloaromadendrene copaene	$C_{15}H_{24}$	nd	nd	0.03 ± 0	0.08 ± 0.03	0.65 ± 0.07	0.51 ± 0.12
A34	1644		$C_{15}H_{24}$	nd	nd	nd	0.06 ± 0.01	0.12 ± 0	nd
A35	1652	β-Panasinsene	$C_{15}H_{24}$	nd	0.14 ± 0	nd	nd	nd	nd
A36	1722	β-Bisabolene	$C_{15}H_{24}$	nd	nd	0.05 ± 0.01	nd	4.13 ± 0.04	nd
A37	1849	β-Farnesene	$C_{15}H_{24}$	nd	0.55 ± 0.02	0.04 ± 0.01	nd	0.04 ± 0	0.59 ± 0.07

166

Table 1. *Cont.*

Codes	RI [1]	Volatile Compounds	Chemical Formula	Concentration for *Litsea cubeba* (%) [2]			Concentration for *Piper cubeba* (%)		
				HD [3]	SDE [3]	SPME [3]	HD [3]	SDE [3]	SPME [3]
Ketones									
B1	987	6-Methyl-5-hepten-2-one	$C_8H_{14}O$	1.2 ± 0.13	2.01 ± 0.02	1.55 ± 0.08	0.39 ± 0.04	0.33 ± 0.01	0.13 ± 0.01
B2	1145	Camphor	$C_{10}H_{16}O$	0.28 ± 0.03	0.22 ± 0.01	0.23 ± 0.09	20.6 ± 0.3	17.59 ± 0.4	18.03 ± 0.67
B3	1183	Pulegone	$C_{10}H_{16}O$	nd	1 ± 0.01	nd	nd	nd	nd
B4	1823	Isoshyobunone	$C_{15}H_{24}O$	nd	nd	nd	nd	0.04 ± 0.01	nd
Alcohols									
C1	1029	Eucalyptol	$C_{10}H_{18}O$	1.43 ± 0.15	2.99 ± 0.03	nd	2.17 ± 0.02	2.53 ± 0.02	nd
C2	1100	Linalool	$C_{10}H_{18}O$	3.74 ± 0.37	nd	3.17 ± 0.04	21.31 ± 0.28	18.57 ± 0.18	14.89 ± 0.17
C3	1107	1,5,7-Octatrien-3-ol, 3,7-dimethyl	$C_{10}H_{16}O$	nd	nd	nd	nd	0.21 ± 0.01	nd
C4	1143	Isopulegol	$C_{10}H_{18}O$	0.1 ± 0.01	0.11 ± 0.01	nd	0.04 ± 0	nd	nd
C5	1164	endo-Borneol	$C_{10}H_{18}O$	0.3 ± 0.03	0.33 ± 0.01	0.07 ± 0	0.6 ± 0.01	0.54 ± 0.01	0.24 ± 0.14
C6	1165	Verbenol	$C_{10}H_{16}O$	nd	0.86 ± 0.01	nd	nd	0.29 ± 0.01	nd
C7	1190	Terpinen-4-ol	$C_{10}H_{18}O$	2.98 ± 0.12	3.19 ± 0.02	0.79 ± 0.05	2.64 ± 0.05	2.64 ± 0.02	1.14 ± 0.02
C8	1218	Carveol	$C_{10}H_{16}O$	nd	0.3 ± 0.03	nd	nd	nd	nd
C9	1228	2,6-Octadien-1-ol, 3,7-dimethyl-	$C_{10}H_{18}O$	0.63 ± 0.06	1.12 ± 0.06	nd	nd	nd	nd
C10	1229	Citronellol	$C_{10}H_{20}O$	nd	nd	nd	0.41 ± 0.1	1.12 ± 0.02	0.4 ± 0.02
C11	1256	Geraniol	$C_{10}H_{18}O$	0.99 ± 0.06	1.66 ± 0.05	nd	0.71 ± 0.01	1.76 ± 0.57	nd
C12	1610	.tau.-Cadinol	$C_{15}H_{26}O$	nd	nd	nd	0.49 ± 0.05	nd	nd
C13	1640	Muurolol	$C_{15}H_{26}O$	nd	nd	nd	nd	0.71 ± 0.01	nd
C14	1648	α-Cadinol	$C_{15}H_{26}O$	nd	nd	nd	0.6 ± 0.03	nd	nd
C15	1655	Nerolidol	$C_{15}H_{26}O$	nd	nd	0.14 ± 0.02	nd	nd	nd
Aldehydes									
D1	1154	Citronellal	$C_{10}H_{18}O$	0.48 ± 0.06	0.62 ± 0.01	0.21 ± 0.01	nd	0.46 ± 0.01	0.29 ± 0.01
D2	1241	2,6-Octadienal, 3,7-dimethyl-, (Z)	$C_{10}H_{16}O$	6.09 ± 0.54	11.32 ± 0.13	0.72 ± 0.04	0.52 ± 0.04	nd	1.1 ± 0.03
D3	1272	Citral	$C_{10}H_{16}O$	7.96 ± 0.72	11.85 ± 0.12	1.09 ± 0.03	0.62 ± 0.02	5.77 ± 0.03	1.74 ± 0.04
D4	1740	2,6,10-Dodecatrienal, 3,7,11-trimethyl-	$C_{15}H_{24}O$	nd	nd	nd	0.58 ± 0.02	0.2 ± 0.01	nd
Esters									
E1	1193	Methyl salicylate	$C_8H_8O_3$	nd	nd	0.04 ± 0.01	nd	nd	nd
E2	1213	Acetic acid, octyl ester	$C_{10}H_{20}O_2$	nd	nd	nd	0.12 ± 0.01	0.17 ± 0.01	0.17 ± 0.01
E3	1285	Bornyl acetate	$C_{12}H_{20}O_2$	nd	nd	nd	0.63 ± 0.1	0.52 ± 0.01	0.55 ± 0.01
E4	1297	Decanoic acid, methyl ester	$C_{11}H_{22}O_2$	0.34 ± 0.02	nd	nd	0.55 ± 0.03	nd	nd
E5	1324	2,6-Octadienoic acid, 3,7-dimethyl-, methyl ester	$C_{11}H_{18}O_2$	nd	nd	nd	nd	0.17 ± 0.01	0.11 ± 0.01
E6	1370	2,6-Octadien-1-ol, 3,7-dimethyl-acetate, (Z)-	$C_{12}H_{20}O_2$	nd	nd	nd	0.07 ± 0	nd	nd
E7	1381	2-Propenoic acid, 3-phenyl-, methyl ester	$C_{10}H_{10}O_2$	nd	nd	nd	0.42 ± 0.01	0.33 ± 0	0.14 ± 0.01
E8	1385	lavandulyl acetate	$C_{12}H_{20}O_2$	nd	nd	nd	nd	0.64 ± 0.01	nd
E9	1387	Geranyl acetate	$C_{12}H_{20}O_2$	nd	nd	nd	0.78 ± 0.02	nd	nd
E10	1661	Dodecanoic acid, methyl ester	$C_{12}H_{26}O_2$	0.71 ± 0.05	nd	nd	nd	nd	nd

Table 1. Cont.

Codes	RI [1]	Volatile Compounds	Chemical Formula	Concentration for *Litsea cubeba* (%) [2]			Concentration for *Piper cubeba* (%)		
				HD [3]	SDE [3]	SPME [3]	HD [3]	SDE [3]	SPME [3]
Acids									
F1	1360	Geranic acid	$C_{10}H_{16}O_2$	nd	nd	0.17 ± 0.02	nd	nd	nd
F2	1391	n-Decanoic acid	$C_{10}H_{20}O_2$	15.64 ± 1.51	9.29 ± 0.45	0.34 ± 0.02	14.85 ± 0.6	nd	nd
F3	1582	Dodecanoic acid	$C_{12}H_{24}O_2$	32.9 ± 0.6	13.27 ± 0.36	1.89 ± 1.03	3.85 ± 0.31	nd	nd
Others									
G1	1053	Cyclopentene, 1-methyl-	C_6H_{10}	nd	0.13 ± 0.01	nd	nd	nd	nd
G2	1197	Estragole	$C_{10}H_{12}O$	nd	nd	0.04 ± 0.02	nd	nd	nd
G3	1223	Cyclohexene, 3,3,5-trimethyl-	C_9H_{16}	nd	0.73 ± 0.03	nd	nd	nd	nd
G4	1285	Anethole	$C_{10}H_{12}O$	0.34 ± 0.03	0.77 ± 0.01	0.6 ± 0.02	nd	nd	nd
G5	1287	Safrole	$C_{10}H_{10}O_2$	nd	nd	nd	nd	nd	nd
G6	1326	Eugenol	$C_{10}H_{12}O_2$	0.12 ± 0.01	0.15 ± 0.07	nd	0.1 ± 0.09	0.23 ± 0.01	nd
G7	1336	1,5,5-Trimethyl-6-methylene-cyclohexene	$C_{10}H_{16}$	nd	nd	nd	nd	0.51 ± 0.06	2.04 ± 0.05
G8	1354	2,6-Octadiene, 2,6-dimethyl-	$C_{10}H_{18}$	nd	nd	nd	0.38 ± 0	0.43 ± 0.03	0.32 ± 0.01
G9	1392	Cyclohexane, 1-ethenyl-1-methyl-2,4-bis(1-methylethenyl)-	$C_{15}H_{24}$	nd	nd	nd	nd	0.44 ± 0.01	nd
G10	1405	Methyleugenol	$C_{11}H_{14}O_2$	nd	nd	nd	0.71 ± 0.03	0.11 ± 0	nd
G11	1439	Ethanone, 1-(2-hydroxy-4-methoxyphenyl)-	$C_9H_{10}O_3$	nd	nd	nd	nd	0.17 ± 0.01	nd
G12	1448	trans-Isoeugenol	$C_{10}H_{12}O_2$	nd	nd	nd	nd	0.23 ± 0.01	0.11 ± 0.01
G13	1485	Naphthalene, decahydro-4a-methyl-1-methylene-7-(1-methylethenyl)-	$C_{15}H_{24}$	nd	nd	nd	0.69 ± 0.01	0.79 ± 0.01	nd
G14	1513	Naphthalene, 1,2,4a,5,6,8a-hexahydro-4,7-dimethyl-1-(1-methylethyl)-	$C_{15}H_{24}$	nd	nd	nd	0.31 ± 0	0.3 ± 0	0.81 ± 0.03
G15	1523	Naphthalene, 1,2,3,5,6,8a-hexahydro-4,7-dimethyl-1-(1-methylethyl)-	$C_{15}H_{24}$	nd	nd	nd	1.59 ± 0.02	1.58 ± 0.01	0.06 ± 0.01
G16	1652	Naphthalene, 1,2,4a,5,6,8a-hexahydro-4,7-dimethyl-1-(1-methylethyl)-	$C_{15}H_{24}$	nd	nd	nd	nd	1.4 ± 0.01	nd
G17	1669	Caryophyllene oxide	$C_{15}H_{24}O$	4.93 ± 0.27	0.25 ± 0.01	1.47 ± 0.09	nd	nd	nd

[1] RI: retention indices. [2] %: relative concentration was expressed by peak area, and data listed were the mean of three assays ± SD (standard deviation). [3] HD: hydrodistillation; SDE: simultaneous distillation and extraction; SPME: solid phase micro-extraction. [4] nd: not detected.

For ketones, 6-methyl-5-hepten-2-one (B1) and camphor (B2) were both detected in all the cubeb berries samples. As irregular terpene, 6-methyl-5-hepten-2-one is probably derivative of carotenoids produced by enzymatic action [26]. Camphor (17.59–20.6%) was detected as the major ketones in all the extracts of *P. cubeba*. The contents of ketones in *P. cubeba* (17.97–20.99%) with different extraction methods were significantly higher than in the *L. cubeba* (1.49–3.24%) by one-way ANOVA (Figure 2).

Alcohols were also present with a high proportion in cubeb fruit, and the contents of alcohols in *P. cubeba* (16.68–28.97%) with different extraction methods were significantly higher than in the *L. cubeba* (4.16–10.56%) by a One-way ANOVA (Figure 2). Terpinen-4-ol (C7) could be detected in all the extracts of *L. cubeba* and *P. cubeba*. Linalool (C2, 14.89–21.31%) was the most abundant volatile and could be found in all the extracts of *P. cubeba*.

Citral was present as the most abundant aldehyde compound in *L. cubeba* extracts. This result is in accordance with the study by Wang et al. [20], who reported that citral was one of the main constituents in the fruit oil of *L. cubeba* extracted by HD. The contents of 3,7-dimethyl-2,6-octadienal in *L. cubeba* was higher than *P. cubeba*. Only 10 esters were identified in the current work, most were relatively high-boiling esters and with lower contents than other chemical families in cubeba berries.

Only two kinds of acids, decanoic acid and dodecanoic acid, were mainly extracted by HD and SDE with longer extraction time and higher temperature. The lack of acids with low volatility in SPME extracts may be caused by the low extraction temperature during the extraction process. Most acids may exist as esters form or have been changed to aldehydes, alcohols, or other secondary metabolites [13]. The contents of acids in *L. cubeba* (2.39–48.55%) with different extraction methods were significantly higher than in the *P. cubeba* (0–18.69%) by one-way ANOVA (Figure 2).

2.3. Aroma-Active Compounds by GC-MS-O

The extracts obtained by SPME were analyzed to assess the aroma-active compounds of the cubeb fruits using GC-O. Table 2 listed the identified aroma-active compounds of the *L. cubeba* and *P. cubeba*. A total of 12 compounds were tentatively found to be the aroma-active compounds at olfactometry port for odor description in GC-O analysis, including eight terpenes, one ketone, two alcohols, and one aldehyde. The odor descriptions of all the aroma-active compounds identified in the volatiles of *L. cubeba* and *P. cubeba* were basically similar to the reported of other fruits, such as blackberry [27], bayberry [17], strawberry [28], orange [29], and gooseberry [30].

Table 2. Principal aroma-active components of cubeb berries determined by SPME-GC-MS-O using DFA.

Codes	Odor	DF [a]	
		Litsea cubeba	Piper cubeba
A1	pine, turpentine	4	2
A4	pine, resin, turpentine	4	6
A6	turpentine, mint, spice	0	2
A10	lemon, citrus, mint	6	6
A12	herb	4	6
A13	turpentine	4	4
A18	herb	0	4
A23	wood, herb	2	2
B2	camphor wood	2	6
C2	flower, lavender	2	8
C8	turpentine, nutmeg	4	4
D3	lemon, fruit	6	4

[a] Sum of times detected by four assessors.

According to the evaluation of the odor and the odor description of the reported, it can be concluded that the flavor of turpentine-like might be caused by α-pinene (A1), β-pinene (A4), α-phellandrene (A6), γ-terpinene (A13), and terpinen-4-ol (C7); the fruity and flower flavor might be due to the presence of D-limonene (A10), citral (D3) with higher level in *L. cubeba*, and linalool (C2) with higher level in *P. cubeba*; the herbal flavor might come from β-ocimene (A12), α-cubebene (A18) identified only in *P. cubeba*, and caryophyllene (A23); camphoraceous flavor might be caused by camphor (C2), which is stronger in *P. cubeba* with higher content than in *L. cubeba*. It is interesting to note that α-cubebene was in very low proportions, also had high detection frequencies in all the extracts of *P. cubeba*. α-cubebene had been detected in other study for cubeb fruit [7].

2.4. Principal Component Analysis (PCA)

Principal component analysis (PCA) is an unsupervised clustering method and could reduce the dimensionality of multivariate data and preserve most of the variance therein [31]. To get a clear distribution of the volatiles with the separation of the samples, PCA was applied to the data presented in Table 1, the first two principal components explained nearly 91% of the total variability of the GC-MS data set between the samples, is shown in Figure 3a. The corresponding loading weight plot, establishing the magnitude of each volatile component (variable), is illustrated in Figure 3b. Figure 3 plots the samples on the coordinate grid defined by the first two principal components and showed that PC1 and PC2 separated the *L. cubeba* samples from the *P. cubeba* samples.

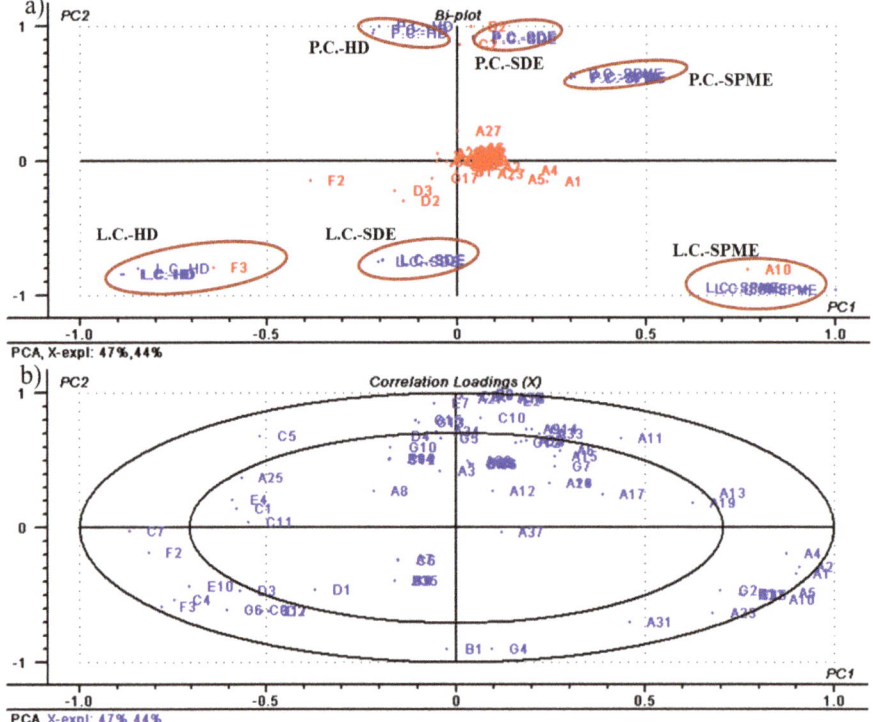

Figure 3. PC1 vs. PC2 scatter plot for variability among the different cubeba samples: (**a**) bio-plots of PC1 & PC2 of the volatile compounds of different extracts (HD, SDE, and SPME) of *L. cubeba* and *P. cubeba*; (**b**) relation between the volatile compounds (loadings).

Principal component 1 (PC1) and PC 2, explained 47% and 44% of the total variance among the sample batches, showed that the cubeb berries discrimination based on varietal volatile profile.

170

The extracts (HD, SDE, and SPME) of *P. cubeba* was projected in positive PC2 and highly associated with α-cubebene (A18, 0.4–0.52%), bicyclosesquiphellandrene (A27, 2.41–5.28%), camphor (B2, 17.59–20.6%), and linalool (C2, 14.89–21.31%). The extracts (HD, SDE, and SPME) of *L. cubeba* were located in the negative PC 2 with high D-limonene (A10, 0–38.89%), 3,7-dimethyl-2,6-octadienal (0.72–11.32%), citral (D3, 1.09–11.85%), and dodecanoic acid (F3) content (Figure 3b).

In conclusion, 12 volatiles were identified as aroma-active compounds in both cubeb berries with mainly 'turpentine-like', 'fruity and flowery', 'herbal', and 'camphoraceous' flavors. Principal component analysis was further applied to the data of GC-MS, which differentiated and discriminated the two ambiguous cubeb berries according to their unique volatile compounds. *Litsea cubeba* was characterized by higher level of D-limonene ('fruit, citrus' note, 0–38.89%), 3,7-dimethyl-2,6-octadienal (0.72–11.32%), citral ('fruit, lemon' note, 1.09–11.85%) and dodecanoic acid (1.89–32.9%). α-Cubebene ('herb' note, 0.4–0.52%) was identified as a marker compound for *Piper cubeba* with higher camphor ('camphoraceous' note, 17.59–20.6%), and linalool ('flower' note, 14.89–21.31%) contents.

3. Materials and Methods

3.1. Materials and Chemicals

Litsea cubeba was collected from Guizhou province (Guiyang, China) and *Piper cubeba* was from Yunnan province of China (Yuxi, China). The collected cubeb samples were kept in a dry and dark place and stored at 4 °C in order to minimize any deteriorative changes to the volatile components of the cubeb berries until their processing. For the precise measurements of GC-MS-O (Agilent Technologies Inc., Santa Clara, CA, USA), cubeb fruit samples were ground to a fine powder using a grinder.

The *n*-alkane standard (C8–C20) was provided by Sigma-Aldrich Chemical Co. (Sigma Chemical Co., St. Louis, MO, USA).

3.2. Hydro-Distillation (HD)

All the air-dried cubeb berries samples (an amount of 100 g each) were subjected to hydro-distillation using a Clevenger-type apparatus to extract essential oil using the reported methods with some modifications [20,32]. The Clevenger-type apparatus consisted of a 2000 mL glass flask, a vertical tube, a condenser, a measuring tube with stopcock, and a return tube. The return tube connected the bottom of the measuring tube to the vertical tube, which combined with the top of the condenser. The flask was filled with 1200 mL of distilled water and heated by an electric heating mantle. The extraction time was 4 h, after which no more essential oil was obtained. The vapor mixture of water–essential oil produced in the flask passed through the condenser and then the distillate was collected. The essential oil in the upper layer of the distillate was dried over anhydrous sodium sulfate (Na_2SO_4) and stored at 4 °C until subsequent GC-MS analysis.

3.3. Simultaneous Distillation Extraction (SDE)

SDE was performed in a modified Lickens–Nickerson apparatus (Chrompack, Netherlands) [33]. A 25 g measure crushed air-dried cubeb berry, with 1.6g sodium chloride and 200 mL distilled water, was placed in a 500 mL flask. The sample and 40 mL of a mixture of pentane–diethyl ether (1:1 *v/v*) solvent placed in another flask were heated up to their boiling points and the temperature conditions were maintained for about 3 h. After cooling to ambient temperature for 10 min, the pentane-diethyl ether extract was dried over anhydrous Na_2SO_4. The extract was kept at 4 °C until subsequent GC-MS analysis [26].

3.4. Optimization of SPME Conditions

A SPME (Supelco, Inc., Bellefonte, PA, USA) fiber (50/30 μm divinylbenzene/carboxen/polydimethylsiloxane; DVB/CAR/PDMS) was used for volatile extraction after the fiber had been conditioned at 270 °C for 1 h. The ground samples were passed through a 20 mesh sieve to achieve uniform particle size. A 1.5 g measure of the sieved cubeb fruits powder was placed in a 20 mL vial with a sealed cap and equilibrated in a laboratory stirrer/hot plate (model PC-420, Corning Inc. Life Science, Acton, MA, USA). Then, a stainless steel needle, housing the SPME fiber, was placed through a hole to expose the fiber in the vial [19]. Three independent extractions were done for each cubeb fruit sample.

To improve the volatile absorption, the SPME parameters were optimized: desorption time (1, 2, 3, and 4 min); incubation time (10, 15, 20, and 25 min); extraction temperature (40, 50, 60, and 70 °C); extraction time (20, 30, 40, and 50 min). For each parameter investigated, the analysis was conducted in triplicate.

3.5. Analysis of Volatiles by GC-MS

7890A gas chromatograph with 5975C mass spectrometer selective detector (Agilent Technologies Inc., Santa Clara, CA, USA) was used, and a DB-5 capillary column (30 m ×0.25 mm × 0.25 µm) was applied for GC-MS. The extraction was injected into the inlet of GC-MS and desorbed at 250 °C for 3 min. The injection port was operated in splitless mode, helium (99.999%) was used as carrier gas at the flow rate of 1.2 mL/min. The initial oven temperature was 40 °C (2 min), ramped at 3 °C min^{-1} to 170 °C (5 min), and then ramped at 10 °C min^{-1} to 260 °C (5 min). Mass detector conditions were performed by EI (electronic impact) mode at 70 eV, source temperature at 230 °C, mass spectra acquisition range of 45–500 amu, scanning rate of 3.18 amu/s. The transfer line temperature was 280 °C. The volatile compounds were identified by comparing the mass spectra with mass-spectral library (NIST 2011), retention index (RI), aroma description, and matching against the published data [34]. Each extract was analyzed in triplicate. Mean data and relative standard deviation (mean ± SD) of volatiles were reported.

3.6. GC-MS-Olfactometry

GC-MS-O was performed by trained panelists on a sniffing port (Sniffer 9000, Brechbühler, Schlieren, Switzerland). The modified method described earlier by Pang et al. [18] was used in this study and the conditions of SPME and GC-MS were the same to the volatile analysis described above.

Four trained panelists take part in the detection frequency analysis (DFA) combined with GC-MS-O for identification aroma-active compounds. The panel consisted of an age from 20 to 35 years (mixed of male and female). The panelists were trained by solutions of artificial odorants and different cubeb berries samples to be familiar with the odor descriptions. In total, eight runs by GC-MS-O were conducted by four assessors (two runs for one person). The judges sniffed the effluent from the mask and recorded the time and odor characteristic of the aroma-active compounds of different cubeb berries samples. When the total detection frequencies were more than twice for the odorants perceived by two different assessors at the sniffing port, the odorants were considered potential aroma-active compounds [18,35].

3.7. Statistical Data Analysis

Significant differences for the volatile constituents among the cubeb berries were determined by one-way analysis of variance (ANOVA) using a SPSS statistics (version 20.0; SPSS, Inc., Chicago, IL, USA). The column figures in the context were plotted using Origin software (version 8.5; Northampton, MA, USA). The Unscrambler v.9.7 (CAMO AS, Trondheim, Norway) software was used for the statistical analysis (PCA) on volatiles.

4. Conclusions

In this study, the aroma compounds of two ambiguous cubeb berries were isolated by HD, SDE, and SPME pretreatment methods in order to fully obtain the complex aroma profiles of the cubeb berries and were analyzed by GC-MS-O combined with PCA. By GC-MS-O analysis, a total of 12 aroma-active compounds were found to play a key role in the characteristic flavor of the cubeb berries. The PCA results clearly indicated that the two ambiguous cubeb berries could be discriminated by the aroma profiles. *Litsea cubeba* was characterized by higher level of D-limonene, citral and dodecanoic acid; *Piper cubeba* was marked with α-cubebene, higher camphor, and linalool. Therefore, using the volatile profile combined with PCA is an appropriate method to discriminate the cubeb berries and assure the related product quality.

Author Contributions: Methodology, J.C.; Software, D.W.; Formal Analysis, S.C.; Writing—Original Draft Preparation, H.C.; Writing—Review & Editing, P.J.W., and D.L.; Supervision, X.Y.

Funding: This research was funded by the Special Science and Technology Program of Xinjiang Uygur Autonomous Region (grant no. 2016A03008-02), Natural Science Foundation of Zhejiang province, China (grant no. LY17C200013).

Conflicts of Interest: The authors declare no conflict of interest.

References

1. Li, Y.J.; Kong, W.J.; Li, M.H.; Liu, H.M.; Zhao, X.; Yang, S.H.; Yang, M.H. *Litsea cubeba* essential oil as the potential natural fumigant: Inhibition of aspergillus flavus and afb1 production in licorice. *Ind. Crops Prod.* **2016**, *80*, 186–193. [CrossRef]
2. Liao, P.C.; Yang, T.S.; Chou, J.C.; Chen, J.; Lee, S.H.; Kuo, Y.H.; Ho, C.L.; Chao, L.K.P. Anti-inflammatory activity of neral and geranial isolated from fruits of *Litsea cubeba* Lour. *J. Funct. Foods* **2015**, *19*, 248–258. [CrossRef]
3. Costa, R.; Salvo, A.; Rotondo, A.; Bartolomeo, G.; Pellizzeri, V.; Saija, Z. Combination of separation and spectroscopic analytical techniques: Application to compositional analysis of a minor citrus species. *Nat. Prod. Res.* **2018**, 1–7. [CrossRef] [PubMed]
4. Salvo, A.; Bruno, M.; La Torre, G.L.; Vadala, R.; Mottese, A. Interdonato lemon from Nizza di Sicilia (Italy): Chemical composition of hexane extract of lemon peel and histochemical investigation. *Nat. Prod. Res.* **2016**, *30*, 1517–1525. [CrossRef] [PubMed]
5. Niwa, A.M.; Marcarini, J.C.; Sartori, D.; Maistro, E.; Mantorani, M.S. Effects of (−)-cubebin (*Piper cubeba*) on cytotoxicity, mutagenicity and expression of p38 MAP kinase and GSTa2 in a hepatoma cell line. *J. Food Compos. Anal.* **2013**, *30*, 1–5. [CrossRef]
6. Perazzo, F.F.; Rodrigues, I.V.; Maistro, E.L.; Souza, S.M.; Nanaykkara, N.P.D.; Bastos, J.K.; Carvalho, J.C.T.; De Souza, G.H.B. Anti-inflammatory and analgesic evaluation of hydroalcoholic extract and fractions from seeds of *Piper cubeba* L. (Piperaceae). *Pharmacogn. J.* **2013**, *5*, 13–16. [CrossRef]
7. Kim, H.J.; Baek, W.S.; Jang, Y.P. Identification of ambiguous cubeb fruit by DART-MS-based fingerprinting combined with principal component analysis. *Food Chem.* **2011**, *129*, 1305–1310. [CrossRef] [PubMed]
8. D'Acampora Zellner, B.; Dugo, P.; Dugo, G. Gas chromatography–olfactometry in food flavour analysis. *J. Chromatogr. A* **2008**, *1186*, 123–143. [CrossRef] [PubMed]
9. Jerkovic, I.; Mastelic, J.; Marijanovic, Z.; Mondello, L. Comparison of hydrodistillation and ultrasonic solvent extraction for the isolation of volatile compounds from two unifloral honeys of *Robinia pseudoacacia* L. and *Castanea sativa* L. *Ultrason. Sonochem.* **2007**, *14*, 750–756. [CrossRef] [PubMed]
10. Cheng, H.; Chen, J.; Li, X.; Pan, J.; Xue, S.J.; Liu, D.; Ye, X. Differentiation of the volatile profiles of Chinese bayberry cultivars during storage by HS-SPME–GC/MS combined with principal component analysis. *Postharvest Biol. Technol.* **2015**, *100*, 59–72. [CrossRef]
11. Majcher, M.; Jeleń, H.H. Comparison of suitability of SPME, SAFE and SDE methods for isolation of flavor compounds from extruded potato snacks. *J. Food Compos. Anal.* **2009**, *22*, 606–612. [CrossRef]
12. Ning, L.; Fu-Ping, Z.; Hai-Tao, C.; Si, Y.L.; Chen, G.; Zhen, Y.S.; Bao, G.S. Identification of volatile components in Chinese Sinkiang fermented camel milk using SAFE, SDE, and HS-SPME-GC/MS. *Food Chem.* **2011**, *129*, 1242–1252. [CrossRef] [PubMed]
13. Du, L.; Li, J.; Li, W.; Li, Y.; Li, T. Characterization of volatile compounds of pu-erh tea using solid-phase microextraction and simultaneous distillation–extraction coupled with gas chromatography–mass spectrometry. *Food Res. Int.* **2014**, *57*, 61–70. [CrossRef]

14. Watkins, P.J.; Rose, G.; Warner, R.D.; Dunshea, F.R.; Pethick, D.W. A comparison of solid-phase microextraction (SPME) with simultaneous distillation–extraction (SDE) for the analysis of volatile compounds in heated beef and sheep fats. *Meat Sci.* **2012**, *91*, 99–107. [CrossRef] [PubMed]
15. Selli, S.; Kelebek, H.; Ayseli, M.T.; Tokbas, H. Characterization of the most aroma-active compounds in cherry tomato by application of the aroma extract dilution analysis. *Food Chem.* **2014**, *165*, 540–546. [CrossRef] [PubMed]
16. Pino, J.A. Odour-active compounds in papaya fruit cv. Red Maradol. *Food Chem.* **2014**, *146*, 120–126. [CrossRef] [PubMed]
17. Kang, W.; Li, Y.; Xu, Y.; Jiang, W.; Tao, Y. Characterization of aroma compounds in Chinese bayberry (*Myrica rubra* Sieb. et Zucc.) by Gas Chromatography Mass Spectrometry (GC-MS) and Olfactometry (GC-O). *J. Food Sci.* **2012**, *77*, C1030–C1035. [CrossRef] [PubMed]
18. Pang, X.; Guo, X.; Qin, Z.; Yao, Y.; Hu, X.; Wu, J. Identification of Aroma-Active compounds in Jiashi Muskmelon Juice by GC-O-MS and OAV Calculation. *J. Agric. Food Chem.* **2012**, *60*, 4179–4185. [CrossRef] [PubMed]
19. Cheng, H.; Chen, J.; Chen, S.; Wu, D.; Liu, D.; Ye, X. Characterization of aroma-active volatiles in three Chinese bayberry (Myrica rubra) cultivars using GC–MS–olfactometry and an electronic nose combined with principal component analysis. *Food Res. Int.* **2015**, *72*, 8–15. [CrossRef]
20. Wang, H.; Liu, Y. Chemical Composition and antibacterial activity of essential oils from different parts of *Litsea cubeba*. *Chem. Biodivers.* **2010**, *7*, 229–235. [CrossRef] [PubMed]
21. Wang, F.; Chang, Z.; Duan, P.; Xu, Y.; Zhang, L.; Miao, J.; Fan, Y. Hydrothermal liquefaction of *Litsea cubeba* seed to produce bio-oils. *Bioresour. Technol.* **2013**, *149*, 509–515. [CrossRef] [PubMed]
22. Díaz-Maroto, M.C.; Pérez-Coello, M.S.; Cabezudo, M.D. Effect of drying method on the volatiles in bay leaf (*Laurus nobilis* L.). *J. Agric. Food Chem.* **2002**, *50*, 4520–4524. [CrossRef] [PubMed]
23. Barra, A.; Baldovini, N.; Loiseau, A.M.; Albino, L.; Lesecq, C.; Cuvelier, L.L. Chemical analysis of French beans (*Phaseolus vulgaris* L.) by headspace solid phase microextraction (HS-SPME) and simultaneous distillation/extraction (SDE). *Food Chem.* **2007**, *101*, 1279–1284. [CrossRef]
24. Boschfuste, J.; Riuaumatell, M.; Guadayol, J.; Caixach, J.; López-Tamames, E.; Buxaderas, S. Volatile profiles of sparkling wines obtained by three extraction methods and gas chromatography–mass spectrometry (GC–MS) analysis. *Food Chem.* **2007**, *105*, 428–435. [CrossRef]
25. Yi, Z.; Feng, T.; Zhuang, H.; Ye, R.; Li, M.; Liu, T. Comparison of different extraction methods in the analysis of volatile compounds in pomegranate juice. *Food Anal. Methods* **2016**, *9*, 2364–2373. [CrossRef]
26. Kraujalyté, V.; Leitner, E.; Venskutonis, P.R. Characterization of Aronia melanocarpa Volatiles by Headspace-Solid-Phase Microextraction (HS-SPME), Simultaneous Distillation/Extraction (SDE), and Gas Chromatography-Olfactometry (GC-O) Methods. *J. Agric. Food Chem.* **2013**, *61*, 4728–4736. [CrossRef] [PubMed]
27. Du, X.; Finn, C.E.; Qian, M.C. Volatile composition and odour-activity value of thornless 'Black Diamond' and 'Marion' blackberries. *Food Chem.* **2010**, *119*, 1127–1134. [CrossRef]
28. Du, X.; Plotto, A.; Baldwin, E. Evaluation of volatiles from two subtropical strawberry cultivars using GC–Olfactometry, GC-MS Odor activity values, and sensory analysis. *J. Agric. Food Chem.* **2011**, *59*, 12569–12577. [CrossRef] [PubMed]
29. Plotto, A.; Margaría, C.A.; Goodner, K.L.; Baldwin, E.A. Odour and flavour thresholds for key aroma components in an orange juice matrix: Esters and miscellaneous compounds. *Flavour Fragr. J.* **2008**, *23*, 398–406. [CrossRef]
30. Hempfling, K.; Fastowski, O.; Kopp, M.; Pour Nikfardjam, M.; Engel, K.H. Analysis and sensory evaluation of gooseberry (*Ribes uva crispa* L.) volatiles. *J. Agric. Food Chem.* **2013**, *61*, 6240–6249. [CrossRef] [PubMed]
31. Lee, S.; Ahn, B. Comparison of volatile components in fermented soybean pastes using simultaneous distillation and extraction (SDE) with sensory characterisation. *Food Chem.* **2009**, *114*, 600–609. [CrossRef]
32. Zhang, Z.; Zeng, D.; Li, G. Study of the volatile profile characteristics of longan during storage by a combination sampling method coupled with GC/MS. *J. Sci. Food Agric.* **2008**, *88*, 1035–1042. [CrossRef]
33. Ceva-Antunes, P.M.N.; Bizzo, H.R.; Alves, S.M.; Alves, S.M.; Antunes, O.A.C. Analysis of volatile Compounds of Taperebá (*Spondias mombin* L.) and Cajá (*Spondias mombin* L.) by Simultaneous Distillation and Extraction (SDE) and Solid Phase Microextraction (SPME). *J. Agric. Food Chem.* **2003**, *51*, 1387–1392. [CrossRef] [PubMed]

34. Molyneux, R.J.; Schieberle, P. Compound Identification: A journal of agricultural and food chemistry perspective. *J. Agric. Food Chem.* **2007**, *55*, 4625–4629. [CrossRef] [PubMed]
35. Cheng, H.; Chen, J.; Chen, S.; Xia, Q.; Liu, D.; Ye, X. Sensory evaluation, physicochemical properties and aroma-active profiles in a diverse collection of Chinese bayberry (*Myrica rubra*) cultivars. *Food Chem.* **2016**, *212*, 374–385. [CrossRef] [PubMed]

Sample Availability: Samples of the compounds are not available from the authors.

© 2018 by the authors. Licensee MDPI, Basel, Switzerland. This article is an open access article distributed under the terms and conditions of the Creative Commons Attribution (CC BY) license (http://creativecommons.org/licenses/by/4.0/).

Communication

Evaluation of Volatile Metabolites Emitted In-Vivo from Cold-Hardy Grapes during Ripening Using SPME and GC-MS: A Proof-of-Concept

Somchai Rice [1,2,3], Devin L. Maurer [3], Anne Fennell [4], Murlidhar Dharmadhikari [1] and Jacek A. Koziel [2,3,*]

1. Midwest Grape and Wine Industry Institute, Iowa State University, Ames, IA 50011, USA; somchai@iastate.edu (S.R.); murli@iastate.edu (M.D.)
2. Interdepartmental Toxicology Graduate Program, Iowa State University, Ames, IA 50011, USA
3. Department of Agricultural and Biosystems Engineering, Iowa State University, Ames, IA 50011, USA; dmaurer@iastate.edu
4. Department of Agronomy, Horticulture and Plant Science, BioSNTR, South Dakota State University, Brookings, SD 57006, USA; anne.fennell@sdstate.edu
* Correspondence: koziel@iastate.edu; Tel.: +1-515-294-4206

Academic Editors: Constantinos K. Zacharis and Paraskevas D. Tzanavaras
Received: 1 January 2019; Accepted: 30 January 2019; Published: 1 February 2019

Abstract: In this research, we propose a novel concept for a non-destructive evaluation of volatiles emitted from ripening grapes using solid-phase microextraction (SPME). This concept is novel to both the traditional vinifera grapes and the cold-hardy cultivars. Our sample models are cold-hardy varieties in the upper Midwest for which many of the basic multiyear grape flavor and wine style data is needed. Non-destructive sampling included a use of polyvinyl fluoride (PVF) chambers temporarily enclosing and concentrating volatiles emitted by a whole cluster of grapes on a vine and a modified 2 mL glass vial for a vacuum-assisted sampling of volatiles from a single grape berry. We used SPME for either sampling in the field or headspace of crushed grapes in the lab and followed with analyses on gas chromatography-mass spectrometry (GC-MS). We have shown that it is feasible to detect volatile organic compounds (VOCs) emitted in-vivo from single grape berries (39 compounds) and whole clusters (44 compounds). Over 110 VOCs were released to headspace from crushed berries. Spatial (vineyard location) and temporal variations in VOC profiles were observed for all four cultivars. However, these changes were not consistent by growing season, by location, within cultivars, or by ripening stage when analyzed by multivariate analyses such as principal component analysis (PCA) and hierarchical cluster analyses (HCA). Research into aroma compounds present in cold-hardy cultivars is essential to the continued growth of the wine industry in cold climates and diversification of agriculture in the upper Midwestern area of the U.S.

Keywords: biogenic emissions; veraison; viticulture; nondestructive analysis; wine aroma; diffusion; grape skin; vacuum-assisted extraction; solid-phase microextraction; VOCs

---

1. Introduction

Understanding the development of flavor and aroma compounds in wine grapes is crucial to winemaking. Grape berry development is characterized by two sigmoidal growth periods. The first growth period is berry formation from fruit set to lag phase. This is followed by berry-ripening from veraison to harvest [1]. Veraison is characterized by a change in color of the berries. During the berry ripening phase, sugar accumulates as measured in Brix. The rapid accumulation of sugar in the berry ripening from veraison onto harvest is well understood [2]. This is contrasted by the relative lack of research on aroma compound accumulation during ripening, especially for cold-hardy grapes. Further understanding of the accumulation of aroma compounds during the ripening phase can inform how viticultural practices can be used to influence wine style.

Interest in research of aroma compounds in wine grapes is high. The prevailing share of published research in wine grapes has been in *V. vinifera* ('Old World', well-established varieties). This is expected since vinifera was cultivated as early as the seventh and fourth millennia BC [3]. For example, it well-known that aroma compounds such as pyrazines contribute to the characteristic aroma in Cabernet Sauvignon and Sauvignon Blanc [4,5]. These aromas can be described as 'grassy,' 'herbaceous', and 'green bell pepper' [6]. The decline in pyrazines in developing wine grapes has been linked to the levels of sunlight reaching the cluster and can be reduced through canopy management if this aroma is undesirable [7]. Terroir has been shown to affect wine aroma in Riesling grown in the Niagara Peninsula [8] and Cabernet Sauvignon in China [4]. Aroma compounds have been characterized in Japanese Shine muscat (*V. labruscana* Bailey and *V. vinifera* L.). Levels of linalool, hexanal, (E)-2-hexenal, hexanol, and (Z)-3-hexene-1-ol, nerol in berry skins and pulp were influenced by storage temperatures post-harvest [9]. Viticultural practices such as the timing of early defoliation have been investigated to determine the effect on Tempranillo wine aroma [10], and enological practices such as effects of pre-fermentation cold soaking on Cabernet Sauvignon grape and wine volatiles [11]. The bulk of the research into the aroma of grapes and wine has been done for vinifera because vinifera has existed longer than hybrid grapes. With the recent introduction of cold-hardy (hybrid) grapes, production of quality wines is possible in cold-climate regions where vinifera cannot thrive.

Since the release of high-quality, cold-hardy, and disease-resistant cultivars from the University of Minnesota, the winemaking industry has grown in cold climates such as the upper Midwest region of the U.S. St. Croix, Frontenac, Marquette, and La Crescent cultivars were developed in 1983, 1996, 2006, and 2002, respectively [12]. Current searches in the journal database Web of Science using keywords and variations of 'Marquette,' 'Frontenac,' 'St. Croix', 'La Crescent', volatile, aroma, cold-hardy, and maturity yield < 50 articles. Canopy management effects on fruit and wine aroma have been investigated in Traminette, an interspecific hybrid of Gewürztraminer in the Eastern U.S. [13]. Volatile compounds from Zuoshanyi, a native red grape variety in northeast China, were characterized with 135 VOCs identified and quantified [14]. Effects of pre-fermentation treatments on wine aroma profile were explored in the cold-hardy cultivar Solaris in Denmark [15].

There is a gap in knowledge, especially in aroma research, with these interspecific, cold-hardy hybrid grapes. Previous work showed a constant decrease in the ratio of alpha-linolenic acid degradation products, *cis*-3-hexenol to *trans*-2-hexenal during ripening of Frontenac and Marquette berries grown in Quebec, through the destructive blending of the berries [16]. Frontenac and Marquette aromas reported at harvest were mainly hexanal, *trans*-2-hexenal, 1-hexanol, *cis*-3-hexenol, hexanoic acid, acetic acid, *beta*-damascenone, and 1-phenylethanol. Marquette had significantly higher levels of linalool, geraniol, and *alpha*-citral [17]. Continuing work in Canada has been done in profiling aroma compounds in Frontenac, Marquette, Marechal Foch, Sabrevois, and St. Croix skin, juice and wine. Terpenes were primarily located in the skin, and the highest concentration was in Marquette. Nonanal, (E,Z)-2,6 nonadienal, *beta*-damascenone, ethyl octanoate, and isoamyl acetate were compounds with the highest odor activity values (OAV) in wines [17]. The OAV for a compound is the ratio between the concentration and the odor detection threshold (ODT) and it could be a useful metric for aroma-imparting compounds. The ODT is the minimal concentration that can be detected by human nose in 50% of the population [18–20]. Earlier research has also shown that the majority of aroma compounds present in grape berries are bound to a sugar moiety within the berry [21].

Various methods of sample preparation have been used to characterize aromas from grapes and wine. Thermal desorption was used to determine volatiles from Solaris wine [15]. Solid phase extraction (SPE) has been used to isolate aroma precursors in Merlot, Gewürztraminer, and Tempranillo grapes and wine [22]. Solid-phase microextraction (SPME) has emerged as one of the preferred methods of sample and sample prep for analysis of volatiles in the grapes and wine. SPME offers the advantages of portability, simplicity, and re-usability in field and laboratory settings. Applications using SPME in the food and beverage industry can be found elsewhere [23]. Gas chromatography (GC) has been extensively used to separate aroma compounds from the complex mix of aromas. GC is often coupled with mass spectrometry (MS) to identify and quantify the separated aroma compounds. These analytical methods have been used, sometimes in combination with other analytical methods, in analysis of volatiles from grapefruit (*Citrus paradise* L.) [24], berry cactus (*Myrtillocactus geometrizans*) [25], Shine Muscat [9], Cabernet Sauvignon [4,26], Zuoshanyi grapes [14], Muscat cultivars [27], Nero d'Avola and Fiano grapes [28], Monastrell wines [29], and selected cold-hardy grape cultivars and wine [16,17] and cold-hardy wines [17,30–34].

A review of SPME use for in-vivo and in-vitro in whole plant and plant organ analysis is found elsewhere [35]. It is clear that there is little research in cold-hardy wine grape cultivars when compared to *V. vinifera*. There is a need for a better understanding of these new cultivars in order to produce quality cold-climate wines that can compete in the world market. To date, this research is the first report of aroma compounds (1) emitted in-vivo from veraison to harvest using two novel sampling methods and collected by SPME and analyzed by GC-MS from Frontenac, Marquette, St. Croix, and La Crescent grape cultivars.

In this research, we propose a novel concept for a non-destructive evaluation of volatiles emitted from ripening grapes using SPME. This concept is novel to both the traditional vinifera grapes and the cold-hardy cultivars. Our models are cold-hardy varieties in the upper Midwest for which many of the basic multiyear grape flavor and wine style data is needed. Research into aroma compounds present in cold-hardy cultivars is essential to the continued growth of the wine industry in cold climates and diversification of agriculture in the upper Midwestern area of the U.S. The need for data is confounded by the small resources available to conduct long-term research.

If proven feasible, the concept of non-destructive analysis of ripening grapes presents a tantalizing possibility to investigate the effects of different viticulture practices throughout the stages of berry ripening on berry aroma. This, in turn, could be used to develop better quality wines. If volatile compounds emitted in-vivo could be identified as developmental biomarkers, portable target VOC detectors could then be developed. These detectors can give vineyards a real-time gauge to guide them in harvesting for flavor.

The main objective was to develop the proof-of-concept for a non-destructive (in-vivo) sampling of volatile compounds from growing and ripening grapes. Specific objectives (1–4) were to (1) develop sampling devices to capture volatiles emitted from a whole cluster and single berry; (2) characterize the volatile compounds emitted in-vivo from four cold-hardy grape cultivars using: (2i) whole cluster analysis, (2ii) single berry analysis; (3) compare volatile compounds emitted in-vivo (objective 1) with crushed berry (i.e., destructive analysis including skin, seeds, and pulp); (4) search for preliminary links between volatile compounds detected (objectives 1 and 2) and selected: (4i) microclimates (Iowa and South Dakota), (4ii) the individual cultivars (i.e., Frontenac, Marquette, St. Croix, and La Crescent), and (4iii) time stages of berry ripening.

The working hypotheses were: (1) aroma compound development from veraison-to-harvest can be detected in-vivo by sampling volatile emissions from ripening grapes (from both a single berry and whole grape cluster) and (2) that the flavor accumulation (i.e., increasing concentration of VOCs) can be correlated with berry ripening in all four cultivars. Testing these hypotheses can potentially translate into improving viticulture practices that lead to timing the harvesting for flavor. This research aims at addressing the gap in knowledge for cold-hardy grape cultivars by cataloging VOCs from Frontenac, Marquette, St. Croix and La Crescent emitted in-vivo and whole crushed berries throughout berry ripening.

2. Results

2.1. Sampling Devices to Capture Volatiles Emitted from a Whole Cluster and Single Berry

In this research, non-destructive and destructive sampling methods for the detection of VOCs emitted from cold-hardy grapes were explored. Non-destructive sampling included (1) a use of polyvinyl fluoride (PVF) chambers temporarily enclosing and concentrating volatiles emitted by a whole cluster of grapes on a vine (Figure 1), and (2) a modified 2 mL glass vial for a vacuum-assisted sampling of volatiles from a single berry (Figure 2).

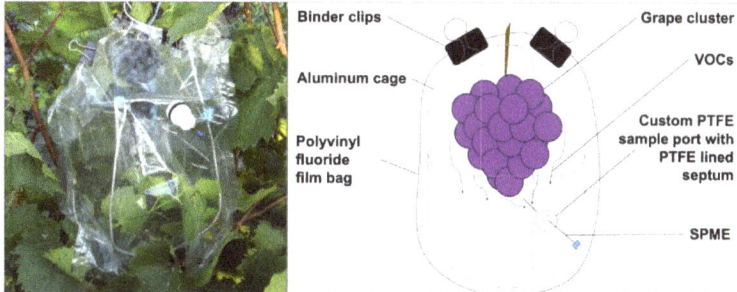

Figure 1. Non-destructive sampling of biogenic volatiles emitted by the whole cluster of grapes on a vine. Schematic of polyvinyl fluoride (PVF) film chambers used for short-term enclosing of growing clusters of cold-hardy grapes during in-vivo sampling of volatile emissions using solid-phase microextraction (SPME). An aluminum wire cage was constructed to hold the PVF chamber spread around and to be secured to the grape vine's training system. The PVF chamber was modified with a custom polytetrafluoroethylene (PTFE) port fitted with 11 mm PTFE lined silicone septa (SPME sampling port).

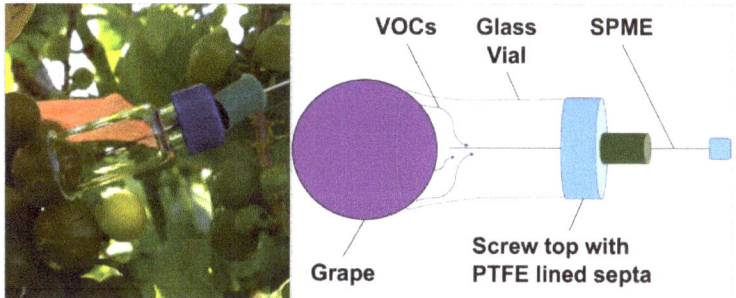

Figure 2. Non-destructive sampling of biogenic volatiles emitted by a modified 2 mL glass vial for a vacuum-assisted sampling of volatiles from a single grape berry. Schematic of a modified screw top 2 mL glass vial with PTFE lined septa used for characterizing in-vivo metabolite emissions from selected cold-hardy grapes. Negative pressure was created with a syringe to hold the sampling device with SPME sealed onto the grape berry surface.

A total of 124 VOCs were identified across all sampling methods, 79 of these VOCs were verified with analytical standards matching retention times and mass spectral data (i.e., using the identification of compounds with Automated Mass Spectral Deconvolution and Identification System (AMDIS) target library search with at least 80% mass spectral match. Target libraries included (a) the 6 libraries that are included with the AMDIS program, (b) an onsite (our laboratory) library created from analysis of pure standards (200+ compounds), (c) NIST11 mass spectral library described in Materials and Methods section on data analysis). A full summary of VOCs identified in Frontenac, Marquette,

St. Croix, and La Crescent berries from South Dakota and Iowa by each sampling method is provided in data paper [36] with known aroma descriptors for pure compounds [37,38]. All PCA biplots are given in Appendix A (Figures A1 and A2). It should be noted that significant changes in volatiles emitted were only observed in Frontenac grapes grown in South Dakota in 2013 as indicated by the variance accounted for in component 1 and 2 in PCA (i.e., greater than 70%). However, due to the exploratory nature of this research, all PCA data is presented is subsequent sections.

2.2. Volatiles Emitted In-Vivo from Four Cold-Hardy Grape Cultivars

PVF chambers were used in 2012 on Frontenac and Marquette in Iowa and South Dakota. Modified glass vials were used in 2013 on Frontenac, Marquette, St. Croix, and La Crescent in South Dakota. Only St. Croix and La Crescent were sampled by modified glass vials in Iowa, limited by funding.

2.2.1. Emissions from Whole Grape Cluster

Forty-four of the total 124 grape VOCs emitted in-vivo were detected by whole grape cluster sampling chambers in Frontenac and Marquette cultivars grown in Iowa and South Dakota, monitored from veraison to harvest. Table 1 presents the VOCs that are characteristic of biogenic emissions from Frontenac and Marquette clusters during the 2012 growing season from Iowa and South Dakota. These volatiles were detected in-vivo from whole grape clusters and determined through interpretation of principal component analysis (PCA) and hierarchical clustering analysis (HCA). A detailed summary of all 124 VOC can be found elsewhere [36]. However, only key representative volatiles from the hierarchical cluster analysis (HCA) are labeled with numbers on PCA biplot figures presented in Results.

Table 1. Whole cluster analysis. Volatiles emitted from Frontenac and Marquette clusters, grown in Iowa (IA) and South Dakota (SD). These VOCs were indicated to be the most representative variable from hierarchical clustering analysis (HCA) after the PCA (JMP Pro 12.0.1, SAS Institute Inc., Cary, NC, USA).

Sample	Cluster	No. of Members [2]	Most Representative Variable [3]	Cluster Proportion of Variation Explained [4]	Total Proportion of Variation Explained [5]
IA Frontenac (0.709) [1]	1	5	Heptanal	0.937	0.173
	3	5	4-Methyl-3-penten-2-one	0.745	0.138
	5	6	Nonanal	0.512	0.114
	2	3	3-Methyl-1-butanol	0.845	0.094
	4	4	1,4-Butanolide	0.576	0.085
	6	2	5-(Hydroxymethyl)-2-furancarboxaldehyde	0.858	0.064
	7	2	Benzophenone	0.548	0.041
SD Frontenac (0.686) [1]	1	7	Toluene	0.712	0.208
	2	5	Nonanal	0.820	0.171
	5	3	3-Phenyl-2-propenal	0.810	0.101
	3	4	4-Methyl-3-penten-2-one	0.592	0.099
	4	3	Acetic acid	0.502	0.063
	6	2	Benzyl alcohol	0.536	0.045
IA Marquette (0.739) [1]	1	10	1-Octanol	0.805	0.310
	2	5	Acetaldehyde	0.814	0.156
	3	3	Methyl ethyl ketone	0.616	0.071
	4	3	1-Hexadecanol	0.505	0.063
	5	3	Acetophenone	0.536	0.062
	6	1	Acetic acid	1.000	0.038
	7	1	2-Ethyl-1-hexanol	1.000	0.038
SD Marquette (0.783) [1]	1	7	Acetone	0.817	0.249
	2	4	4-Methyl-3-penten-2-one	0.799	0.139
	6	4	Decane	0.797	0.139
	3	2	1-Pentanol	1.000	0.087
	4	3	2-Ethyl-1-hexanol	0.532	0.069
	7	2	1-Hexadecanol	0.654	0.057
	5	1	Indene	1.000	0.043

[1] The total proportion of variation explained by all the cluster components. [2] The number of variables in the cluster. [3] The cluster variable that has the largest squared correlation with its cluster component. [4] The cluster's proportion of variance explained by the first principal component amount the variables in the cluster, based only on variables within the cluster. [5] The overall proportion of variance explained by the cluster component, using only the variables within each cluster to calculate the first principal component.

Frontenac

3-Methyl-1-butanol and heptanal were emitted and detected in 2012 Iowa Frontenac grapes at veraison. At harvest, 1,4-butanolide was detected (Figure A1). Nonanal, benzyl alcohol, and toluene were emitted and detected in 2012 South Dakota Frontenac grapes (Figure 3) at veraison. Only one compound, i.e., acetic acid, was associated with harvest time. Other compounds were detected (e.g., 2-methyl-3-penten-2-one and 3-phenyl-2-propenal) but were not indicated to be the most representative compounds from HCA. Results are also presented in this manner throughout the manuscript.

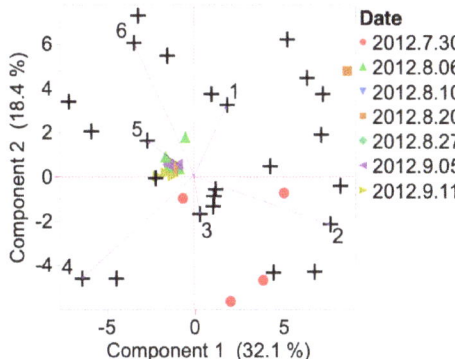

Figure 3. Evolution of VOCs emitted from whole clusters of 2012 Frontenac grapes grown in South Dakota from veraison to harvest. Key to most representative volatiles from HCA, shown as vectors from the origin and read clockwise: 1 = 2-Methyl-3-penten-2-one, 2 = Nonanal, 3 = Benzyl alcohol, 4 = Toluene, 5 = Acetic acid, 6 = 3-Phenyl-2-propenal.

Marquette

2012 Iowa Marquette did not have a 'representative' VOC at veraison, as indicated by HCA, and replicate samples had high variability (i.e., unevenly distributed between 2 quadrants of the PCA biplot). By harvest, 1-hexadecanol and methyl ethyl ketone had developed. Similarly, 2012 South Dakota Marquette VOCs emitted at veraison (Figure 4) did not have 'representative' VOC at veraison. At harvest, indene was the representative VOC emitted.

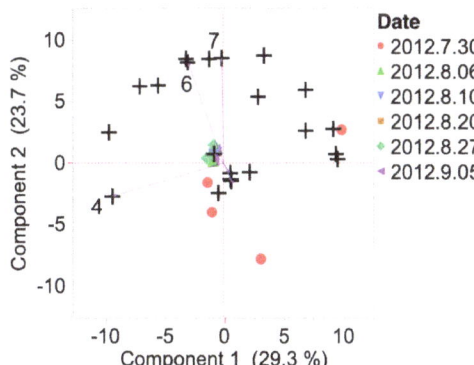

Figure 4. Evolution of VOCs emitted from whole clusters of 2012 Marquette grapes grown in South Dakota from veraison to harvest. Key to most representative volatiles from HCA, shown as vectors from the origin and read clockwise: 1 = 1-Hexadecanol, 2 = 2-Ethyl-1-hexanol, 3 = 1-Pentanol, 4 = Acetone, 5 = Indene, 6 = Decane, 7 = 4-Methyl-3-penten-2-one.

2.2.2. Emissions from Single Berries

Thirty-nine VOCs emitted in-vivo were also detected by modified glass vial (vacuum assisted) method in Frontenac, Marquette, St. Croix, and La Crescent cultivars grown in Iowa and South Dakota. Table 2 presents the VOCs that are characteristic of these 4 cold-hardy cultivars during the 2013 growing season in Iowa and South Dakota, detected in-vivo from single berries, and determined through multivariate statistical analysis previously discussed.

Table 2. 'Characteristic' VOCs emitted from single berries of Frontenac, Marquette, St. Croix, and La Crescent grapes grown in Iowa and South Dakota. These VOCs were indicated to be the most representative variable from hierarchical clustering analysis after the PCA (JMP Pro 12.0.1, SAS Institute Inc., Cary, NC, USA).

Sample	Cluster	No. of Members [2]	Most Representative Variable [3]	Cluster Proportion of Variation Explained [4]	Total Proportion of Variation Explained [5]
SD Frontenac (0.750) [1]	1	2	Palmitic acid	0.830	0.415
	2	2	Acetic acid	0.669	0.334
SD Marquette (0.870) [1]	1	7	1,4-Butanolide	0.878	0.473
	2	4	Ethyl octanoate	0.883	0.272
	3	2	2-Ethyl-1-hexanol	0.819	0.126
IA St. Croix (0.896) [1]	1	4	3-Methyl indole	1.000	0.500
	2	3	Acetic acid	0.722	0.271
	3	1	Benzyl alcohol	1.000	0.125
SD St. Croix (0.855) [1]	1	3	Nonanal	0.802	0.241
	3	3	Diacetone alcohol	0.713	0.214
	2	2	1,4-Butanolide	1.000	0.200
	4	1	5-(Hydroxymethyl)-2-furancarboxaldehyde	1.000	0.100
	5	1	Ethyl acetate	1.000	0.100
IA La Crescent (0.909) [1]	1	28	2-Ethyl-1-hexanol	0.973	0.757
	2	4	3-Methyl indole	0.772	0.086
	4	2	Ethanol	0.636	0.035
	3	2	2-Phenylethanol	0.556	0.031
SD La Crescent (0.936) [1]	1	3	2-Phenylethanol	1.000	0.300
	3	3	Diacetone alcohol	0.853	0.256
	4	2	Acetic acid	0.976	0.195
	2	2	6-Methyl-5-hepten-2-one	0.926	0.185

[1] The total proportion of variation explained by all the cluster components. [2] The number of variables in the cluster. [3] The cluster variable that has the largest squared correlation with its cluster component. [4] The cluster's proportion of variance explained by the first principal component amount the variables in the cluster, based only on variables within the cluster. [5] The overall proportion of variance explained by the cluster component, using only the variables within each cluster to calculate the first principal component.

Frontenac

In-vivo detection of VOCs by modified glass vial did not identify a key representative compound in 2013 South Dakota Frontenac grapes at veraison (Figure 5). At harvest, palmitic acid was emitted and detected in these berries.

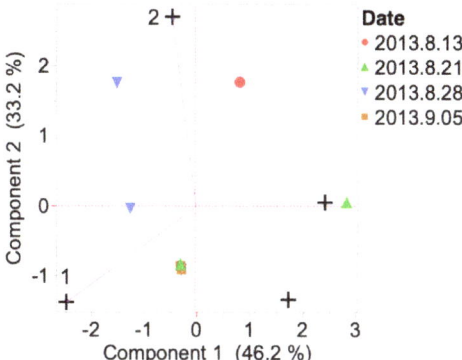

Figure 5. Evolution of VOCs emitted from single berry from veraison to harvest of 2013 Frontenac grapes grown in South Dakota. Key to most representative volatiles from HCA, shown as vectors from the origin and read clockwise: 1 = Palmitic acid, 2 = Acetic acid.

Marquette

VOCs detected by modified glass vial emitted from 2013 Marquette grown in South Dakota generally did not vary during berry development. The variability decreased between the replicate samples, indicated by less spread between the data points as the berries developed. Aromas from berries grown in South Dakota during the 2013 growing season (Figure 6) can be characterized from 3 VOCs. These compounds were 2-ethyl-1-hexanol, ethyl octanoate and 1,4-butanolide.

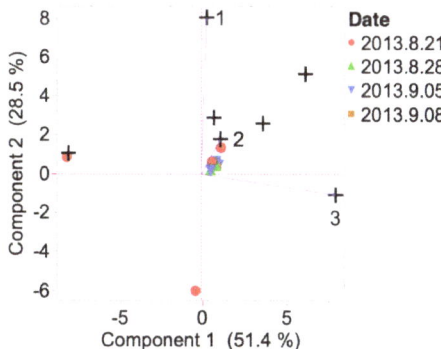

Figure 6. Evolution of VOCs emitted from single berry from veraison to harvest of 2013 Marquette grapes grown in South Dakota. Key to most representative volatiles from HCA, shown as vectors from the origin and read clockwise: 1 = Ethyl octanoate, 2 = 2-Ethyl-1-hexanol, 3 = 1,4-Butanolide.

St. Croix

Statistical analysis of VOCs detected from 2013 St. Croix grown in Iowa at veraison and harvest determined 3 important compounds. These compounds were 3-methyl indole, benzyl alcohol, and acetic acid. Decreased variability between replicate samples was observed as the berries ripened, although no strong associations were noticed between these compounds and berry development. Compounds emitted and detected in 2013 St. Croix from South Dakota (Figure 7) were 1,4-butanolide, 5-(hydroxymethyl)-2-furancarboxaldehyde, ethyl acetate, and nonanal. Of the 5 key VOCs detected in 2013 South Dakota St. Croix at veraison, nonanal was most associated with development at harvest.

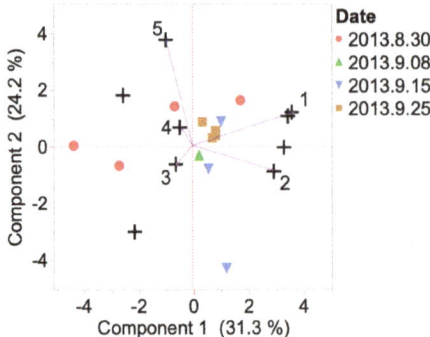

Figure 7. Evolution of VOCs emitted from single berry from veraison to harvest of 2013 St. Croix grapes grown in South Dakota. Key to most representative volatiles from HCA, shown as vectors from the origin and read clockwise: 1 = Nonanal, 2 = Diacetone alcohol, 3 = 1,4-Butanolide, 4 = 5-(Hydroxymethyl)-1-furancarboxaldehyde, 5 = Ethyl acetate.

La Crescent

VOCs detected by modified glass vial emitted from 2013 La Crescent grown in Iowa were highly variable at veraison. A characteristic compound (i.e., 3-methyl-indole) was determined to be present at veraison. By harvest, octanal was present but not statistically representative. La Crescent berries from 2013 grown in South Dakota (Figure 8) were highly variable between replicate samples. Compounds emitted included 2-phenylethanol, 6-methyl-5-hepten-2-one, and acetic acid. By harvest, 1,4-butanolide was present but not statistically representative.

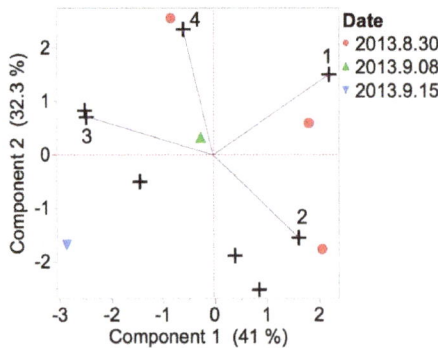

Figure 8. Evolution of VOCs emitted from single berry from veraison to harvest of 2013 St. Croix grapes grown in South Dakota. Key to most representative volatiles from HCA, shown as vectors from the origin and read clockwise: 1 = 2-Phenylethanol, 3 = Diacetone alcohol, 4 = Acetic acid, 2 = 6-Methyl-5-hepten-2-one.

2.3. Destructive Sampling

117 grape VOCs were detected by destructive analysis (i.e., crushed berries) in Frontenac, Marquette, St. Croix, and La Crescent cultivars grown in Iowa and South Dakota. The sample matrix included skins, pulp, and seeds. Crushed berry analysis was used in 2012 on Frontenac and Marquette cultivars grown in South Dakota, and all 4 cultivars in 2013. A freezer malfunction in resulted in the loss of Iowa 2012 berries stored for crushed berry analysis. Table 3 presents the VOCs that are characteristic of these 4 cold-hardy cultivars during the 2012 and 2013 growing seasons in Iowa and South Dakota, detected in whole, crushed berries and determined through multivariate statistical analysis previously discussed.

Table 3. 'Characteristic' VOCs emitted from crushed berries of Frontenac, Marquette, St. Croix, and La Crescent grapes grown in Iowa and South Dakota. These VOCs were indicated to be the most representative variable from hierarchical clustering analysis after the PCA (JMP Pro 12.0.1, SAS Institute Inc., Cary, NC, USA).

Sample	Cluster	No. of Members [B]	Most Representative Variable [C]	Cluster Proportion of Variation Explained [D]	Total Proportion of Variation Explained [E]
IA Frontenac (0.803) [A]	1	7	3-Methyl-1-butanol	0.993	0.257
	2	9	Cyclohexanol	0.770	0.257
	4	4	Isoamyl acetate	0.745	0.110
	3	4	Isovaleraldehyde	0.622	0.092
	5	3	Toluene	0.774	0.086
SD Frontenac (0.627) [A]	2	8	Styrene	0.636	0.083
	10	7	Acetaldehyde	0.546	0.063
	5	6	2-Octanone	0.608	0.060
	1	6	Acetone	0.602	0.059
	6	5	1-Hexanol	0.645	0.053
	4	5	Nonane	0.624	0.051
	8	4	Ethyl hexanoate	0.732	0.048
	3	4	Ethyl palmitate	0.694	0.045
	7	3	Hexanoic acid	0.740	0.036
	9	4	N-benzyl-2-phenethylamine	0.509	0.033
	12	2	2-Methyl-1-propanol	0.953	0.031
	11	3	Benzoic acid, methyl ester	0.560	0.028
	14	2	Isophorone	0.569	0.019
	13	2	Octanal	0.524	0.017
IA Marquette (0.863) [A]	1	9	Hexanal	0.925	0.347
	3	5	Isoamyl acetate	0.878	0.183
	2	4	Styrene	0.776	0.129
	5	2	Ethanol	0.933	0.078
	6	2	Benzophenone	0.813	0.068
	4	2	Allyl alcohol	0.703	0.059
SD Marquette (0.654) [A]	7	7	Acetaldehyde	0.617	0.062
	6	6	Methyl ethyl ketone	0.639	0.055
	3	5	Decane	0.760	0.054
	19	5	Nonanal	0.667	0.048
	1	4	Styrene	0.777	0.044
	5	5	Amyl acetate	0.621	0.044
	4	4	(E)-2-Hexenoic acid	0.704	0.040
	10	4	Cyclohexanol	0.692	0.040
	9	5	Octanal	0.480	0.034
	2	3	1-Pentanol	0.673	0.029
	8	2	Nonane	0.966	0.028
	18	3	Valeraldehyde	0.630	0.027
	11	3	1-Heptanol	0.629	0.027
	14	4	beta-Damascenone	0.470	0.027
	12	4	Allyl alcohol	0.435	0.025
	13	2	p-Cymene	0.835	0.024
	16	2	Methyl disulfide	0.635	0.018
	15	1	beta-Cyclocitral	1.000	0.014
	17	1	Nerol acetate	1.000	0.014
IA St. Croix (0.772) [A]	1	9	Formic acid, octyl ester	0.832	0.150
	4	8	Ethyl decanoate	0.901	0.144
	2	7	Isobutyraldehyde	0.674	0.094
	3	5	Aspirin methyl ester	0.813	0.081
	5	3	Benzeneacetaldehyde	0.858	0.052
	10	3	Ethanol	0.771	0.046
	8	3	Methacrolein	0.682	0.041
	12	2	Isoamyl acetate	0.841	0.034
	6	2	1-Butanol	0.790	0.032
	7	3	Ethyl butyrate	0.493	0.030
	9	2	1-Hexanol	0.649	0.026
	11	2	beta-Damascenone	0.576	0.023
	13	1	Valeraldehyde	1.000	0.020

Table 3. Cont.

Sample	Cluster	No. of Members [B]	Most Representative Variable [C]	Cluster Proportion of Variation Explained [D]	Total Proportion of Variation Explained [E]
SD St. Croix (0.692) [A]	2	8	Acetophenone	0.685	0.081
	3	6	Linalool	0.786	0.069
	6	6	Benzaldehyde	0.727	0.064
	7	6	Methyl salicylate	0.681	0.060
	5	6	Cyclohexanol	0.662	0.058
	4	6	2-Heptanone	0.627	0.055
	1	5	2-Phenylethanol	0.728	0.054
	10	5	1-Pentanol	0.581	0.043
	8	5	Benzyl alcohol	0.557	0.041
	11	3	Safrol	0.855	0.038
	9	3	Benzoic acid, methyl ester	0.797	0.035
	12	3	Ethyl acetate	0.622	0.027
	14	2	Aspirin methyl ester	0.840	0.025
	13	2	Propionaldehyde	0.797	0.023
	15	2	N-Benzyl-2-phenethylamine	0.628	0.018
IA La Crescent (0.699) [A]	1	11	beta-Cyclocitral	0.682	0.121
	2	8	beta-Pinene	0.836	0.108
	3	9	Ethyl butyrate	0.710	0.103
	8	4	p-Cymene	0.663	0.043
	9	4	Propanoic acid	0.648	0.042
	12	3	1-Hexanol	0.792	0.038
	6	3	Nerol acetate	0.776	0.038
	4	4	Methacrolein	0.565	0.036
	5	4	Beta-damascenone	0.521	0.034
	7	3	(+)-4-Carene	0.650	0.031
	11	3	Valeric acid	0.641	0.031
	13	3	3-Methyl-1-butanol	0.638	0.031
	10	2	Acetic acid	0.845	0.027
	14	1	Propyl-benzene	1.000	0.016
SD La Crescent (0.741) [A]	3	8	Allyl alcohol	0.845	0.086
	2	8	beta-Pinene	0.837	0.085
	1	7	Toluene	0.691	0.061
	11	5	Isoamyl acetate	0.915	0.058
	6	6	Isophorone	0.637	0.048
	7	6	Ethyl butyrate	0.567	0.043
	8	5	Hexanal	0.669	0.042
	4	5	Benzaldehyde	0.657	0.042
	13	5	Styrene	0.618	0.039
	15	4	Carbon disulfide	0.771	0.039
	9	3	Ethyl vinyl ketone	1.000	0.038
	5	3	Camphene	0.900	0.034
	18	3	Linalyl acetate	0.806	0.031
	17	3	Geraniol	0.730	0.028
	10	2	Furfural	0.908	0.023
	12	3	Isobutyraldehyde	0.499	0.019
	16	2	2-Ethyl-1-hexanol	0.500	0.013
	14	1	Propyl-benzene	1.000	0.013

[A] The total proportion of variation explained by all the cluster components. [B] The number of variables in the cluster. [C] The cluster variable that has the largest squared correlation with its cluster component. [D] The cluster's proportion of variance explained by the first principal component amount the variables in the cluster, based only on variables within the cluster. [E] The overall proportion of variance explained by the cluster component, using only the variables within each cluster to calculate the first principal component.

2.3.1. Frontenac

VOCs detected after crushing the berries of Frontenac grapes from the 2013 growing season in Iowa were isovaleraldehyde and isoamyl acetate at veraison. At harvest, VOCs were cyclohexanol and) and 3-methyl-1-butanol, as shown in Figure A1. VOCs detected after crushing berries of Frontenac grapes from the 2012 growing season in South Dakota were acetaldehyde and 1-hexanol at veraison in 2012. Frontenac grapes from the 2013 growing season in South Dakota was associated with ethyl hexanoate. VOCs detected after crushing berries of Frontenac grapes from the 2012 growing season at harvest in South Dakota were associated with alkane and styrene. In 2013 at harvest, however, compounds emitted were acetone, ethyl palmitate, hexanoic acid, 2-methyl-1-propanol, and 2-octanone in Frontenac grapes in South Dakota, Figure 9.

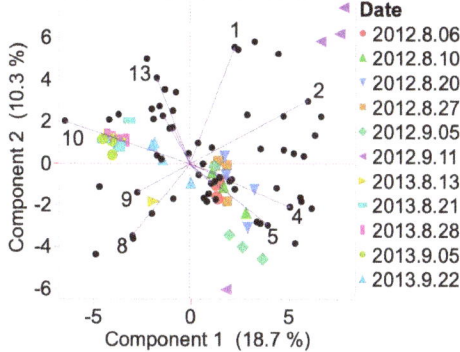

Figure 9. Evolution of VOCs emitted from crushed berries from veraison to harvest of 2012 and 2013 Frontenac grapes grown in South Dakota. Key to most representative volatiles from HCA, shown as vectors from the origin and read clockwise: 1 = Nonane, 2 = Styrene, 3 = Octanal, 4 = Acetaldehyde, 5 = 1-Hexanol, 6 = Benzoic acid, methyl ester, 7 = Isophorone, 8 = Ethyl hexanoate, 9 = N-benzyl-2-phenethylamine, 10 = Acetone, 11 = Ethyl palmitate, 12 = Hexanoic acid, 13 = 2-Methyl-1-propanol, 14 = 2-Octanone.

2.3.2. Marquette

Compounds emitted from Marquette grapes from the 2013 growing season in Iowa were formic acid, octyl ester at veraison. By harvest, 2013 Marquette grapes emitted benzophenone, hexanal, and isoamyl acetate, Figure A1. In the 2012 South Dakota growing season, compounds such as cyclohexanol and (E)-2-hexenoic acid were most associated with Marquette berries at veraison. By harvest, these compounds shifted to styrene, beta-cyclocitral, and nonanal (Figure 10).

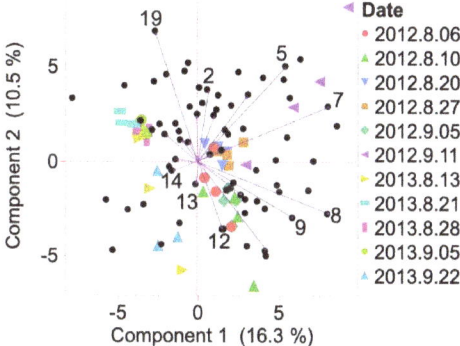

Figure 10. Evolution of VOCs emitted from crushed berries from veraison to harvest of 2012 and 2013 Marquette grapes grown in South Dakota. Key to most representative volatiles from HCA, shown as vectors from the origin and read clockwise: 1 = 1-Heptanol, 2 = Amyl acetate, 3 = Methyl ethyl ketone, 4 = Decane, 5 = Styrene, 6 = beta-Cyclocitral, 7 = Nonanal, 8 = Acetaldehyde, 9 = Valeraldehyde, 10 = Octanal, 11 = Cyclohexanol, 12 = (E)-2-hexenoic acid, 13 = Methyl disulfide, 14 = Nonane, 15 = Allyl alcohol, 16 = beta-Damascenone, 17 = Nerol acetate, 18 = p-cymene, 19 = 1-Pentanol.

2.3.3. St. Croix

VOCs from crushed St. Croix grapes from the 2013 Iowa growing season changed from benzene acetaldehyde, isobutyraldehyde, ethyl butyrate, 1-Hexanol, beta-Damascenone, valeraldehyde, ethyl decanoate, methacrolein, 1-butanol, aspirin methyl ester at veraison to formic acid, and octyl ester and isoamyl acetate at harvest (Figure A1). VOCs from crushed St. Croix grapes from the 2013 growing season in South Dakota changed from benzyl alcohol, benzaldehyde, and N-benzyl-2-phenethylamine at veraison to ethyl acetate, methyl salicylate, safrol, propionaldehyde, 2-phenylethanol, 1-Pentanol, 2-heptanone, benzoic acid, and methyl ester at harvest (Figure 11).

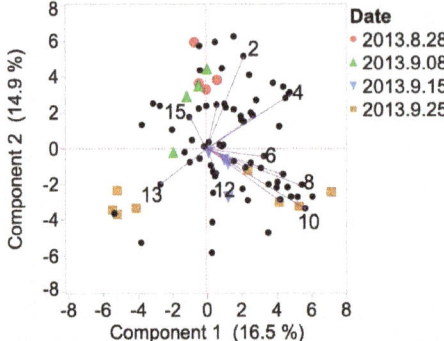

Figure 11. Evolution of VOCs emitted from crushed berries from veraison to harvest of 2013 St. Croix grapes grown in South Dakota. Key to most representative volatiles from HCA, shown as vectors from the origin and read clockwise: 1 = Benzyl alcohol, 2 = Benzaldehyde, 3 = Octanal, 4 = Acetophenone, 5 = Linalool, 6 = Ethyl acetate, 7 = Methyl salicylate, 8 = Safrol, 9 = Propionaldehyde, 10 = 2-Phenylethanol, 11 = 1-Pentanol, 12 = 2-Heptanone, 13 = Benzoic acid, methyl ester, 14 = Aspirin methyl ester, 15 = N-benzyl-2-phenethylamine.

2.3.4. La Crescent

VOCs from La Crescent berries from the 2013 Iowa growing season changed from propanoic acid, ethyl butyrate, 3-methyl-1-butanol, beta-cyclocitral at veraison to p-cymene, beta-damascenone, 1-hexanol, and beta-pinene at harvest (Figure A1). In the 2013 South Dakota growing season, La Crescent VOCs from crushed berries changed from isoamyl acetate, linalyl acetate, 2-ethyl-1-hexanol, geraniol, isophorone, and allyl alcohol at veraison to ethyl butyrate, propyl-benzene, and styrene at harvest (Figure 12).

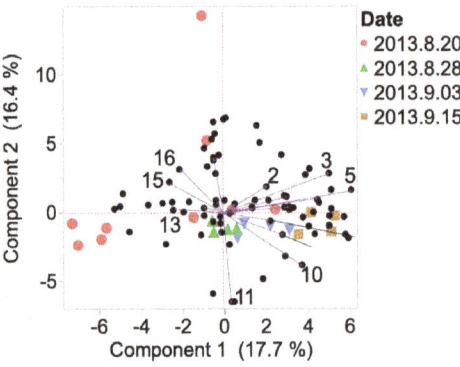

Figure 12. Evolution of VOCs emitted from crushed berries from veraison to harvest of 2012 and 2013 La Crescent grapes grown in South Dakota. Key to most representative volatiles from HCA, shown as vectors from the origin and read clockwise: 1 = Furfural, 2 = Isobutyraldehyde, 3 = Benzaldehyde, 4 = Ethyl vinyl ketone, 5 = Toluene, 6 = Camphene, 7 = Ethyl butyrate, 8 = Propyl-Benzene, 9 = Styrene, 10 = Hexanal, 11 = beta-Pinene, 12 = Isoamyl acetate, 13 = Carbon disulfide, 14 = Linalyl acetate, 15 = 2-Ethyl-1-hexanol, 16 = Geraniol, 17 = Isophorone, 18 = Allyl alcohol.

3. Discussion

The effect of vineyard practices on grape and wine aroma merit study. The claims to a regions' wines by sensory attributes need to be scientifically correlated to bolster the local economies. Otherwise consumers are inundated with marketing claims. This study attempted to compare microclimates of Iowa and South Dakota during 3 months of the growing season, over 2 years. The Iowa plot is in USDA plant hardiness zone 5a [39]. In comparison, the South Dakota plot is in USDA plant hardiness zone 4b [39]. These metrics were obtained from the National Oceanic and Atmospheric Administration, formerly the National Climatic Data Center [40].

Preliminary analyses using ANOVA were also completed (shown in 'ANOVA in-vivo' spreadsheet, Supplementary Materials). Type 1 sum of squares analysis indicated statistical significance of method, cultivar, date and time, method and cultivar interaction, and method and site and cultivar interactions (shown in Table A1). Post-hoc Tukey HSD test is (Table A2) shows differences within a method, cultivar, date and time, method and cultivar interaction, and method and site and cultivar interactions. PCA analyses were used to determine key volatile compounds emitted. Although significant differences were noted using ANOVA and Tukey HSD for total VOCs emitted, the more detailed analysis with PCA (focused on individual VOCs) accounted for less than 70% variance. This was the case for all but one case (Figure 5) of Frontenac grown in South Dakota in 2013. Statistical analysis using PCA indicated that Frontenac and Marquette were most similar in total VOC emission profile (i.e., clustered around the origin). St. Croix cultivars had a higher positive correlation with the first principal component. Seventeen VOCs with correlation ≥ 0.300 are listed in a section on statistical analysis. Any differences in the soil and microclimate of these two sites affected overall VOCs emitted from La Crescent and St. Croix cultivars during this research. It is cautioned that these differences could also be affected by the genetics of the cold-hardy hybrids. Frontenac and Marquette share similar parentage [41].

The advantages of the modified glass vials over PVF film chambers are its compact design for field sampling, reusability, reduced background contamination from glass, and isolation of VOCs emitted from a single berry. Vacuum-assisted headspace SPME sampling has been used in carefully controlled laboratory settings, to successfully achieve shorter sampling times at lower sampling temperature with good sensitivity and precision to extract polychlorinated biphenyls (PCB) from water [42]. This novel sampling device was the logical next step to isolate VOCs emitted from grape berries during development. This sampling technique is comparable to a viticulturalist 'smelling' a grape, and detecting only the volatile compounds emitted through the grape skin. These VOCs are recognized as "free" aroma compounds not bound to a sugar moiety within the berry [43]. This could allow for monitoring of VOCs to measure berry ripeness by instrumental methods. Grapes get softer as they develop, and some cultivars are prone to slip skin (i.e., the grape skin slips easily from the fruit pulp). A disadvantage in using modified glass vials for grape sampling in this research is that increased vacuum was needed as the grapes developed and softened, and sometimes broke the grape skin, more often in the St. Croix cultivar. Another confounding element could be the presence and interference of volatile compounds on the grape skin but not produced by the grape (i.e., pesticide residues, naturally occurring yeasts and molds).

Non-destructive, sampling of VOCs emitted in-vivo from cold-hardy grapes was conducted using 2 methods. PVF film sampling chambers with custom SPME sampling port was used to monitor whole cluster VOC emissions. Modified glass vials supported SPME sampling of individual berries. For comparison to both non-destructive methods, a random 5 berry sample was collected, crushed, and analyzed under controlled laboratory conditions. Statistical analysis using PCA indicated that sampling by PVF chambers and modified glass vials detected similar VOC emission profiles across all 4 cultivars. There was 1 outlier from the glass vial method, indicating a higher than average concentration of styrene in La Crescent grapes. This data could provide evidence of styrene as a product of 2-phenylethanol synthesis from yeast cells [44] (p. 309) present during sampling. It should be noted that 2-phenylethanol variable is positively correlated with principal component 2, orthogonal to styrene. It is expected to have more VOCs detected at higher relative concentrations in crushed berry analysis because of the release of juices and volatiles bound within the berry skin and pulp. Research is warranted to compare headspace SPME analyses of crushed berries with conventional analytical methods such as liquid-liquid extraction [45].

Berry VOCs were sampled within a 3-month growing period each year for 2 years. VOC profile in 2012, sampled by PVF chambers and crushed berries were similar in profile. Data points from the PCA fall close to the origin throughout the year in 2012, not shown. This indicated that VOCs collected via PVF chambers did not show noticeable changes from veraison to harvest. VOC profile in 2013, sampled by modified glass vials and crushed berries were similar in profile (points near the origin) until the end of August, with the exception of the outlier on 14 August 2013, Figure 3. In La Crescent and St. Croix cultivars grown in Iowa, there is a movement towards higher than average VOC emission on 24 August 2016, Figure 4. In Iowa, VOC development was still trending above average at harvest on 3 August 2016 for St. Croix and on 29 September 2016 for La Crescent, not shown. VOCs emitted from Frontenac, Marquette, and St Croix cultivars grown in South Dakota started to develop and deviate from average later than Iowa on 29 August 2016, Figure 5. VOCs emitted from La Crescent grown in South Dakota start to trend above average on 3 September 2016, Figure 6. VOC emissions returned to average levels between 29 August and 5 September (harvest) in South Dakota Frontenac berries, Figure 7. The same decreasing trend was observed in South Dakota Marquette between 5 September and 8 September 2016, not shown. Similar to Iowa, the increased VOCs emitted from South Dakota St. Croix and La Crescent do not decline by harvest, not shown.

Differences in microclimate of Iowa and South Dakota plots did not affect VOC emissions from 4 cold-hardy grape cultivars. Little difference in VOC emissions is expected from Marquette and Frontenac because of a shared pedigree. Greater changes in VOC emissions was observed between destructive crushed berry analysis and non-destructive in-vivo analysis methods, but not within the non-destructive methods. In Iowa and South Dakota plots, VOCs emitted from St. Croix and La Crescent cultivars continued to change from veraison through harvest. VOCs emitted in-vivo from Frontenac and Marquette cultivars in South Dakota started to decline 8 days and 3 days before harvest, respectively. More research is warranted in order to make recommendations to viticulturists regarding ideal harvest time for maximum aromas in the cold-hardy grapes. Linking correlations between viticultural practices can enhance the quality of wines for new cold-climate cultivars.

Several improvements to the proposed in-vivo sampling are warranted. Addition of internal standard (IS) [46], for example a small vial with a membrane for controlled emission of IS during sampling (e.g., inside a PVF bag) would to ensure that sampling temperature and SPME fiber variables are controlled. This information would help to normalize sampling variables in field conditions and potentially help with data quality. Secondly, IS addition would enable quantification of volatiles.

4. Materials and Methods

4.1. Overview

A detailed description of Materials and Methods is provided elsewhere [36]. Briefly, below are the summaries of particular approaches used. Research vineyards were located at South Dakota State University (SDSU, Brookings, SD, USA) and Iowa State University (ISU, Ames, IA, USA). Grape clusters were randomly selected, and volatiles from the same clusters were sampled from veraison to harvest. Veraison is defined as when half of the clusters have changed to their ripe color and is shown as the first time point in Results. Collection of volatiles from whole clusters and single berries was completed in 2012 and 2013 seasons, respectively. Berry chemistry data (i.e., Brix, pH, ambient temperatures, and titratable acidity (IA only)) is provided in Supplementary Material. Volatiles from crushed berries were collected at the same time as in-vivo sampling for both growing seasons. A SPME (65 μm polydimethylsiloxane (PDMS)/divinylbenzene (DVB)) fibers were used for on-site sampling at vineyards and for headspace extraction from crushed berries. No internal standard was used. However, trip blanks (i.e., ambient air samples collected at each vineyard) and sampling vial blanks (for destructive sampling) were used to account for potential interfering volatiles. Four replicates (vines) were sampled per site and cultivar at each time point.

4.2. In-Vivo Sampling of Volatiles from a Whole Cluster of Grapes

Sampling chambers (~5 L volume) for the non-destructive collection of in-vivo volatiles were made from a PVF film and held firm with clean aluminum wire cage framing. Preconditioned (cleaned) PVF chambers were fitted with custom sampling ports for insertion of SPME needles. Typical sampling time was 30 min.

4.3. In-Vivo Sampling of Volatiles from a Single Grape

A standard 2 mL glass vials were modified by removing flat bottoms (Fisher Scientific, Waltham, MA, USA) at a glass shop. The edges of were flared and rounded. A half hole septa was added to the screw top to support the SPME needle. The SPME fiber was placed through the septa prior to sampling. After assembly and placement of the vial apparatus on the individual berry (Figure S2), 5 mL of air was pulled from the vial using a syringe. Care was taken not to disturb the SPME fiber with the syringe needle. The resultant vacuum held the apparatus in place (i.e., sealed by suction into berry surface) while the SPME fiber was exposed for vacuum-assisted VOC sampling. The single berry sampling vials were cleaned prior to each sampling by rinsing in deionized water and oven baked overnight at 107 °C. Cleaned vials were transported in an aluminum lined box. PTFE screw tops were replaced after each sampling.

4.4. Destructive Sampling

Berries were collected from each cultivar on the same day and time of in-vivo sampling. Five berries were collected from clusters adjacent to the cluster tagged for in-vivo sampling (i.e., from the same vine but a different cluster than in-vivo sampled berries). Collected berries were frozen prior to analysis and stored in a −20 °C freezer. Berries collected in South Dakota were also frozen and shipped on ice overnight for analysis in Iowa. Frozen berries were hand-crushed in the lab, placed into 20 mL amber screw top vials (Wheaton, Millville, NJ) with PTFE/silicone septa. A CTC CombiPal (LEAP Technologies, Carrboro, NC, USA) was used for automated SPME sampling. Briefly, the vials were agitated and heated to 50 °C for 10 min, followed by 30 min agitated headspace sampling using 65 µm PDMS/DVB SPME fiber. The fiber was thermally desorbed under a flow of helium prior to each sample exposure. These sampling parameters were determined, not shown.

4.5. Data Acquisition and Analysis

A custom multidimensional GC was used (Microanalytics, a part of Volatile Analysis Corporation, Round Rock, TX, USA), built on a standard Agilent 6890 platform (Agilent Technologies, Santa Clara, CA, USA). System automation and data acquisition software were MultiTrax (Microanalytics, Round Rock, TX, USA) and ChemStation (Agilent Technologies, Santa Clara, CA, USA). Chromatography was performed on two capillary columns connected in series. The first column was 5% phenyl polysilphenylene-siloxane (30 m × 0.53 mm inner diameter × 0.5 µm thickness, Trajan Scientific, Austin, TX, USA) with a fixed restrictor pre-column. The second polar column was bonded polyethylene glycol in a Sol-Gel matrix (30 m × 0.53 mm inner diameter × 0.5 µm thickness, Trajan Scientific, Austin, TX, USA). The midpoint between the two columns was maintained at a constant pressure of 0.39 atm by a pneumatic switch. In this research, all effluent from the first column was directed into the 2nd analytical column, i.e., no heartcutting was performed. The instrument was also equipped with a flame ionization detector (FID). Flow to the FID can also directed at the midpoint, but FID was not utilized in this research. True multidimensional analyses were not performed, i.e., the system was used in full heartcut mode, meaning separation was performed on both columns in series. Effluent from the second polar column was simultaneously directed to a single quadrupole MS (Model 5973N, Agilent Technologies, Santa Clara, CA, USA) and an olfactometry (sniff) port (Microanalytics, Round Rock, TX,

USA) via an open spit interface at atmospheric pressure. The sniff port is equipped with a purge flow controller and supplied with humidified air at 0.54 atm. Flow to the MS and sniff port is determined by fixed restrictor columns, 1 part to MS and 3 to sniff port. Olfactometry was not utilized in the research. The GC inlet was operated in splitless mode at 250 °C. GC oven parameters start with an initial temperature of 40 °C, held for 3.0 min, followed by a 7 °C per min ramp to 240 °C, held for 8.43 min. Total run time was 40 min. Carrier gas is ultra-high purity (UHP) helium (99.999%, Airgas, Des Moines, IA, USA). Temperature of the sniff port and MS transfer lines were 240 °C and 280 °C, respectively. MS full scan range was set from 34 m/z to 350 m/z. Scans were collected in electron ionization (EI) mode with an ionization energy of 70 eV. MS heated zones for quadrupole and source were 150 °C and 230 °C, respectively. Daily tuning of the MS was performed with perfluorotributylamine (PFTBA) before each analysis.

Identification of compounds was performed using Automated Mass Spectral Deconvolution and Identification System (AMDIS) target library search with at least 80% mass spectral match. Target libraries included (a) the 6 libraries that are included with the AMDIS program, (b) an onsite library created from analysis of pure standards (200+ compounds), (c) NIST11 mass spectral library. Analysis of variance (ANOVA) was performed using XLSTAT 2016.04.33113 (Addinsoft, New York, NY, USA). The effects of cultivar, site, sampling time, and sampling methods and their interactions on volatiles emitted were analyzed using ANOVA (with confidence interval of 95% and the tolerance of 0.0001) followed by post-hoc (Tukey honestly significant difference, HSD) test. Multivariate analysis was performed using JMP Pro 12.0.1 (SAS Institute Inc., Cary, NC, USA).

5. Conclusions

We have shown that is feasible to detect VOCs emitted in-vivo from single grape berries (39 compounds) and whole clusters (44 compounds). Over 110 VOCs were released to headspace from crushed berries. Spatial (vineyard location) and temporal variations in VOC profiles were observed for all four cultivars. However, these changes were not consistent by growing season, by location, within cultivars, by ripening stage when analyzed by multivariate analyses such principal component analysis (PCA) and hierarchical cluster analyses (HCA). Research into aroma compounds present in cold-hardy cultivars is essential to the continued growth of the wine industry in cold climates and diversification of agriculture in the upper Midwestern area of the U.S.

Supplementary Materials: The full list of biogenic volatiles emitted from four cold-hardy grape cultivars during ripening is available online at [36]. In addition, analysis of variance (ANOVA) is provided in 'ANOVA in vivo' spreadsheet. Berry chemistry data is provided in 'Berry Chemistry 2012 and 2013 data' spreadsheet.

Author Contributions: Conceptualization, J.A.K., S.R. and M.D.; methodology, S.R.; validation, S.R., J.A.K.; formal analysis, S.R.; investigation, S.R., D.L.M.; resources, D.L.M., S.R., A.F., J.A.K., and M.D.; data curation, S.R. and J.A.K.; writing—original draft preparation, S.R.; writing—review and editing, S.R., A.F., J.K., and M.D.; visualization, S.R.; supervision, J.A.K. and M.D.; project administration, J.A.K. and M.D.; funding acquisition, M.D. and J.A.K.

Funding: This research was funded by the United States Department of Agriculture's Special Crops Research Initiative Program of the National Institute for Food and Agriculture, [grant number 2011-51181-30850], titled "Northern grapes: integrating viticulture, winemaking, and marketing of new cold-hardy cultivars supporting new and growing rural wineries." In addition, this project was partially supported by the Iowa Agriculture and Home Economics Experiment Station, Ames, Iowa. [Project no. IOW05556] (Future Challenges in Animal Production Systems: Seeking Solutions through Focused Facilitation) is sponsored by Hatch Act and State of Iowa funds and USDA National Institute of Food and Agriculture, Hatch project [Project no. SD00H449-12] and SD Agriculture Experiment Station.

Acknowledgments: The authors are thankful to Jason Vallone for his support with sample collection and laboratory analyses.

Conflicts of Interest: The authors declare no conflict of interest. The funders had no role in the design of the study; in the collection, analyses, or interpretation of data; in the writing of the manuscript, or in the decision to publish the results.

Appendix A

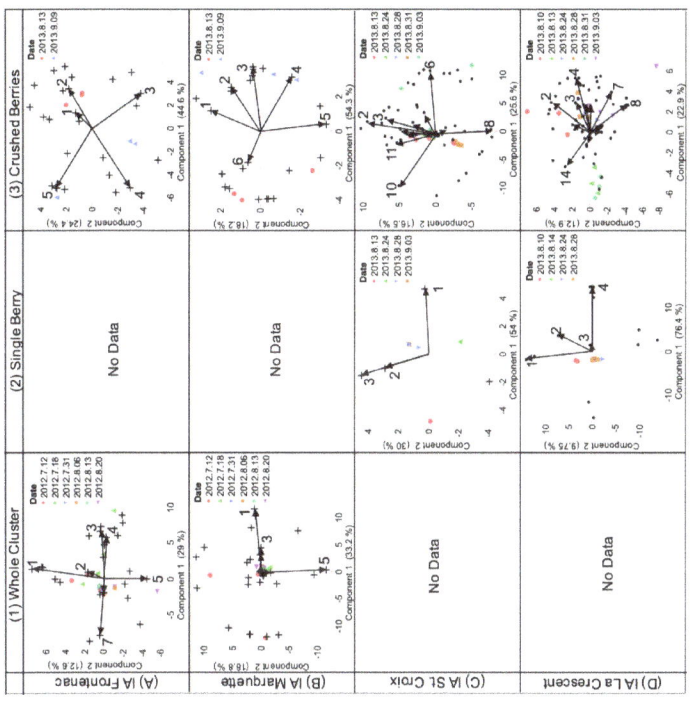

Figure A1. Results of a PCA on VOCs emitted from Iowa cold-hardy grapes by sampling methods (1) whole cluster, (2) single berry, (3) crushed berries, and cultivars (**A**) Frontenac, (**B**) Marquette, (**C**) St. Croix, (**D**) La Crescent. These plots show the relationships of grape maturity during ripening (Date) to each other and the associations among the most representative variable from cluster analysis. Key: 1A: 1 = 3-Methyl-1-butanol, 2 = 5-(Hydroxymethyl)-2-furancarboxaldehyde, 3 = Nonanal, 4 = 4-Methyl-3-penten-2-one, 5 = 1,4-Butanolide, 6 = Benzophenone, 7 = Heptanal; 1B: 1 = 1-Octanol, 5 = Acetaldehyde, 3 = Methyl ethyl ketone, 2 = 1-Hexadecanol,

6 = Acetophenone, 4 = Acetic acid, 7 = 2-Ethyl-1-hexanol; 2C: 1 = 3-Methyl indole, 2 = Benzyl alcohol, 3 = Acetic Acid; 2D: 1 = 3-methyl indole, 2 = Ethanol, 3 = 2-Phenylethanol, 4 = 2-ethyl-1-hexanol; 3A: 1 = Isovaleraldehyde, 2 = Isoamyl acetate, 3 = Toluene, 4 = Cyclohexanol, 5 = 3-Methyl-1-butanol; 3B: 1 = Benzophenone, 2 = Ethanol, 3 = Hexanal, 4 = Isoamyl acetate, 5 = Styrene, 6 = Allyl alcohol; 3C: 1 = Benzene acetaldehyde, 2 = Isobutyraldehyde, 3 = Ethanol, 4 = Ethyl butyrate, 5 = 1-Hexanol, 6 = Formic acid, octyl ester, 7 = beta-Damascenone, 8 = Isoamyl acetate, 9 = Valeraldehyde, 10 = Ethyl decanoate, 11 = Methacrolein, 12 = 1-Butanol, 13 = Aspirin methyl ester; 3D: 1 = Propanoic acid, 2 = Ethyl butyrate, 3 = 3-Methyl-1-butanol, 4 = beta-Cyclocitral, 5 = p-Cymene, 6 = beta-Damascenone, 7 = 1-Hexanol, 8 = beta-Pinene, 9 = Nerol acetate, 10 = Propyl-benzene, 11 = Acetic acid, 12 = (+)-4-Carene, 13 = Valeric acid, 14 = Methacrolein.

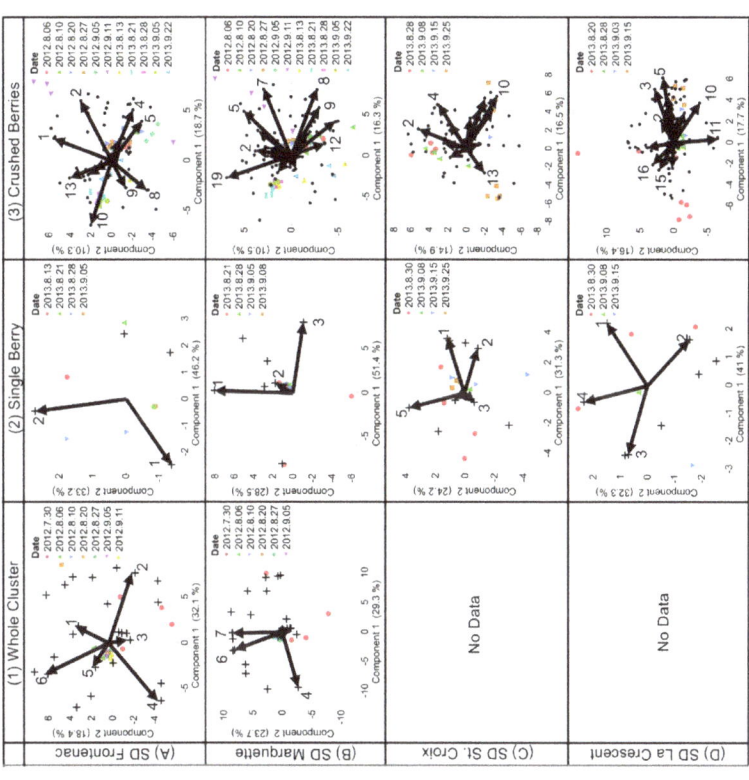

Figure A2. Results of a PCA on VOCs emitted from South Dakota cold-hardy grapes by sampling methods (1) whole cluster, (2) single berry, (3) crushed berries; and cultivars (**A**) Frontenac, (**B**) Marquette, (**C**) St. Croix, (**D**) La Crescent. These plots show the relationships of grape maturity during ripening (Date) to each other and the associations among the most representative variable from cluster analysis. Key: 1A: 1 = 2-Methyl-3-penten-2-one, 2 = Nonanal, 3 = Benzyl alcohol, 4 = Toluene,

5 = Acetic acid, 6 = 3-Phenyl-2-propenal; 1B: 1 = 1-Hexadecanol, 2 = 2-Ethyl-1-Hexanol, 3 = 1-Pentanol, 4 = Acetone, 5 = Indene, 6 = Decane, 7 = 4-methyl-3-penten-2-one; 2A: 1 = Palmitic acid, 2 = Acetic acid; 2B: 1 = Ethyl octanoate, 2 = 2-Ethyl-1-hexanol, 3 = 1,4-Butanolide; 2C: 1 = Nonanal, 2 = Diacetone alcohol, 3 = 1,4-Butanolide, 4 = 5-(Hydroxymethyl)-1-furancarboxaldehyde, 5 = Ethyl acetate; 2D: 1 = 2-Phenylethanol, 3 = Diacetone alcohol, 4 = Acetic acid, 2 = 6-Methyl-5-hepten-2-one; 3A: 1 = Nonane, 2 = Styrene, 3 = Octanal, 4 = Acetaldehyde, 5 = 1-Hexanol, 6 = Benzoic acid, methyl ester, 7 = Isophorone, 8 = Ethyl hexanoate, 9 = N-benzyl-2-phenethylamine, 10 = Acetone, 11 = Ethyl palmitate, 12 = Hexanoic acid, 13 = 2-Methyl-1-propanol, 14 = 2-Octanone; 3B: 1 = 1-Heptanol, 2 = Amyl acetate, 3 = Methyl ethyl ketone, 4 = Decane, 5 = Styrene, 6 = beta-Cyclocitral, 7 = Nonanal, 8 = Acetaldehyde, 9 = Valeraldehyde, 10 = Octanal, 11 = Cyclohexanol, 12 = (E)-2-hexenoic acid, 13 = Methyl disulfide, 14 = Nonane, 15 = Allyl alcohol, 16 = beta-Damascenone, 17 = Nerol acetate, 18 = p-Cymene, 19 = 1-Pentanol; 3C: 1 = Benzyl alcohol, 2 = Benzaldehyde, 3 = Octanal, 4 = Acetophenone, 5 = Linalool, 6 = Ethyl acetate, 7 = Methyl salicylate, 8 = Safrol, 9 = Propionaldehyde, 10 = 2-Phenylethanol, 11 = 1-Pentanol, 12 = 2-Heptanone, 13 = Benzoic acid, methyl ester, 14 = Aspirin methyl ester, 15 = N-benzyl-2-phenethylamine; 3D: 1 = Furfural, 2 = Isobutyraldehyde, 3 = Benzaldehyde, 4 = Ethyl vinyl ketone, 5 = Toluene, 6 = Camphene, 7 = Ethyl butyrate, 8 = Propyl-benzene, 9 = Styrene, 10 = Hexanal, 11 = beta-Pinene, 12 = Isoamyl acetate, 13 = Carbon disulfide, 14 = Linalyl acetate, 15 = 2-Ethyl-1-hexanol, 16 = Geraniol, 17 = Isophorone, 18 = Allyl alcohol.

Table A1. Type I sum of squares analyses.

Source	DF	Sum of Squares	Mean Squares	F	Pr > F
Method	3	1.88×10^{16}	6.26×10^{15}	122.794	<0.0001
Site	1	6.92×10^{10}	6.92×10^{10}	0.001	0.971
Cultivar	3	1.91×10^{16}	6.37×10^{15}	124.900	<0.0001
Date/Time	31	1.14×10^{16}	3.67×10^{14}	7.204	<0.0001
Method*Site	2	6.54×10^{14}	3.27×10^{14}	6.416	0.002
Method*Cultivar	4	6.43×10^{15}	1.61×10^{15}	31.532	<0.0001
Method*Date/Time	14	1.45×10^{15}	1.04×10^{14}	2.035	0.016
Site*Cultivar	3	7.36×10^{14}	2.45×10^{14}	4.815	0.003
Site*Date/Time	2	2.24×10^{14}	1.12×10^{14}	2.202	0.113
Cultivar*Date/Time	24	2.37×10^{15}	9.87×10^{13}	1.937	0.007
Method*Site*Cultivar	1	8.96×10^{14}	8.96×10^{14}	17.588	<0.0001
Method*Site*Date/Time	0	0.00×10^{00}			
Method*Cultivar*Date/Time	7	6.53×10^{13}	9.33×10^{12}	0.183	0.989

Table A2. Results of Tukey HSD test.

Category	Groups
Method-GV*Cultivar-La Crescent*Date/Time-9/8/2013	A
Method-GV*Cultivar-La Crescent*Date/Time-9/15/2013	A
Method-GV*Cultivar-La Crescent*Date/Time-8/30/2013	A B
Method-GV*Cultivar-St. Croix*Date/Time-9/15/2013	A B C
Method-GV*Cultivar-St. Croix*Date/Time-9/8/2013	A B C
Method-GV*Cultivar-St. Croix*Date/Time-8/30/2013	A B C D
Method-GV*Cultivar-St. Croix*Date/Time-9/25/2013	A B C D E
Method-CB*Cultivar-Frontenac*Date/Time-8/13/2013	A B C D E F
Method-CB*Cultivar-Marquette*Date/Time-8/13/2013	A B C D E F
Method-CB*Cultivar-La Crescent*Date/Time-8/13/2013	A B C D E F
Method-CB*Cultivar-Frontenac*Date/Time-9/9/2013	A B C D E F
Method-GV*Cultivar-St. Croix*Date/Time-8/29/2013	A B C D E F
Method-PVF*Cultivar-Frontenac*Date/Time-8/10/2012	A B C D E F
Method-PVF*Cultivar-Frontenac*Date/Time-8/27/2012	A B C D E F
Method-PVF*Cultivar-Marquette*Date/Time-8/13/2012	A B C D E F
Method-*PVF*Cultivar-Marquette*Date/Time-7/31/2012	A B C D E F
Method-*PVF*Cultivar-Marquette*Date/Time-8/06/2012	A B C D E F
Method-PVF*Cultivar-Marquette*Date/Time-9/5/2012	A B C D E F
Method-PVF*Cultivar-Frontenac*Date/Time-9/5/2012	A B C D E F
Method-PVF*Cultivar-Marquette*Date/Time-9/11/2012	A B C D E F
Method-PVF*Cultivar-Frontenac*Date/Time-8/6/2012	A B C D E F
Method-PVF*Cultivar-Marquette*Date/Time-8/6/2012	A B C D E F
Method-*PVF*Cultivar-Marquette*Date/Time-8/6/2012	A B C D E F
Method-PVF*Cultivar-Frontenac*Date/Time-8/13/2012	A B C D E F
Method-PVF*Cultivar-Marquette*Date/Time-8/10/2012	A B C D E F
Method-PVF*Cultivar-Marquette*Date/Time-8/20/2012	A B C D E F
Method-PVF*Cultivar-Marquette*Date/Time-8/27/2012	A B C D E F
Method-PVF*Cultivar-Frontenac*Date/Time-7/31/2012	A B C D E F
Method-CB*Cultivar-Marquette*Date/Time-9/9/2013	A B C D E F
Method-*PVF*Cultivar-Marquette*Date/Time-8/6/2012	A B C D E F
Method-PVF*Cultivar-Frontenac*Date/Time-8/20/2012	A B C D E F
Method-PVF*Cultivar-Marquette*Date/Time-7/18/2012	A B C D E F
Method-PVF*Cultivar-Frontenac*Date/Time-7/12/2012	A B C D E F
Method-PVF*Cultivar-Marquette*Date/Time-7/31/2012	A B C D E F
Method-PVF*Cultivar-Marquette*Date/Time-7/30/2012	A B C D E F
Method-PVF*Cultivar-Frontenac*Date/Time-7/18/2012	A B C D E F
Method-PVF*Cultivar-Frontenac*Date/Time-7/30/2012	A B C D E F
Method-CB*Cultivar-La Crescent*Date/Time-8/31/2013	A B C D E F
Method-PVF*Cultivar-Marquette*Date/Time-7/12/2012	A B C D E F
Method-GV*Cultivar-La Crescent*Date/Time-8/29/2013	A B C D E F
Method-CB*Cultivar-Frontenac*Date/Time-8/21/2013	A B C D E F
Method-CB*Cultivar-Frontenac*Date/Time-9/22/2013	A B C D E F
Method-CB*Cultivar-Frontenac*Date/Time-8/28/2013	A B C D E F
Method-CB*Cultivar-Frontenac*Date/Time-9/5/2013	A B C D E F
Method-CB*Cultivar-Frontenac*Date/Time-9/13/2013	A B C D E F
Method-CB*Cultivar-Marquette*Date/Time-8/21/2013	A B C D E F
Method-GV*Cultivar-Frontenac*Date/Time-9/5/2013	A B C D E F G
Method-GV*Cultivar-Frontenac*Date/Time-8/21/2013	A B C D E F G
Method-GV*Cultivar-Frontenac*Date/Time-9/13/2013	A B C D E F G
Method-GV*Cultivar-Marquette*Date/Time-9/13/2013	A B C D E F G
Method-GV*Cultivar-Marquette*Date/Time-9/8/2013	A B C D E F G
Method-GV*Cultivar-Marquette*Date/Time-9/5/2013	A B C D E F G

Table A2. Cont.

Category	Groups
Method-GV*Cultivar-Marquette*Date/Time-8/21/2013	A B C D E F G
Method-CB*Cultivar-Frontenac*Date/Time-9/5/2012	A B C D E F G
Method-CB*Cultivar-Frontenac*Date/Time-8/6/2012	A B C D E F G H
Method-CB*Cultivar-Marquette*Date/Time-9/22/2013	A B C D E F G H
Method-CB*Cultivar-Frontenac*Date/Time-8/27/2012	A B C D E F G H
Method-CB*Cultivar-Marquette*Date/Time-9/5/2013	A B C D E F G H
Method-GV*Cultivar-St. Croix*Date/Time-9/3/2013	A B C D E F G H
Method-GV*Cultivar-St. Croix*Date/Time-8/24/2013	A B C D E F G H
Method-CB*Cultivar-Marquette*Date/Time-8/28/2013	A B C D E F G H
Method-CB*Cultivar-Frontenac*Date/Time-8/10/2012	A B C D E F G H I
Method-CB*Cultivar-Marquette*Date/Time-9/13/2013	A B C D E F G H I
Method-CB*Cultivar-Frontenac*Date/Time-8/20/2012	A B C D E F G H I
Method-CB*Cultivar-St. Croix*Date/Time-8/29/2013	B C D E F G H I J
Method-CB*Cultivar-Marquette*Date/Time-8/6/2012	C D E F G H I J
Method-CB*Cultivar-Frontenac*Date/Time-9/11/2012	C D E F G H I J
Method-CB*Cultivar-Marquette*Date/Time-8/10/2012	C D E F G H I J
Method-CB*Cultivar-St. Croix*Date/Time-8/31/2013	C D E F G H I J
Method-GV*Cultivar-La Crescent*Date/Time-8/24/2013	C D E F G H I J K
Method-GV*Cultivar-La Crescent*Date/Time-8/10/2013	C D E F G H I J K
Method-CB*Cultivar-La Crescent*Date/Time-8/10/2013	C D E F G H I J K
Method-CB*Cultivar-St. Croix*Date/Time-9/8/2013	C D E F G H I J K L
Method-GV*Cultivar-Frontenac*Date/Time-8/29/2013	C D E F G H I J K L
Method-GV*Cultivar-Marquette*Date/Time-8/29/2013	C D E F G H I J K L
Method-CB*Cultivar-St. Croix*Date/Time-9/25/2013	D E F G H I J K L M
Method-CB*Cultivar-St. Croix*Date/Time-8/13/2013	E F G H I J K L M
Method-CB*Cultivar-Marquette*Date/Time-9/5/2012	F G H I J K L M
Method-CB*Cultivar-Marquette*Date/Time-8/20/2012	F G H I J K L M
Method-CB*Cultivar-Marquette*Date/Time-9/11/2012	F G H I J K L M
Method-CB*Cultivar-St. Croix*Date/Time-8/24/2013	F G H I J K L M N
Method-CB*Cultivar-Marquette*Date/Time-8/27/2012	F G H I J K L M N
Method-CB*Cultivar-La Crescent*Date/Time-8/29/2013	G H I J K L M N O
Method-CB*Cultivar-La Crescent*Date/Time-9/20/2013	H I J K L M N O
Method-CB*Cultivar-La Crescent*Date/Time-8/28/2013	I J K L M N O
Method-CB*Cultivar-St. Croix*Date/Time-9/3/2013	J K L M N O
Method-GV*Cultivar-La Crescent*Date/Time-8/14/2013	K L M N O
Method-CB*Cultivar-St. Croix*Date/Time-9/15/2013	L M N O
Method-CB*Cultivar-La Crescent*Date/Time-8/24/2013	M N O
Method-CB*Cultivar-La Crescent*Date/Time-9/3/2013	N O
Method-CB*Cultivar-La Crescent*Date/Time-9/15/2013	O

Note: Tukey's d critical value: 6.125. Categories not sharing a group letter are significantly different (p value ≤ 0.05).

References

1. Coombe, B.G.; McCarthy, M.B. Dynamics of grape berry growth and physiology of ripening. *Aust. J. Grape Wine Res.* **2000**, *6*, 131–135. [CrossRef]
2. Robinson, S.P.; Davis, C. Molecular biology of grape berry ripening. *Aust. J. Grape Wine Res.* **2000**, *6*, 169–174. [CrossRef]
3. Terral, J.F.; Tabard, E.; Bouby, L.; Ivorra, S.; Pastor, T.; Figueiral, I.; Picq, S.; Chevance, J.B.; Jung, C.; Fabre, L.; et al. Evolution and history of grapevine (*Vitis vinifera*) under domestication: New morphometric perspectives to understand seed domestication syndrome and reveal origins of ancient European cultivars. *Ann. Bot.* **2010**, *105*, 443–456. [CrossRef] [PubMed]
4. Xu, X.Q.; Liu, B.; Zhu, B.Q.; Lan, Y.B.; Gao, Y.; Wang, D.; Reeves, M.J.; Duan, C.Q. Differences in volatile profiles of Cabernet Sauvignon grapes grown in two distinct regions of China and their responses to weather conditions. *Plant Physiol. Biochem.* **2015**, *89*, 123–133. [CrossRef] [PubMed]
5. Hashizume, K.; Samuta, T. Grape maturity and light exposure affect berry methoxypyrazine concentration quality. *Am. J. Enol. Viticult.* **1999**, *50*, 194–198.
6. Cai, L.; Koziel, J.A.; O'Neal, M. Determination of characteristic odorants from *Harmonia axyridis* beetles using in-vivo solid-phase microextraction and multidimensional gas chromatography-mass spectrometry—Olfactometry. *J. Chromatogr. A* **2007**, *1147*, 66–78. [CrossRef] [PubMed]
7. Conde, C.; Silva, P.; Fontes, N.; Dias, A.; Tavares, R.; Sousa, M.; Agasse, A.; Delrot, S.; Geros, H. Biochemical changes throughout grape berry development and fruit and wine quality. *Food* **2007**, 1–22.
8. Wilwerth, J.J.; Reynolds, A.G.; Lesschaeve, I. Sensory analysis of Riesling wines from different sub-appellation in the Niagara Peninsula in Ontario. *Am. J. Enol. Viticult.* **2015**, *66*, 279–293. [CrossRef]

9. Matsumoto, H.; Ikoma, Y. Effect of postharvest temperature on the muscat flavor and aroma volatile content in the berries of 'Shine Muscat' (*Vitis labruscana* Baily × *V. vinifera* L.). *Postharvest. Biol. Tec.* **2016**, *112*, 256–265. [CrossRef]
10. Diago, M.P.; Vilanova, M.; Tardaguila, J. Effects of timing of early defoliation (manual and mechanical) on the aroma attributes of Tempranillo (*Vitis vinifera* L.) wines. *Am. J. Enol. Viticult.* **2010**, *61*, 382–391.
11. Gardner, D.; Zoecklein, B.W. Electronic nose analysis on the effect of prefermentation cold soak of *Vitis vinifera* L. cv. Cabernet Sauvignon grape and wine volatiles. *Am. J. Enol. Vitic.* **2009**, *62*, 387A. [CrossRef]
12. University of Minnesota, Minnesota Hardy. Available online: http://mnhardy.umn.edu/varieties/fruit/grapes (accessed on 26 December 2018).
13. Skinkis, P.A.; Bordelon, B.P.; Butz, E.M. Effects of sunlight exposure on berry and wine monoterpenes and sensory characteristics of Traminette. *Am. J. Enol. Viticult.* **2010**, *61*, 147–156.
14. Liu, B.; Xu, X.Q.; Cai, J.; Lan, Y.B.; Zhu, B.Q.; Wang, J. The free and enzyme-released volatile compounds of distinctive *Vitis amurensis* var. Zuoshanyi grapes in China. *Eur. Food Res. Technol.* **2015**, *240*, 985–997. [CrossRef]
15. Zhang, S.; Petersen, M.A.; Liu, J.; Toldam-Andersen, T.B. Influence of pre-fermentation treatments on wine volatile and sensory profile of the new disease tolerant cultivar Solaris. *Molecules* **2015**, *20*, 21609–21625. [CrossRef] [PubMed]
16. Pedneault, K.; Dorais, M.; Angers, P. Flavor of cold-hardy grapes: Impact of berry maturity and environmental conditions. *J. Agric. Food. Chem.* **2013**, *64*, 10418–10438. [CrossRef] [PubMed]
17. Slegers, A.; Angers, P.; Ouillet, E.; Truchon, T.; Pedneault, K. Volatile compounds from grape skin, juice and wine from five interspecific hybrid grape cultivars grown in Quebec (Canada) for wine production. *Molecules* **2015**, *20*, 10980–11016. [CrossRef] [PubMed]
18. Rice, S.; Koziel, J.A. The relationship between chemical concentration and odor activity value explains the inconsistency in making a comprehensive surrogate scent training tool representative of illicit drugs. *Forensic. Sci. Int.* **2015**, *257*, 257–270. [CrossRef] [PubMed]
19. Rice, S.; Koziel, J.A. Odor impact of volatiles emitted from marijuana, cocaine, heroin and their surrogate scents. *Data in Brief* **2015**, *5*, 653–706. [CrossRef] [PubMed]
20. Rice, S.; Koziel, J.A. Characterizing the smell of marijuana by odor impact of volatile compounds: An application of simultaneous chemical and sensory analysis. *PLoS ONE* **2015**, *10*, e0144160. [CrossRef] [PubMed]
21. Hjelmeland, A.K.; Ebeler, S.E. Glycosidically bound volatile aroma compounds in grapes and wine; A review. *Am. J. Enol. Vitic.* **2015**, *66*, 1–11. [CrossRef]
22. Hernandez-Orte, P.; Concejero, B.; Astrain, J.; Lacau, B.; Cacho, J.; Ferreira, V. Influence of viticulture practices on grape aroma precursors and their relation with wine aroma. *J. Sci. Food Agric.* **2015**, *95*, 688–701. [CrossRef] [PubMed]
23. Pawliszyn, J. *Handbook of Solid Phase Microextraction*; Chemical Industry Press of China: Beijing, China, 2009.
24. Flamini, G.; Cioni, P.L. Odour gradients and patterns in volatile emissions of different plant parts and developing fruits of grapefruit (*Citrus paradise* L.). *Food Chem.* **2010**, *120*, 984–992. [CrossRef]
25. Vasquez-Cruz, M.A.; Jimenez-Garcia, S.N.; Torres-Pacheco, I.; Guzman-Maldonado, S.H.; Guevara-Gonzalez, R.G.; Miranda-Lopez, R. Effect of maturity stage and storage on flavor compounds and sensory description of Berrycatus (*Myrtillocactus geometrizans*). *J. Food Sci.* **2012**, *77*, C366–C373. [CrossRef] [PubMed]
26. Cai, J.; Zhu, B.; Wang, Y.; Lu, L.; Lan, Y.; Reeves, M.J.; Duan, C. Influence of pre-fermentation cold maceration treatment on aroma compounds of Cabernet Sauvignon wines fermented in different industrial scale fermenters. *Food Chem.* **2014**, *154*, 217–229. [CrossRef] [PubMed]
27. Battilana, J.; Emanuelli, F.; Gambino, G.; Gribaudo, I.; Gasperi, F.; Boss, P.K.; Grando, M.S. Functional effect of grapevine 1-deoxy-D-xylulose 5-phosphate synthase substitution K284N on Muscat flavour formation. *J. Exp. Bot.* **2011**, *62*, 5497–5508. [CrossRef] [PubMed]
28. Esti, M.; Tamborra, P. Influence of winemaking techniques on aroma precursors. *Anal. Chim. Acta* **2006**, *563*, 173–179. [CrossRef]
29. Pardo-Garcia, A.I.; Serrano de la Hoz, K.; Zalacain, A.; Alonso, G.L.; Salinas, M.R. Effect of vine foliar treatments on the varietal aroma of Monastrell wines. *Food Chem.* **2014**, *163*, 258–266. [CrossRef]
30. Mansfield, A.K.; Reineccius, G. Identifying characteristic volatiles of Frontenac wine by stir bar sorptive extraction, GCO/FID, and GCO/MS. In *Proceedings of the American Society of Enology and Viticulture*, Seattle, WA, USA, 22–24 June 2005.

31. Mansfield, A.K.; Schirle-Keller, J.; Reineccius, G.A. Identification of odor-impact compounds in red table wines produced from Frontenac grapes. *Am. J. Enol. Vitic.* **2011**, *62*, 169–176. [CrossRef]
32. Cai, L.; Rice, S.; Koziel, J.A.; Dharmadhikari, M. Development of an automated method for aroma analysis of red wines from cold-hardy grapes using simultaneous solid-phase microextraction-multidimensional gas chromatography-mass spectrometry-olfactometry. *Separations* **2017**, *4*, 24. [CrossRef]
33. Rice, S.; Lutt, N.; Koziel, J.A.; Dharmadhikari, M.; Fennell, A. Determination of selected aromas in Marquette and Frontenac wine using headspace-SPME coupled with GC-MS and simultaneous olfactometry. *Separations* **2018**, *5*, 20. [CrossRef]
34. Rice, S.; Tursumbayeva, M.; Clark, M.; Greenlee, D.; Dharmadhikari, M.; Fennell, A.; Koziel, J.A. Effects of harvest time on aroma of white wines made from cold-hardy Brianna and Frontenac gris grapes using headspace solid-phase microextraction and gas-chromatography-mass-spectrometry-olfactometry. *Foods* **2019**, *8*, 29. [CrossRef] [PubMed]
35. Zhu, F.; Xu, J.; Ke, Y.; Huang, S.; Zeng, F.; Luan, T.; Ouyang, G. Applications of in vivo and in vitro solid-phase microextraction techniques in plant analysis: A review. *Anal. Chim. Acta* **2013**, *794*, 1–14. [CrossRef] [PubMed]
36. Rice, S.; Maurer, D.L.; Fennell, A.; Dharmadhikari, M.; Koziel, J.A. Biogenic volatiles emitted from four cold-hardy grape cultivars during ripening. *Data* **2019**, *4*, 22. [CrossRef]
37. Acree, T.; Arn, H. Flavornet and Human Odor Space. Available online: http://www.flavornet.org (accessed on 20 August 2018).
38. The Good Scents Company Information System. Available online: http://www.thegoodscentscompany.com/ (accessed on 20 August 2018).
39. United States Department of Agriculture, Agricultural Research Service. Available online: https://planthardiness.ars.usda.gov/PHZMWeb/ (accessed on 26 December 2018).
40. National Oceanic and Atmospheric Administration, National Centers for Environmental Information. Available online: https://www.ncdc.noaa.gov (accessed on 19 December 2018).
41. Chateau Stripmine. Available online: http://chateaustripmine.info/Parentage/Marquette.gif (accessed on 26 December 2018).
42. Yiantzi, E.; Kalogerakis, N.; Psillakis, E. Design and testing of a new sampler for simplified vacuum-assisted headspace solid-phase microextraction. *Anal. Chim. Acta* **2016**, *927*, 46–54. [CrossRef] [PubMed]
43. Maicias, S.; Mateo, J.J. Hydrolysis of terpenyl glycosides in grape juice and other fruit juices: A review. *Appl. Microbiol. Biotechnol.* **2005**, *67*, 322–335. [CrossRef] [PubMed]
44. Jackson, R.S. *Wine Science Principles and Applications*; Elsevier: Amsterdam, The Netherlands, 2008; ISBN 978-0-12-373646-8.
45. Rapid determination of the aromatic compounds methyl-anthranilate, 2′-aminoacetophenone and furaneol by GC-MS: Method validation and characterization of grape derivatives. *Food Res. Intl.* **2018**, *107*, 613–618. [CrossRef] [PubMed]
46. Risticevic, S.; Chen, Y.; Kudlejova, L.; Vatinno, R.; Baltensperger, B.; Stuff, J.R.; Hein, D.; Pawliszyn, J. Protocol for the development of automated high-throughput SPME–GC methods for the analysis of volatile and semivolatile constituents in wine samples. *Nat. Protoc.* **2010**, *5*, 162–176. [CrossRef] [PubMed]

Sample Availability: Samples of the compounds are not available from the authors.

© 2019 by the authors. Licensee MDPI, Basel, Switzerland. This article is an open access article distributed under the terms and conditions of the Creative Commons Attribution (CC BY) license (http://creativecommons.org/licenses/by/4.0/).

Article

Analysis of Volatile Compounds in Pears by HS-SPME-GC×GC-TOFMS

Chenchen Wang [1,2], Wenjun Zhang [1,2], Huidong Li [1,2], Jiangsheng Mao [1,2], Changying Guo [1,2], Ruiyan Ding [1,2], Ying Wang [3], Liping Fang [1,2], Zilei Chen [1,2,4,*] and Guosheng Yang [5,*]

1. Institution of Quality Standard and Testing Technology for Agro-Product, Shandong Academy of Agricultural Science, Jinan 250100, China; wangchenchen0826@163.com (C.W.); zipingguozhang@163.com (W.Z.); lihuidong8066@163.com (H.L.); maojiangsheng@163.com (J.M.); cyguo808@163.com (C.G.); zengding-1978@163.com (R.D.); lpfang922@163.com (L.F.)
2. Shandong Provincial Key Laboratory of Testing Technology for Food Quality and Security, Jinan 250100, China
3. Department of Bioengineering, Qilu University of Technology, Jinan 250353, China; hanyan1226@126.com
4. College of Life Science, Shandong Normal University, Jinan 250014, China
5. School of Chemistry and Chemical Engineering, Shandong University, Jinan 250100, China
* Correspondence: czl7274@163.com (Z.C.); gsyang@sdu.edu.cn (G.Y.); Tel.: +86-531-66659036 (Z.C.); +86-531-88364464 (G.Y.)

Academic Editors: Constantinos K. Zacharis and Paraskevas D. Tzanavaras
Received: 15 April 2019; Accepted: 6 May 2019; Published: 9 May 2019

Abstract: Aroma plays an important role in fruit quality and varies among different fruit cultivars. In this study, a sensitive and accurate method based on headspace solid-phase microextraction (HS-SPME) coupled with comprehensive two-dimensional gas chromatography time-of-flight mass spectrometry (GC×GC-TOFMS) was developed to comprehensively compare aroma components of five pear cultivars. In total, 241 volatile compounds were identified and the predominant volatile compounds were esters (101 compounds), followed by alcohols (20 compounds) and aldehydes (28 compounds). The longyuanyangli has the highest relative concentration (838.12 ng/g), while the Packham has the lowest (208.45 ng/g). This study provides a practical method for pear aroma analysis using SPME and GC×GC-TOFMS.

Keywords: pears; HS-SPME; volatile compounds; GC×GC-TOFMS

1. Introduction

Pear (*Pyrus spp.*, Rosaceae) is a popular fruit and is extensively grown in China. In 2017, based on the FAO Statistical Database, 16,527,694 tonnes of pears were produced in China taking up 68.38% of the total pear production in the world [1]. The pear cultivars mainly belong to 4 types, *P. communis* L., *P. pyrifolia* (Burm.) Nakai, *P. ussuriensis* Max. and *P. bretschneideri* Redh [2]. The *P. sinkiangensis* Yu. was reported as the fifth pear category.

Fruit aroma is one of the most important factors contributing to the overall flavor and consumer preference [2]. Therefore, several studies have investigated the aroma components of different pear cultivars [2–7]. The aroma compounds of pears are complicated and vary among pear cultivars. Most Occidental pears have intense aromas and juicy texture, whereas *P. bretschneideri* cultivars are characterized by their faint odor and crisp texture [2]. Investigation of pear aromas has focused on composition changes among pear cultivars [3–7], storage conditions influences [8–12] and postharvest treatment [11,13]. Low temperature conditioning [12], calcium treatment [13], ultralow oxygen environment [10,11] and 1-methylcyclopropene (1-MCP) treatment [11] are external factors that affect pear aroma formation and emission. Volatile compounds of pears include esters, aldehydes, alcohols, ketones and hydrocarbons. Esters are the major volatile components in *P. ussuriensis* and *P. communis*, while in *P. pyrifolia* aldehydes are the dominant volatile compounds, followed by alcohols and esters [3–5]. C_6 compounds (C_6 aldehydes and C_6 alcohols) that were reported to be significant components in fruits [14–16] were also detected in pears [3,4].

Volatiles emitted from pear fruits have been studied by SPME and gas chromatography-mass spectrometry (GC-MS) in recent years. SPME makes great contributions to volatile compounds analysis [17]. However, investigations using GC-MS only identified a small quantity of volatile compounds. In comparison with one-dimensional gas chromatography (1D-GC), comprehensive two-dimensional gas chromatography (GC×GC) can provide significant signal enhancement and a 5-fold to 15-fold improvement in peak capacity [18,19]. GC×GC-TOFMS has been applied to volatiles identification such as wines [20], cloud waters [21] and green teas [22], but pears are not included.

In this research, HS-SPME and GC×GC-TOFMS were used to analyse volatile compounds in five pear cultivars. Packham's Triumph, Docteur Jules Guyot, Clapp's Favorite and Starkrimson are four occidental pears, which are introduced from abroad. Longyuanyangli is a hybrid variety that has intense aroma. Therefore in this study, the aromas of five pear cultivars are comprehensively investigated.

2. Results and Discussion

2.1. Optimization of the Modulation Period

Modulation period is a parameter of crucial importance in GC×GC-TOFMS analysis. Modulator is responsible to trap and refocus the components from the 1D-column and transfer them to the 2D column for further separation [18]. The modulator makes the GC×GC possible. For aroma components analysis, the modulation period was set as 2 s, 3 s, and 4 s. The shorter the modulator period is, the narrower chromatography band and the higher peak capacity are obtained. As shown in Figure 1, when the modulation period was 3 s, the retention time of 2-methylnaphthalene and 1-methylnaphthalene was 1185 s, 2.00 s and 1212 s, 2.08 s, respectively. When the modulation period was 2 s, the retention time of 2-methylnaphthalene and 1-methylnaphthalene was 1186 s, 0.00 s and 1214 s, 0.08 s, respectively. The chromatographic peak of 2-methylnaphthalene was divided into two parts, which has a significant impact on the quantification process. The peak of 1-methylnaphthalene was at the very bottom in the chromatography. In order to ensure the peak shape of the volatile compounds and the accuracy of the quantitative analysis, the modulation period was set for 3 s with a 0.6 s hot pulse time.

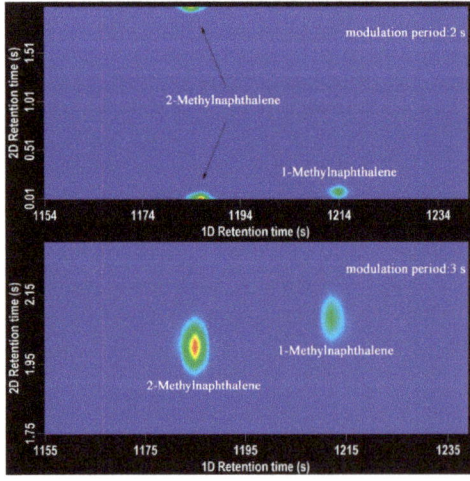

Figure 1. The 2D chromatography of 2-methylnaphthalene and 1-methylnaphthalene. (top-modulation period 2 s, bottom-modulation period 3 s).

2.2. SPME Fibre Selection

Pear aromas are extremely complex and may comprise hundreds of constitutes of different physical and chemical properties. Therefore, four SPME fibres coated with different stationary phases were compared for extraction of volatile compounds. They are 100 μm PDMS (nonpolar), 85 μm PA (polar), 65 μm PDMS/PVB (bipolar) and 50/30 μm DVB/CAR/PDMS (bipolar). The aroma components of a same yali pear (*P. bertschneideri* Reld) were analyzed for fibres comparison. The aroma extraction process was repeated for three times to guarantee the accuracy of the results. Figure 2 illustrates the peak numbers and the average peak areas for volatile compounds extracted from yali pear using different SPME fibres. A total of 146 and 163 volatile compounds were identified using 65 μm PDMS/PVB and 50/30 μm DVB/CAR/PDMS, respectively. The fewest compounds were extracted by 85 μm PA fibre. In contrast to PDMS/PVB fibre, the use of DVB/CAR/PDMS fibre can obtain higher peak areas. The peak numbers of different classes obtained using four SPME fibres were shown in Table S1. These results indicated that the 50/30 μm DVB/CAR/PDMS fibre was the optimum for extracting volatile compounds from pears. Therefore, 50/30 μm DVB/CAR/PDMS fibre was selected for aroma extraction in this study.

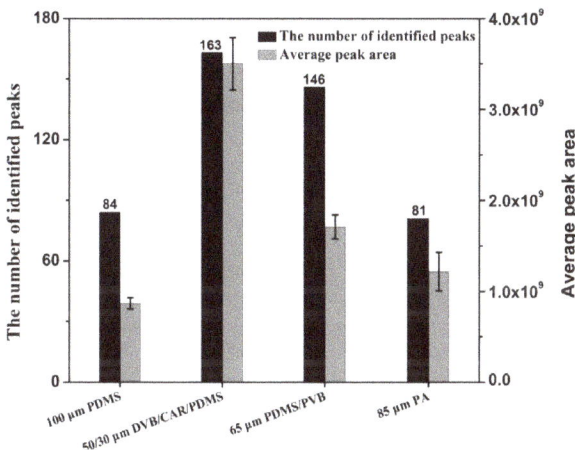

Figure 2. Comparison of aroma amounts and peak areas in pears using four different SPME fibres.

2.3. Volatile Compounds

The 2D chromatography of five pears obtained after HS-SPME-GC×GC-TOFMS analysis is shown in Figures S1–S5. The colour gradient reflects the intensity of the TOFMS signal from low (blue) to high (red). In this study, 241 volatile compounds were tentatively identified, including 101 esters, 30 alkenes, 12 alkanes, 19 arenes, 28 aldehydes, 8 ketones, 20 alcohols and 23 others compounds. The volatile compounds amounts show great variation in different pear cultivars and ranged from 67 compounds in Packham to 160 compounds in longyuanyangli. The number of chemical classes of each pear is shown in Figure 3. Figure 4 shows that the percentage contents of volatile compounds in pears are of large differences. Esters are the dominant aromas in pears, followed by alcohols and aldehydes. Table S2 summarizes the volatile compounds detected in five pear cultivars. In this study, the retention time of *n*-alkanes (C_5–C_{20}) was obtained and the retention index of each volatile compound was calculated. The ChromaTOF-GC uses the Van den Dool and Kratz equation for the calculation of the retention index. The equation is:

$$RI_a = \left(\frac{RT_a - RT_n}{RT_N - RT_n}\right)100(N - n) + 100RT_n$$

RI_a: the retention index of the compound of interest; a: the compound of interest; n: the carbon number of the lower normal alkane; N: the carbon number of the higher normal alkane; RT: the retention time.

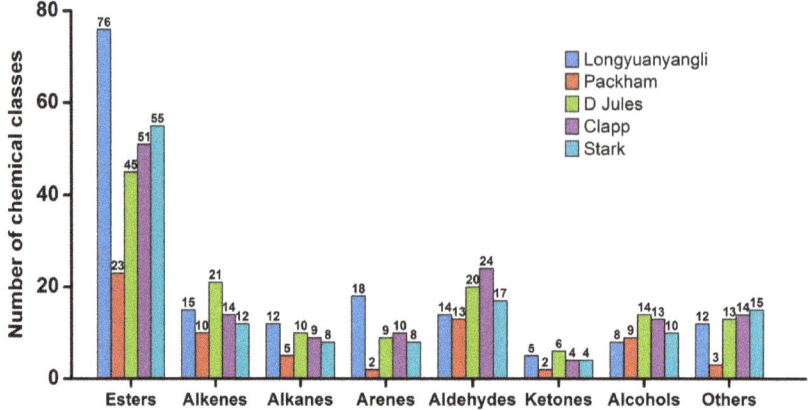

Figure 3. The number of chemical classes in pears.

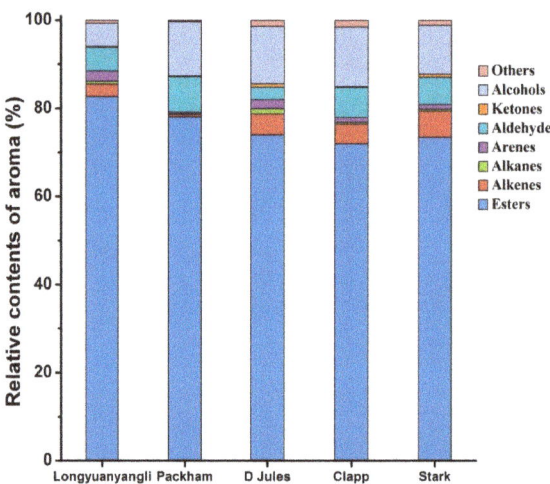

Figure 4. Relative contents of volatile compounds in five pears.

2.3.1. Esters

Esters are the dominant compounds in pears. A total of 76 esters were identified in longyuanyangli. Followed by Stark (55), Clapp (51), D Jules (45), and Packham, who had the fewest esters (23) (Figure 3). The longyuanyangli has the highest concentration of esters (692.72 ng/g, 82.65%), while the Packham has the lowest (162.66 ng/g, 78.03%) (Table S2). Acetates with high concentrations were the major ester constituents, including methyl acetate, ethyl Acetate, n-propyl acetate, butyl acetate, pentyl acetate, hexyl acetate and heptyl acetate (Table S2). Sulfur-containing compounds, ethyl 3-(methylsulfanyl)propanoate, 3-(methylthio)propyl acetate and ethyl 3-(methylthio)-(E)-2-propenoate were also detected in this study. Sulfur-containing compounds have been reported to have originated from methionine and cysteine [3] and provided the juicy, fresh aroma to many fruits [23]. Methyl (E,Z)-2,4-decadienoate and ethyl (E,Z)-2,4-decadienoate are two esters that have a pear-like smell and are major volatile compounds existed in Bartlett [10] and Beurre Bosc [3]. In this study, methyl (E,Z)-2,4-decadienoate and ethyl (E,Z)-2,4-decadienoate were also detected in longyuanyangli and Packham, but the contents (less than 0.53 ng/g) were very low. Furthermore, long-chain aliphatic acid esters such as methyl tetradecanoate, methyl hexadecanoate, methyl (Z)-9-octadecenoate were also detected in this study.

2.3.2. Alcohols

Alcohols were the second dominant volatile compounds in the five pear cultivars (Figure 4). Alcohols account for 5.33 percent (44.67 ng/g) in longyuanyangli to 13.48 percent (51.19 ng/g) in Clapp. Ethanol was the primary alcohol compounds with the concentrations ranging from 12.34 ng/g in Packham to 27.35 ng/g in D Jules. 1-Hexanol and trans-2-hexen-1-ol were reported in many fruits and regarded as C_6 alcohols. 1-Hexanol was found in the range of 9.84 ng/g and 26.97 ng/g, but trans-2-hexen-1-ol was only detected in Packham in the concentration of 0.50 ng/g (Table S2). In addition, 1-butanol and 1-heptanol are straight-chain alcohols, which existed in all pears. Linalool, which possesses a floral and citrus-like aroma [22,24], was identified in D Jules cultivar for the first time. Citronellol was identified in D Jules, Clapp and Stark with low concentration.

2.3.3. Aldehydes and Ketones

The largest number of aldehydes compounds (24) was detected in Clapp, and Packham contained the fewest aldehydes (13) (Figure 3). The concentrations of aldehydic compounds were relatively low in all pears other than acetaldehyde, hexanal and (E)-2-hexenal. Acetaldehyde was detected in high concentrations ranged from 4.81 ng/g in D Jules to 23.71 ng/g in longyuanyangli. C_6 compounds (C_6 aldehydes and C_6 alcohols) are regarded as green leaf volatiles and contribute to the herbaceous odour in fruits [14–16]. In this study, hexanal, (E)-2-hexenal and 1-hexanol are dominant C_6 compounds. The concentration of hexanal ranged from 2.07 ng/g in D Jules to 12.46 ng/g in longyuanyangli and the (E)-2-hexenal concentration changed from 1.60 ng/g in D Jules to 6.27 ng/g in longyuanyangli. In addition, benzaldehyde was detected in all pears. Benzaldehyde has an almond-like smell and has been previously isolated from green teas [22], lychee [24], and apricot [16]. Figure 4 shows that small proportion ketones were detected. In total, eight ketones were found, but 6-methyl-5-heptene-2-one was the only ketone that presented in all pears. It has been reported that 6-methyl-5-heptene-2-one was present in higher amount in the peel compared to the flesh and was a degradation product of lycopene [16] or α-farnesene [25,26]. In addition, 6-methyl-5-heptene-2-one possesses fatty, green, citrus odour [27] and is a common ketone existed in many fruits [16,24,28,29].

2.3.4. Hydrocarbons

Although 12 alkanes were identified, the relative contents were very low. A series of n-alkanes (C_{13}–C_{16}) existed in all five pears. Alkenes account for 0.56–5.88% in total volatile compounds (Figure 4). A total of 30 alkenes were detected, comprising aliphatic alkenes (5), aromatic alkenes (6) and terpenes (19). Styrene was previously identified in Chinese white pear [5]. In addition to styrene, aromatic alkenes identified in this study comprise 1-propenylbenzene, 1-ethenyl-3-ethylbenzene, 1-ethenyl-4-ethylbenzene, 1,4-dethenyl benzene, 1,3-diethenylbenzene. Terpenes which play important role in fruit flavors have been identified in pears even if their contents were much lower than other compounds. Among these terpenes, β-myrcene (Grassy, piney), (Z)-β-ocimene (floral, citrusy) and limonene (citrusy) are three monoterpenes [30]. Furthermore, β-myrcene has been previously identified in mango [31], apricot [16] and lychee [24]. In this study, four isomers of farnesene were detected for the first time, including (E)-β-farnesene, (Z,E)-α-farnesene, α-farnesene and (Z,Z)-α-farnesene. α-farnesene is the only alkenes found in five pear cultivars and accounted for the highest proportion in alkenes. (E)-β-farnesene, (Z,E)-α-farnesene and (Z,Z)-α-farnesene were also identified in four pear cultivars other than Packham. (E)-γ-bisabolene, (Z)-γ-bisabolene and α-humulene are major volatile compounds in carrots [32,33]. It is the first time that (E)-γ-bisabolene and (Z)-γ-bisabolene were found in pears. The two isomers existed in all pears apart from Packham. The α-humulene contributing to the woody smell [32] was identified in Packham and Clapp. Other terpenes such as α-cubenene, copaene, α-muurolene, (+)-δ-cadinene, cis-calamenene, α-calacorene also play important role in pear aroma.

In contrast to other chemical classes of volatiles, arenes are minor components of total volatiles. A total of 19 arenes were identified in this study. In addition to benzene and benzene homologous compounds, polycyclic aromatic hydrocarbon 2-methylnaphthalene, 1-methylnaphthalene, naphthalene were also found (Table S2).

2.3.5. Others

Pears have high concentrations of esters, alcohols, aldehydes and alkenes which are of great significance on fruit aroma. Nevertheless, other compounds (23 compounds) such as benzonitrile, 2-pentylfuran, estragole and sesquirosefuran also contribute to the overall flavor of pears and account for 0.26–1.53% of the total volatiles. Among these volatiles, three volatile acids were identified, including acetic acid, thioacetic acid and (E)-3-octenoic acid. Eucalyptol, which is characterized by a fresh, camphoraceous, cool odour [34] was detected in longyuanyangli in the concentration of 0.89 ng/g. Sesquirosefuran, a natural constituent existed in essential oils [35,36], was detected in D Jules, Clapp and Stark. Previous study has shown the existence of estragole in *Pyrus ussuriensis* cultivars [4], but its isomers anethole and (Z)-1-methoxy-4-(prop-1-en-1-yl)benzene were also identified in this study. Additionally, 2-pentylfuran was found in five pear cultivars and was perceived as having a fruity, green, earthy and vegetable-like smell [22]. Benzonitrile which was identified in *Pyrus ussuriensis* cultivars [4] was also indentified in longyuanyangli and D Jules. Other compounds such as phenol, (Z)-rose oxide and (E)-rose oxide were also detected in this study. Various aroma components and concentration difference determine the overall flavor properties of pears.

2.4. Cluster Analysis (CA)

The cluster analysis based on concentrations of identified volatile compounds was performed using the SPSS Statistical 19.0 software. The dendrogram (Figure 5) shows that two main groups are distinguished. Longyuanyangli, which has the maximum aroma numbers and the highest concentrations, is separated from other pear cultivars. Packham, D Jules, Clapp and Stark constitute the second group. They are four Occidental pears, which are introduced from abroad. Figure 5 shows that the D Jules and Stark have the slightest differences compared with other cultivars. Many factors affect the volatile compounds composition of the fruits. In this study, the volatile compositions of pears were found to be considerably different.

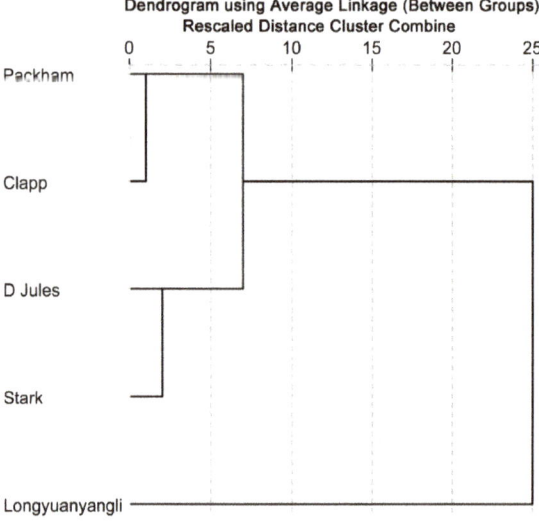

Figure 5. Dendrogram obtained from cluster analysis based on the identified volatile compounds.

3. Materials and Methods

3.1. Materials

Five pear cultivars were prepared for analysis. The detailed information of the pears is shown in Table 1. The conventional indicators such as growing period, external morphology and skin color were used to judge the maturity of each cultivar. All fruits were stored at 1 °C before experiments. For each cultivar in this study, after-ripening process was necessary to enhance the flavor and taste. Samples were placed at room temperature before experiments (approximately five days). The 4 SPME fibres (100 μm PDMS, 85 μm PA, 65 μm PDMS/PVB and 50/30 μm DVB/CAR/PDMS) were supplied by Supelco (Bellefonte, PA, USA). The length of the fibre coating is 1 cm. The internal standard 2-nonanone (>99%) was obtained from Dr. Ehrensorfer (Germany).

Table 1. The cultivars, abbreviation, producing area and sampling time of 5 pears.

Pear Cultivars	Abbreviation	Producing Area	Sampling Time
Longyuanyangli	longyuanyangli	Qiqihaer city of Heilongjiang province	07 September 2018
Packham's Triumph	Packham	Weihai city of Shandong Province	27 September 2018
Docteur Jules Guyot	D Jules	Yantai city of Shandong Province	01 August 2018
Clapp's Favorite	Clapp	Yantai city of Shandong Province	07 August 2018
Starkrimson	Stark	Yantai city of Shandong Province	01 August 2018

3.2. Volatiles Extraction

HS-SPME was used for volatile compounds extraction. A 50/30 μm DVB/CAR/PDMS SPME fibre was used in this study. Fibres were activated according to the conditioning guidelines before the first use. The core of each pear was removed, while the peel was reserved. The skin and flesh of each pear was cut into cubes (0.5 cm × 0.5 cm × 0.5 cm). For each extraction, 6.0 g of sample were placed into a 15-mL screw-cap vial. Prior to sealing of the vials, 5 μL of 10 μg/mL 2-nonanone was added as internal standard, and was standing for 10 min. Then each vial was placed in a constant-temperature controller at 40 °C for 40 min. Finally, the SPME fibre was immediately inserted into the GC injector for desorption at 270 °C for 2 min in the split mode of 10:1.

3.3. GC×GC-TOFMS Conditions

The volatile compounds analysis was performed with an Agilent 7890B gas chromatography equipped with a Pegasus 4D-C time-of-flight mass spectrometric detector. A Rxi-5MS column (30 m × 250 μm × 0.25 μm) was used as the first-dimension (1D) column, and a Rxi-17Sil MS column (2 m × 250 μm × 0.25 μm) was used as the second-dimension (2D) column. Helium was used as the carrier gas at a constant flow of 1.4 mL/min. The front inlet and the transfer line temperature were 270 °C and 280 °C, respectively. The oven temperature programme conditions were as follows: initial temperature was 40 °C for 2 min, rose at 5 °C/min up to 200 °C, then ramped to 280 °C at 20 °C/min and hold for 2 min. The secondary oven temperature was kept at 5 °C above the GC oven temperature throughout the chromatographic run. The modulator temperature was offset by 15 °C in relative to the secondary oven temperature. The modulation period was set for 3 s with a 0.6 s hot pulse time.

The MS parameters were as follows: acquisition delay 60 s, acquisition rate 100 (spectra/s), the acquisition voltage 1450 V, electron energy −70 V, ion source 250 °C. Mass spectra were acquired in the m/z range 35–550 amu.

3.4. Data Processing and Statistical Analysis

LECO ChromaTOF Version 4.73.3.0 software (Leco Corporation, St. Joseph, MO, USA) was used for instrument control, data acquisition and processing. Each chromatograph peak was compared to National Institute of Standards and Technology (NIST2017) library and the minimum similarity is 800.

The area of the base peak was used for quantification. The quantitative analysis for aroma components was carried out by internal standard method using 2-nonanone as an internal standard. Therefore, the concentration of each volatile compound was normalized to that of 2-nonanone. The formula for volatile compounds quantification is as follows:

$$C_a = \frac{\frac{PA_a}{PA_{is}} \times C_{is} \times 5 \; \mu L}{m}$$

C_a: the concentration of aroma components (ng/g); PA_a: peak area of aroma components; PA_{is}: peak area of internal standard; C_{is}: the concentration of internal standard (g/mL); m: mass of sample (g). The concentration of the 2-nonanone was 10 µg/mL and the mass of sample was 6.0 g. Data are means ± SD of three replications. Cluster analysis (CA) was performed using the SPSS Statistical 19.0.

4. Conclusions

In conclusion, the combination of SPME and GC×GC-TOFMS has improved the analysis of pear volatile compounds. The 50/30 µm DVB/CAR/PDMS SPME fibre exhibited obvious advantages for volatile compounds extraction. A total of 241 compounds were identified in five pear cultivars, which are primarily esters, alcohols, aldehydes and alkenes. Volatile compounds such as sesquirosefuran and anethole are reported for the first time in pears. Evaluation of aromas at the germplasm level will facilitate breeding efforts and improve sensory quality of fruits. This research will contribute to further studies related to volatile compounds analysis.

Supplementary Materials: The following are available online, Figure S1: 2D chromatogram (total ion chromatography) of Longyuanyangli, Figure S2: 2D chromatogram (total ion chromatography) of Packham, Figure S3: 2D chromatogram (total ion chromatography) of D Jules, Figure S4: 2D chromatogram (total ion chromatography) of Clapp, Figure S5: 2D chromatogram (total ion chromatography) of Stark, Table S1: The peak numbers of different classes obtained using 4 SPME fibres, Table S2: Concentrations of volatile compounds in pear cultivars.

Author Contributions: C.W. and W.Z. conceived and designed the experiments; C.W. performed the experiments; C.W. and Y.W. analyzed the data; R.D. and L.F. contributed materials; H.L. and J.M. and C.G. helped the use of the GC×GC-TOFMS; C.W. wrote the paper; Z.C. and G.Y. helped to revise the manuscript. All authors read and approved the final manuscript.

Funding: This research was supported by National Pear Industry Technology System (CARS-28-23) and Agricultural Science and Technology Innovation Project of Shandong Academy of Agricultural Sciences (CXGC2017A03).

Conflicts of Interest: The authors declare no conflict of interest.

References

1. FAO Statistical Database. Available online: http://www.fao.org/home/en/ (accessed on 20 December 2018).
2. Rapparini, F.; Predieris, S. Pear fruit volatiles. *Hort. Rev.* **2003**, *28*, 237–324.
3. Chen, Y.Y.; Yin, H.; Wu, X.; Shi, X.J.; Qi, K.J.; Zhang, S.L. Comparative analysis of the volatile organic compounds in mature fruits of 12 Occidental pear (*Pyrus communis* L.) cultivars. *Sci. Hortic.* **2018**, *240*, 239–248. [CrossRef]
4. Qin, G.H.; Tao, S.T.; Cao, Y.F.; Wu, J.Y.; Zhang, H.P.; Huang, W.J.; Zhang, S.L. Evaluation of the volatile profile of 33 *Pyrus ussuriensis* cultivars by HS-SPME with GC-MS. *Food Chem.* **2012**, *134*, 2367–2382. [CrossRef]
5. Yi, X.K.; Liu, G.F.; Rana, M.M.; Zhu, L.W.; Jiang, S.L.; Huang, Y.F.; Lu, W.M.; Wei, S. Volatile profiling of two pear genotypes with different potential for white pear aroma improvement. *Sci. Hortic.* **2016**, *209*, 221–228. [CrossRef]
6. Takeoka, G.R.; Buttery, R.G.; Flath, R.A. Volatile constituents of asian pear (*Pyrus serotina*). *J. Agric. Food chem.* **1992**, *40*, 1925–1929. [CrossRef]
7. Katayama, H.; Ohe, M.; Sugawara, E. Diversity of odor-active compounds from local cultivars and wild accessions of Iwateyamanashi (*Pyrus ussuriensis* var. *aromatica*) revealed by aroma extract dilution analysis (AEDA). *Breed. Sci.* **2013**, *63*, 86–95. [CrossRef] [PubMed]

8. Chen, J.L.; Wu, J.H.; Wang, Q.; Deng, H.; Hu, X.S. Changes in the volatile compounds and chemical and physical properties of Kuerle fragrant pear (*Pyrus serotina* Reld) during Storage. *J. Agric. Food Chem.* **2006**, *54*, 8842–8847. [CrossRef] [PubMed]
9. Chen, J.L.; Yan, S.; Feng, Z.; Xiao, L.; Hu, X.S. Changes in the volatile compounds and chemical and physical properties of Yali pear (*Pyrus bertschneideri* Reld) during storage. *Food Chem.* **2006**, *97*, 248–255. [CrossRef]
10. Zlatić, E.; Zadnik, V.; Fellman, J.; Demšar, L.; Hribar, J.; Čejić, Ž.; Vidrih, R. Comparative analysis of aroma compounds in 'Bartlett' pear in relation to harvest date, storage conditions, and shelf-life. *Postharvest Biol. Technol.* **2016**, *117*, 71–80. [CrossRef]
11. Hendges, M.V.; Neuwald, D.A.; Steffens, C.A.; Vidrih, R.; Zlatić, E.; do Amarante, C.V.T. 1-MCP and storage conditions on the ripening and production of aromatic compounds in Conference and Alexander Lucas pears harvested at different maturity stages. *Postharvest Biol. Technol.* **2018**, *146*, 18–25. [CrossRef]
12. Zhou, X.; Dong, L.; Li, R.; Zhou, Q.; Wang, J.W.; Ji, S.J. Low temperature conditioning prevents loss of aroma-related esters from 'Nanguo' pears during ripening at room temperature. *Postharvest Biol. Technol.* **2015**, *100*, 23–32. [CrossRef]
13. Wei, S.W.; Qin, G.H.; Zhang, H.P.; Tao, S.T.; Wu, J.; Wang, S.M.; Zhang, S.L. Calcium treatments promote the aroma volatiles emission of pear (*Pyrus ussuriensis* 'Nanguoli') fruit during post-harvest ripening process. *Sci. Hortic.* **2017**, *215*, 102–111. [CrossRef]
14. Wang, Y.J.; Yang, C.X.; Li, S.H.; Yang, L.; Wang, Y.N.; Zhao, J.B.; Jiang, Q. Volatile characteristics of 50 peaches and nectarines evaluated by HP-SPME with GC-MS. *Food Chem.* **2009**, *116*, 356–364. [CrossRef]
15. Yang, C.X.; Wang, Y.J.; Liang, Z.C.; Fan, P.G.; Wu, B.H.; Yang, L.; Wang, Y.N.; Li, S.H. Volatiles of grape berries evaluated at the germplasm level by headspace-SPME with GC-MS. *Food Chem.* **2009**, *114*, 1106–1114. [CrossRef]
16. Gokbulut, I.; Karabulut, I. SPME-GC-MS detection of volatile compounds in apricot varieties. *Food Chem.* **2012**, *132*, 1098–1102. [CrossRef]
17. Xu, C.H.; Chen, G.S.; Xiong, Z.H.; Fan, Y.X.; Wang, X.C.; Liu, Y. Applications of solid-phase microextraction in food analysis. *TrAC-Trends Anal. Chem.* **2016**, *80*, 12–29. [CrossRef]
18. Murray, J.A. Qualitative and quantitative approaches in comprehensive two-dimensional gas chromatography. *J. Chromatogr. A* **2012**, *1261*, 58–68. [CrossRef]
19. Freye, C.E.; Bahaghighat, H.D.; Synovec, R.E. Comprehensive two-dimensional gas chromatography using partial modulation via a pulsed flow valve with a short modulation period. *Talanta* **2018**, *177*, 142–149. [CrossRef]
20. Welke, J.E.; Zanus, M.; Lazzarotto, M.; Pulgati, F.H.; Zini, C.A. Main differences between volatiles of sparkling and base wines accessed through comprehensive two dimensional gas chromatography with time-of-flight mass spectrometric detection and chemometric tools. *Food Chem.* **2014**, *164*, 427–437.
21. Lebedev, A.T.; Polyakova, O.V.; Mazur, D.M.; Artaev, V.B.; Canet, I.; Lallement, A.; Vaitilingom, M.; Deguillaume, L.; Delort, A.M. Detection of semi-volatile compounds in cloud waters by GC×GC-TOF-MS. Evidence of phenols and phthalates as priority pollutants. *Environ. Pollut.* **2018**, *241*, 616–625. [CrossRef]
22. Zhu, Y.; Lv, H.P.; Shao, C.Y.; Kang, S.; Zhang, Y.; Guo, L.; Dai, W.D.; Tan, J.F.; Peng, Q.H.; Lin, Z. Identification of key odorants responsible for chestnut-like aroma quality of green teas. *Food Res. Int.* **2018**, *108*, 74–82. [CrossRef]
23. Cannon, R.J.; Ho, C.T. Volatile sulfur compounds in tropical fruits. *J. Food Drug Anal.* **2018**, *26*, 445–468. [CrossRef]
24. Feng, S.; Huang, M.Y.; Crane, J.H.; Wang, Y. Characterization of key aroma-active compounds in lychee (*Litchi chinensis* Sonn.). *J. Food Drug Anal.* **2018**, *26*, 497–503. [CrossRef]
25. Mir, N.A.; Beaudry, R. Effect of superficial scald suppression by diphenylamine application on volatile evolution by stored cortland apple fruit. *J. Agric. Food Chem.* **1999**, *47*, 7–11. [CrossRef]
26. Hui, W.; Niu, J.P.; Xu, X.Y.; Guan, J.P. Evidence supporting the involvement of MHO in the formation of superficial scald in 'Dangshansuli' pears. *Postharvest Biol. Technol.* **2016**, *121*, 43–50. [CrossRef]
27. The Good Scents Company Information System. Available online: http://www.thegoodscentscompany.com/index.html (accessed on 5 January 2019).
28. Cheng, H.; Chen, J.L.; Li, X.; Pan, J.X.; Xue, S.J.; Liu, D.H.; Ye, X.Q. Differentiation of the volatile profiles of Chinese bayberry cultivars during storage by HS-SPME–GC/MS combined with principal component analysis. *Postharvest Biol. Technol.* **2015**, *100*, 59–72. [CrossRef]

29. Mannucci, A.; Serra, A.; Remorini, D.; Castagna, A.; Mele, M.; Scartazza, A.; Ranieri, A. Aroma profile of Fuji apples treated with gelatin edible coating during their storage. *LWT Food Sci. Technol.* **2017**, *85*, 28–36. [CrossRef]
30. Amanpour, A.; Guclu, G.; Kelebek, H.; Selli, S. Characterization of key aroma compounds in fresh and roasted terebinth fruits using aroma extract dilution analysis and GC–MS-Olfactometry. *Microchem. J.* **2019**, *145*, 96–104. [CrossRef]
31. Zakaria, S.R.; Saim, N.; Osman, R.; Abdul Haiyee, Z.; Juahir, H. Combination of sensory, chromatographic, and chemometrics analysis of volatile organic compounds for the discrimination of authentic and unauthentic harumanis mangoes. *Molecules* **2018**, *23*, 2365. [CrossRef]
32. Kjeldsen, F.; Christensen, L.P.; Edelenbos, M. Changes in volatile compounds of carrots (*Daucus carota* L.) during refrigerated and frozen storage. *J. Agric. Food Chem.* **2003**, *51*, 5400–5407. [CrossRef]
33. Kjeldsen, F.; Christensen, L.P.; Edelenbos, M. Quantitative analysis of aroma compounds in carrot (*Daucus carota* L.) cultivars by capillary gas chromatography using large-volume injection technique. *J. Agric. Food Chem.* **2001**, *49*, 4342–4348. [CrossRef] [PubMed]
34. Fariña, L.; Boido, E.; Carrau, F.; Versini, G.; Dellacassa, E. Terpene compounds as possible precursors of 1,8-cineole in red grapes and wines. *J. Agric. Food Chem.* **2005**, *53*, 1633–1636. [CrossRef] [PubMed]
35. Heuskin, S.; Godin, B.; Leroy, P.; Capella, Q.; Wathelet, J.P.; Verheggen, F.; Haubruge, E.; Lognay, G. Fast gas chromatography characterisation of purified semiochemicals from essential oils of *Matricaria chamomilla* L. (Asteraceae) and *Nepeta cataria* L. (Lamiaceae). *J. Chromatogr. A* **2009**, *1216*, 2768–2775. [CrossRef] [PubMed]
36. Jalali-Heravi, M.; Parastar, H.; Sereshti, H. Development of a method for analysis of Iranian damask rose oil: combination of gas chromatography-mass spectrometry with Chemometric techniques. *Anal. Chim. Acta* **2008**, *623*, 11–21. [CrossRef] [PubMed]

Sample Availability: Not available.

© 2019 by the authors. Licensee MDPI, Basel, Switzerland. This article is an open access article distributed under the terms and conditions of the Creative Commons Attribution (CC BY) license (http://creativecommons.org/licenses/by/4.0/).

Article

HS-SPME Analysis of True Lavender (*Lavandula angustifolia* Mill.) Leaves Treated by Various Drying Methods

Jacek Łyczko [1,*], Klaudiusz Jałoszyński [2], Mariusz Surma [2], Klaudia Masztalerz [2] and Antoni Szumny [1]

[1] Faculty of Biotechnology and Food Science, Wrocław University of Environmental and Life Sciences, Norwida 25, 50-375 Wrocław, Poland; antoni.szumny@upwr.edu.pl
[2] Institute of Agricultural Engineering, Wrocław University of Environmental and Life Sciences, Chełmońskiego 37-41, 51-630 Wrocław, Poland; klaudiusz.jaloszynski@upwr.edu.pl (K.J.); mariusz.surma@upwr.edu.pl (M.S.); klaudia.urbanska@upwr.edu.pl (K.M.)
* Correspondence: jacek.lyczko@upwr.edu.pl; Tel.: +48-71-320-51-47

Academic Editor: Constantinos K. Zacharis
Received: 31 January 2019; Accepted: 18 February 2019; Published: 20 February 2019

Abstract: True lavender (*Lavandula angustifolia* Mill.) is a widely used flavoring and medicinal plant, which strong aroma is mainly composed of linalool and linalyl acetate. The most valuable parts of the plant are the flowers, however leaves are also abundant in volatile constituents. One of the main factors responsible for its quality is the preservation procedure, which usually comes down to a drying process. For this reason an attempt to verify the influence of various drying methods (convective drying, vacuum-microwave drying and combined convection pre-drying with vacuum-microwave finishing drying) on the quality of true lavender leaves was carried out by determination of the volatile constituents profile by solid-phase microextraction (SPME) coupled with GC-MS technique. Total essential oil (EO) content was also verified. The study has revealed that the optimal drying method is strongly dependent on the purpose of the product. For flavoring properties convective drying at 60 °C is the most optimal method, while the best for preserving the highest amount of EO is vacuum-microwave drying at 480 W. Furthermore, SPME analysis had shown that drying may increase the value of true lavender leaves by significantly affecting the linalool to linalyl acetate to camphor ratio in the volatile profile.

Keywords: essential oil; drying; SPME; true lavender; volatile constituents

1. Introduction

Lavandula angustifolia Mill. (also named *Lavandula officinalis* Chaix)—the true lavender—is a essential oil-bearing plant known worldwide, which history of usage starts in Greek and Roman times and last up to this day. The entire genus belongs to the large *Lamiacae* family, which is mostly native to the Mediterranean region, however true lavender is a commonly growing plant in England, Europe, North America and Australia. The most valuable part of the plant are flowers due to their much higher essential oil content than leaves, and a favorable linalool to linalyl aceate to camphor ratio [1].

Nowadays due to the well-recognizable aroma lavender plants or their derivatives find applications in numerous ways, like in perfumery, cosmetics and household products, antimicrobial agents, food fragrance and flavor improvement or as food preservatives [1–3]. Furthermore, the essential oil obtained from lavender is an interesting object for trials considering biological activity and even in medicinal trials. Some studies and overviews from recent years mention the anti-aging, analgesic, nuroprotective, sedative or anticancer activities of lavender essential oil [2–9].

These various lavender essential oil applications are due to their unique chemical composition, rich in monoterpenes, sesquiterpenes, sesquiterpenoids, aliphatic compounds and especially an abundance of monoterpenoids [10], with linalool and linalyl acetate highlighted as main flower components [1,11,12]. In the case of the leaves the main essential oil constituents are eucalyptol (1,8-cineole), camphor and borneol [13–15].

As the main factors affecting the quality of the essential oils obtained from essential oils-bearing plants, plant chemotype, growing conditions and location, fertilizers used, time of harvesting and post-harvest treatment (including preservation method) are mentioned [2,11]. Among those factors, the preservation method has the most significant influence, where the most common one for plants rich in essential oils is drying [16–18]. Drying of essential oil-bearing plants allows one to obtain sustainable products with guaranteed quality, although it may cause also considerable losses of valuable constituents—mainly affecting the volatile constituents [17]. Furthermore, the color of the raw material may be strongly influenced by drying [16].

The traditional and natural method of drying uses solar radiation, however nowadays convective drying (CD), which uses flows of the hot air [17,18], is the most common drying method used in natural products treatment. Nevertheless other techniques like freeze-drying, infrared drying, vacuum-microwave drying (VMD), spray drying or a combination of convective pre-drying with vacuum-microwave finishing drying (CPD-VMFD) are lately the objects of numerous investigations regarding natural products drying [18]. Unfortunately in case of drying the true lavender leaves only single factors were investigated. Interest in this topic is due to the necessity to find an optimal drying method for specific raw materials. In addition, not only a specific technique, but also its parameters, like drying time, temperature or pressure have a significant influence on the quality of the obtained products [19–22]. Overall the most important are air velocity and temperature—for plants the most suitable temperature is one between 50 °C and 60 °C [16].

The objective of this study was to determine the volatile profile composition and compound quantity of true lavender leaves and the influence of three drying methods (CD, VMD, CPD-VMFD) applied with various parameters. The study was done by a solid-phase microextraction (SPME) coupled with gas chromatography mass spectrometry technique (GC-MS). Also the total essential oil content was validated by using a hydrodistillation extraction technique.

2. Results and Discussion

2.1. Drying Kinetics

Figure 1 shows changes with time of the moisture ratio (MR) of leaf samples dehydrated by VMD at three magnetron powers (240, 360 and 480 W, Figure 1a), CD at temperatures in the range of 50 to 70 °C (Figure 1b), and combined (CPD-VMD) drying consisting of CD at 60 °C and VMD at a magnetron power of 480 W (Figure 1c). The drying times, together with the maximum temperatures, the final moisture content and the constants of the Page model are listed in Table 1.

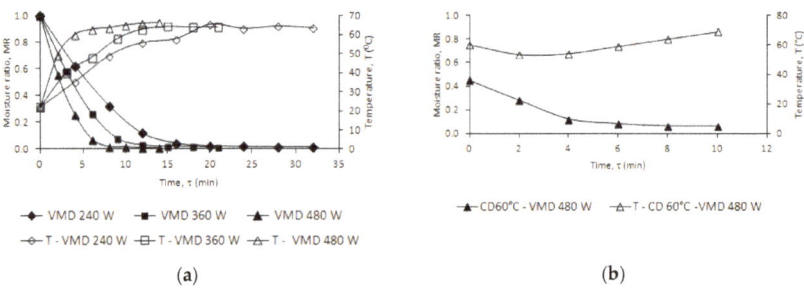

Figure 1. Cont.

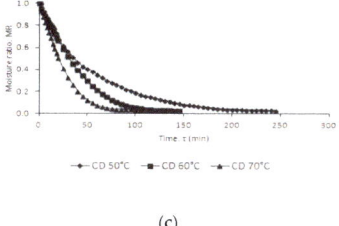

(c)

Figure 1. (a) Drying kinetics of true lavender leaves samples processed using VMD at magnetron powers 240, 360 and 480 W; (b) Drying kinetics of true lavender leaves samples processed using CD at temperatures of 50, 60 and 70 °C; (c) Drying kinetics of true lavender leaves samples processed using VMFD at 480 W after CPD at temperature 60 °C.

Table 1. Final moisture content (M_{fwb}), maximum temperature of the sample T, convective drying time (τ), vacuum microwave drying time (τ_1), and constants A, k and n of the modified Page model describing the drying kinetics.

Drying Conditions	A	Constants K	n	R^2	RMSE	τ	τ_1	T (°C)	M_{fwb} (%)
CD 50 °C	1.000	0.0201	0.953	0.9984	0.0125	245	-	50	7.18
CD 60 °C	1.000	0.0125	1.173	0.9991	0.0104	145	-	60	7.09
CD 70 °C	1.000	0.0202	1.150	0.9983	0.0156	135	-	70	7.42
VMD 240 W	1.000	0.0736	1.328	0.9989	0.0127	-	32	64	6.78
VMD 360 W	1.000	0.1205	1.358	0.9991	0.0104	-	21	65	6.90
VMD 480 W	1.000	0.2339	1.300	0.9991	0.0111	-	14	66	6.87
CPD 60°C + VMFD 480 W	0.449	0.2895	0.893	0.9982	0.0155	60	10	64	7.02

The Page model can be successfully used to describe the drying kinetics of the true lavender leaves dehydrated by the CD, VMD and CPD-VMD methods, characterized by high values of the determination coefficient ($R^2 > 0.99$) and low RMSE values (<0.05). A good adaptation of the applied Page model for description of the drying kinetics can be found in many earlier publications of dill leaves, chanterelle and oyster mushrooms [23–25].

In the case of CD increasing the drying air temperature from 50 to 70 °C decreased the time of drying from 245 to 135 min, respectively. In VMD drying, radical reductions in the total drying time have been observed: the time was shortened from 32 to 14 min with a power change from 240 to 480 W. This radical reduction in the total drying time of VMD compared to CD is a result of the conventional water diffusion occurring, according to Fick's law, that is supported by a pressure diffusion mechanism of the Darcy type [26]. Combined CPD and VMFD using 480 W, shortened the drying time of leaves almost 18-fold compared to CD at 50 °C. The use of CD and 480 W power caused a drop in the material temperature during VMD by 4 °C for leaves and 2 °C for flowers in reference to VMD 480 W. This condition is caused by the molecular distribution of water particles inside the dried CD and the distribution of water particles has an effect on the generation of heat energy production under microwave radiation during VMD [21,27,28]. Energy consumption during the CD of plant materials is much lower than in VMD [29,30]. In industrial conditions, the best solution is a combined drying process consisting of CPD and VMFD. The CD is very effective at the beginning of the drying process (the largest loss of water occurs during that phase) and VMD at the final stage of drying (removal of water strongly bound to the cellular structure of the material being dried) [18,27,28]. The final choice of recommended drying process should be related to the aspects of the dried material (volatile composition and sensory attributes) [27,31].

2.2. Volatile Constituents Profile of Fresh True Lavender Leaves

HS-SPME analysis coupled with the GC-MS technique had revealed one hundred and four peaks (one as a two compound mixture) recognized as volatile constituents, of which only one hundred of them could be identified (the mass spectra of unidentified constituents are available in supplementary materials). Volatile constituents of true lavender leaves are listed in Table 2. Among them nineteen compounds were qualified as monoterpene hydrocarbons, twenty-six as oxygenated monoterpenes, twenty-four as sesquiterpene hydrocarbons, nine as oxygenated sesquiterpenes, ten as esters and eleven as others.

Table 2. Volatile constituents of fresh true lavender leaves.

Compound	RT (min)	RI_lit [1]	RI_lit [2]	RI_exp [3]	Content [%] [4]
1-Penten-3-ol	2.407	-	684	686	Tr [5]
(Z)-3-Hexenal	3.755	797	810	808	0.23 ± 0.14
(E)-2-Hexenal	4.765	846	854	857	0.33 ± 0.17
(Z)-3-Hexen-1-ol	4.821	850	857	859	1.75 ± 0.35
1-Hexanol	5.087	863	868	871	0.32 ± 0.09
(E,E)-2,4-Hexadienal	6.113	909	911	913	0.15 ± 0.09
5.5-Dimethyl-1-vinylbicyclo[2.1.1]hexane	6.380	-	921	924	tr
Tricyclene	6.479	921	926	928	0.17 ± 0.03
α-Thujene	6.591	924	930	932	0.12 ± 0.05
α-Pinene	6.788	932	939	940	0.30 ± 0.07
Camphene	7.209	946	954	955	0.92 ± 0.19
3,7,7-Trimethyl-1.3.5-cycloheptatriene	7.840	-	972	976	tr
Sabinene	7.911	696	976	978	0.11 ± 0.03
1-Octen-3-ol	8.038	974	979	982	0.72 ± 0.06
3-Octanone	8.260	979	986	988	0.22 ± 0.03
β-Myrcene	8.415	988	991	993	0.52 ± 0.22
Mesitylene	8.512	994	995	996	tr
n-Decane	8.681	1000	1000	1000	0.19 ± 0.04
α-Phellandrene	8.850	1002	1005	1007	0.49 ± 0.28
3-Carene	9.031	1008	1011	1013	1.60 ± 0.66
m-Cymene	9.397	1020	1024	1026	2.58 ± 0.33
p-Cymene	9.482	1022	1030	1028	4.81 ± 0.52
Limonene	9.634	1024	1030	1033	3.42 ± 1.16
Eucalyptol	9.692	1026	1031	1035	7.28 ± 1.06
β-cis-Ocimene	9.902	1032	1038	1042	0.16 ± 0.03
β-trans-Ocimene	10.240	1044	1050	1053	0.14 ± 0.04
γ-Terpinene	10.605	1054	1059	1063	0.11 ± 0.03
trans-Sabinene hydrate	10.886	1065	1070	1071	0.23 ± 0.05
cis-Linalool oxide	11.041	1067	1074	1076	0.13 ± 0.03
unknown	11.167	-	-	1079	tr
m-Cymenene	11.419	1082	1085	1086	0.50 ± 0.04
p-Mentha-2.4(8)-diene	11.519	1085	1088	1089	0.34 ± 0.10
p-Cymenene	11.602	1089	1091	1091	0.30 ± 0.03
Camphenone	11.840	1095	1096	1097	0.26 ± 0.02
Linalool	11.953	1095	1096	1100	0.42 ± 0.03
1.3.8-p-Menthatriene	12.206	1108	1110	1108	0.10 ± 0.02
1-Octen-3-ol acetate	12.360	1110	1112	1114	3.80 ± 0.52
cis-p-Menth-2-en-1-ol	12.556	1118	1121	1120	0.18 ± 0.04
trans-p-Mentha-2.8-dien-1-ol	12.724	1119	1122	1125	0.64 ± 0.19
cis-p-Mentha-2.8-dien-1-ol	13.173	1133	1137	1139	0.26 ± 0.03
trans-p-Menth-2-en-1-ol	13.327	1136	1140	1144	0.49 ± 0.08
Camphor	13.496	1141	1146	1149	2.09 ± 0.29
Tetrahydrolavandulol	13.960	1157	1161	1162	0.48 ± 0.09
Borneol + Lavandulol	14.240	1165	1169	1170	4.66 ± 0.69
Melilotal	14.450	1179	1182	1176	tr

Table 2. Cont.

Compound	RT (min)	Retention Indeces (RI)			Content [%] [4]
		RI_lit [1]	RI_lit [2]	RI_exp [3]	
Terpinen-4-ol	14.631	1174	1177	1181	0.59 ± 0.07
m-Cymen-8-ol	14.774	1176	1179	1184	2.09 ± 0.25
p-Cymen-8-ol	14.914	1179	1182	1188	4.09 ± 0.67
α-Terpineol	15.082	1186	1189	1193	0.31 ± 0.06
Myrtenol	15.278	1194	1195	1198	0.20 ± 0.14
trans-Piperitol	15.671	1207	1208	1210	0.65 ± 0.07
cis-Carveol	16.035	1215	1217	1222	0.37 ± 0.07
(Z)-Ocimenone	16.159	1226	1229	1226	0.26 ± 0.07
exo-Fenchyl acetate	16.356	1229	1232	1232	0.49 ± 0.04
cis-Verbenol	16.623	1237	1244	1240	tr
Cumin aldehyde	16.748	1238	1241	1244	1.92 ± 0.59
Carvone	16.874	1246	1243	1247	1.08 ± 0.28
Geraniol	17.055	1249	1252	1253	0.33 ± 0.29
Linalyl acetate	17.263	1254	1257	1259	2.21 ± 0.73
Geranial	17.529	1264	1267	1267	0.10 ± 0.08
trans-Carvone oxide	18.021	1273	1276	1281	0.33 ± 0.07
Bornyl acetate	18.301	1284	1285	1288	5.57 ± 0.82
Lavandulyl acetate	18.428	1288	1290	1292	1.72 ± 0.25
Terpinen-4-ol acetate	18.761	1299	1299	1301	0.18 ± 0.02
unknown	19.124	-	-	1314	0.61 ± 0.09
Myrtenyl acetate	19.435	1324	1326	1326	0.16 ± 0.06
δ-Elemene	19.749	1335	1337	1337	tr
α-Terpinyl acetate	20.036	1346	1349	1347	0.26 ± 0.08
α-Cubebene	20.179	1348	1351	1351	tr
α-Longipinene	20.351	1350	1352	1357	0.18 ± 0.01
unknown	20.465	-	-	1361	0.31 ± 0.05
Silphiperfola-4.7(14)-diene	20.578	1358	1362	1365	tr
Neryl acetate	20.748	1359	1364	1371	0.26 ± 0.06
α-Copaene	21.134	1374	1376	1383	0.14 ± 0.02
Geranyl acetate	21.248	1379	1381	1387	0.49 ± 0.11
α-Bourbonene	21.375	1387	1388	1391	tr
unknown	21.461	1394	1396	1394	tr
β-Longipinene	21.634	1400	1400	1399	0.26 ± 0.06
Sesquithujene	21.833	1405	1405	1409	tr
α-Cedrene	22.049	1410	1411	1420	1.01 ± 0.25
(E)-Caryophyllene	22.176	1417	1419	1427	6.11 ± 1.48
α-Bergamotene	22.506	1432	1435	1443	0.87 ± 0.33
Cadina-3.5-diene	22.745	-	1458	1455	1.12 ± 0.40
(E)-β-Farnesene	22.889	1454	1457	1462	1.35 ± 0.38
cis-Muurola-4(15).5-diene	23.084	1465	1466	1472	1.44 ± 0.45
4-epi-α-Acoradiene	23.155	1474	1475	1475	0.19 ± 0.00
Germacrene D	23.441	1484	1481	1489	0.58 ± 0.17
β-Himachalene	23.629	1500	1500	1498	tr
unknown	23.741	1502	-	1505	tr
α-Bulnesene	23.840	1509	1509	1511	0.92 ± 0.16
γ-Cadinene	24.023	1513	1513	1523	10.53 ± 1.51
cis-Calamenene	24.149	1528	1529	1531	0.65 ± 0.07
10-epi-Cubebol	24.290	1533	1535	1540	0.11 ± 0.05
α-Cadinene	24.402	1537	1538	1547	0.12 ± 0.02
Cadala-1(10).3.8-triene	24.473	-	1555	1552	tr
trans-Cadinene ether	24.669	1557	-	1564	0.35 ± 0.09
unknown	24.851	-	-	1576	0.13 ± 0.05
Spathulenol	24.950	1577	1578	1582	0.25 ± 0.05
Caryophyllene oxide	25.158	1582	1583	1595	3.31 ± 0.18
1-epi-Cubenol	25.552	1627	1628	1628	0.56 ± 0.03
τ-Cadinol	25.860	1635	1340	1656	2.04 ± 0.55
unknown	26.056	-	-	1673	0.11 ± 0.03
14-Hydroxy-4.5-dihydrocaryophyllene	26.407	1706	1706	1706	0.21 ± 0.11
unknown	26.911	1760	1761	1764	0.23 ± 0.02

[1] Retention indices according to Adams [32]; [2] Retention indices according to NIST14 database; [3] Relative retention indices calculated against n-alkanes; [4] % calculated from TIC data; [5] tr. < 0.1%.

The main headspace volatile constituents of the examined true lavender leaves samples were p-cymen-8-ol (4.09% ± 0.67), a mixture of borneol and lavandulol (4.66% ± 0.69), o-cymene (4.81% ± 0.52), bornyl acetate (5.57% ± 0.82), (E)-caryophyllene (6.11% ± 1.48), eucalyptol (7.28% ± 1.06) and γ-cadinene (10.53 ± 1.51). In less amounts cumin aldehyde (1.92% ± 0.59), τ-cadinol (2.04% ± 0.55), m-cymen-8-ol (2.09% ± 0.25), camphor (2.09% ± 0.92), p-cymene (2.58% ± 0.33), caryophyllene oxide (3.31% ± 0.18), limonene (3.42% ± 1.16) and 1-octen-3-pl acetate (3.80% ± 0.52), which have a significant influence on true lavender leaves' fragrance quality, were identified. The most characteristic and valuable constituents for true lavender (flowers), linalool and linalyl acetate, represented 0.42% ± 0.03 and 2.21% ± 0.73 of the total amount of volatile constituents, respectively.

Similar findings were reported in recent studies where eucalyptol (8.50% and 31.9%), borneol (15.21% and 24%), camphor (2.00% and 16.1%), cumin aldehyde (0.50% and 2.2%) were identified as main volatile components of a true lavender leaves sample [33,34]. Also, one of these studies, by Hassanpouraghdam et al. [34] pointed out low amounts or even a lack of linalool (0.7%) and linalyl acetate. This result is contrary to the one obtained in this study, however it may be related to the slightly different plant chemotype or due to the fact that in Hassanpouraghdam's study leaves essential oil was analyzed, not headspace volatiles. Nurzyńska-Wierdak and Zawiślak [35] have identified linalool and linalyl acetate in a similar ratio (1:5), and furthermore they also found higher amounts of γ-cadinene (3.4 ± 0.1) and caryophyllene oxide (7.2% ± 0.2).

Unfortunately, there is a lack of reports in literature including HS-SPME analysis of true lavender leaves volatile constituents. Most of available ones takes as study object lavender flowers or whole aerial parts of the plant, where linalool and linalyl acetate dominate in the chromatographic profile of the volatile constituents [36–38]. Torabbeigi and Aberoomand Azar [39] reported high amounts of eucalyptol (41.37%), camphor (15.83%), borneol (12.32%), α-pinene (4.66%), and γ-cadinene (1.07%) found by HS-SPME analysis of true lavender samples. At the same time they did not find any traces of linalool or linalyl acetate, suggesting that the major part of their samples were lavender leaves.

2.3. Effect of the Drying Methods on the Quantity of True Lavender Leaves Volatile Constituents

In the fresh true lavender leaves cultivated in Poland used in this study the content of essential oil was 3.082 g per 100 g^{-1} of DW. Overall this essential oil yield is high in comparison to previously reported ones, as Mirahmadi and Norouzi [40] obtained just 2.34% of essential oil from true lavender. Moreover, Milojević et al. [41] report the essential oil yield in sage and eucalyptus leaves ranges from 2% up to 2.87%. Changes of essential oil content, the concentration of sixteen major volatile constituents and linalool caused by the various drying methods are shown in Table 3.

In the case of essential oil content all applied drying methods significantly affected the raw material. The most efficient method was VMD 480 W (1.302 g per 100 g^{-1}), followed by VMD 240 W (1.075 g 100 g^{-1}), CD 70 °C (0.992 g per 100 g^{-1}) and CPD-VMFD (0.921 g per 100 g^{-1}) which were in overlapping significant groups. The percent recovery of essential oil in these methods were as follows 42.26%, 34.87%, 32.19% and 29.87%, in comparison to the amount of essential oil obtained from fresh sample. The less efficient drying method was CD 50°C, with a 19.06% recovery. The ratios of percent recovery between fresh sample and ones subjected to drying are presented in Figure 2. Baydar and Erbaş [42], Figiel et al. [19], Ghasemi et al. [43] found as well that due to the applied drying method or its parameters the decrease in essential oil yield of green plant parts may range as high as three to five times. Furthermore, Politowicz et al. [24] and Nöfer et al. [27], in the case of mushroom drying, observed similar effects to the ones found in this study.

Table 3. Variability of major volatile constituents, linalool, and total essential oil of true lavender leaves caused by various drying methods.

Compound	Fresh	Drying Method						
		CD 50 °C	CD 60 °C	CD 70 °C	CPD-VMFD	VMD 240 W	VMD 360 W	VMD 480 W
		Content [%] [1]						
p-Cymene	2.58 [a]	2.73 [c]	2.72 [c]	1.76 [d]	2.01 [d,e]	2.15 [e]	2.27 [e]	3.50 [b]
o-Cymene	4.81 [a]	6.26 [c]	5.65 [d]	3.05 [f]	3.69 [e]	4.62 [g]	4.73 [g]	8.08 [b]
Limonene	3.42 [a]	3.27 [f]	3.47 [f]	1.31 [e]	3.08 [c,f]	1.97 [d,e]	2.41 [c,d]	6.99 [b]
Eucalyptol	7.28 [a]	5.01 [b,c]	3.71 [d]	5.12 [b]	3.98 [c,d]	3.25 [d]	3.74 [d]	3.44 [d]
1-Octen-3-ol. acetate	3.80 [a]	2.70 [d,e]	2.82 [d,e]	4.23 [c]	6.22 [b]	2.10 [e]	4.42 [c]	3.68 [c,d]
Camphor	2.09 [a]	2.32 [b]	1.40 [d]	1.89 [c]	0.40 [f]	1.12 [e]	0.34 [f]	0.39 [f]
Borneol + Lavandulol	4.66 [a]	7.63 [b]	5.75 [d]	6.07 [d]	1.37 [e]	4.78 [c]	1.46 [e]	1.35 [e]
m-Cymen-8-ol	2.09 [a]	3.18 [c]	2.48 [d]	3.67 [b]	0.02 [e]	2.74 [d]	0.08 [e]	0.07 [e]
p-Cymen-8-ol	4.09 [a]	6.31 [c]	6.05 [c]	7.17 [b]	1.10 [d]	6.07 [c]	1.04 [d]	0.89 [d]
Cumin aldehyde	1.92 [a]	3.59 [c]	3.74 [c]	4.48 [b]	0.59 [d]	4.30 [b]	0.66 [d]	0.59 [d]
Linalyl acetate	2.21 [a]	3.46 [d]	11.06 [b]	4.23 [d]	1.60 [e]	5.29 [c]	1.66 [e]	1.75 [e]
Bornyl acetate	5.57 [a]	3.54 [c]	2.36 [e]	4.04 [b]	0.07 [f]	3.07 [d]	0.07 [f]	0.14 [f]
Caryophyllene <(E)->	6.11 [a]	2.11 [d]	2.78 [c]	3.51 [f]	6.38 [b]	4.85 [e]	5.28 [e]	3.98 [f]
γ-Cadinene	10.53 [a]	3.67 [f]	4.43 [e,f]	4.74 [e]	8.48 [c,d]	7.80 [d]	9.20 [c]	5.88 [b]
Caryophyllene oxide	3.31 [a]	2.43 [c]	1.63 [b]	2.12 [b,c]	2.24 [b,c]	2.11 [b,c]	2.47 [c]	1.89 [b,c]
τ-Cadinol	2.04 [a]	1.62 [d]	1.44 [d]	2.78 [c]	2.74 [c]	3.37 [b,c]	3.85 [b]	1.74 [d]
Σ	66.51 [a]	59.83 [b]	61.42 [b]	60.17 [b]	43.97 [c]	59.59 [b]	43.68 [c]	42.47 [c]
Linalool	0.38 [a]	4.32 [c]	6.33 [b]	4.71 [c]	0.76 [a,d]	6.62 [b]	0.73 [a,d]	1.16 [d]
TOTAL essential oil [mL 100g^{-1} dw] [2]	3.082 [a]	0.588 [e]	0.726 [f]	0.992 [c,d]	0.921 [b,c]	1.075 [d]	0.881 [b]	1.302 [g]

[1] Values followed by the same letter within a row are not significantly different ($p > 0.05$, Duncan's test); [2] Values obtained from steam distillation in Deryng apparatus.

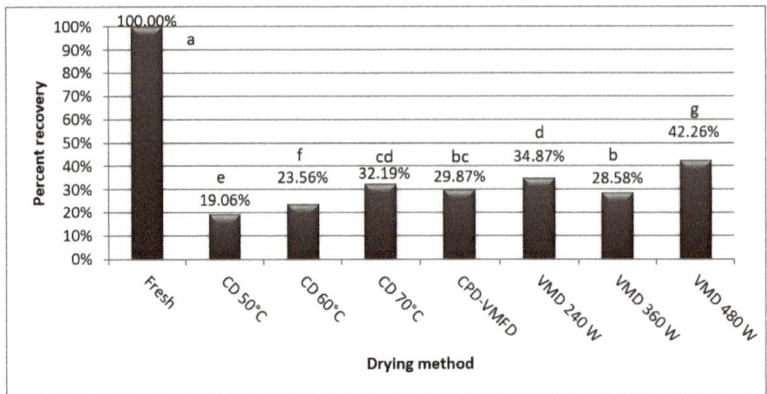

Figure 2. Percent recovery of essential oil of true lavender leaves after applying various drying methods.

The total essential oil content results are not equivalent to the content of sixteen major constituents. In fresh true lavender leaves, sixteen major constituents accounting for 66.51% of the total volatile constituents and changes caused by all drying methods were significantly distinct. Less differences were observed in the CD and CPD-VMFD methods (5.09–6.68 percentage points) and the highest were observed for the VMD method (22.54–24.04 percentage points). Further, some results in the case of particular constituents among the sixteen major ones are worth underlining. Again, all drying methods had a significant influence on a particular constituent share of total volatile constituents. The most interesting was the increase a share of linalyl acetate (even up to 11.06% of the share in the CD 60 °C method) along with the decrease of camphor share (down to 1.40%) at the same time. Also the share of linalool, the main aroma compound for true lavender, increased significantly after all drying treatments, except for VMD 240 W and VMD 360 W. These results suggest that applying drying, mainly CD, for true lavender leaves, may improve the characteristics for use in flavoring, in accordance with Kim and Lee [44] and Da Porto and Decorti [45], who report that the high ratio of linalool and linalyl acetate to camphor ratio is an important quality marker for lavender fragrance. Similar changes after applying drying were obtained by Śmigielski et al. [10]. Nevertheless, if the aim is to preserve as much essential oil as possible, the VMD methods would be more applicable. Very poor results, both in case of total essential oil and major volatile constituents, were obtained after the CPD-VMFD method, what is in contradiction with results obtained by Szumny et al. [20] for rosemary drying (*R. officinalis*), however the taxonomic differences between rosemary and true lavender should be considered.

3. Materials and Methods

3.1. Plant Material

The drying process was carried out on true lavender cultivated in Poland (Kawon-Hurt Nowak Sp.j. Company, Gostyń, Poland). The initial moisture content of material was 2.7 kg per kg of dry weight. The drying processes were stopped after no further change in weights was observed. Moisture content of samples was determined using a vacuum dryer (SPT-200. ZEAMIL Horyzont, Krakow, Poland).

3.2. Drying Methods

3.2.1. Convective Drying (CD)

CD was performed using the equipment designed and constructed at the Institute of Agricultural Engineering (Wrocław University of Environmental and Life Sciences, Wrocław, Poland). Samples were placed in the container (d = 100 mm) and dried at 50 °C. 60 °C and 70 °C—all with an air velocity of 0.5 ms^{-1}.

3.2.2. Vacuum-Microwave Drying (VMD)

VMD was performed on samples with a SM 200 dryer (Plazmatronika, Wrocław, Poland). The dryer was equipped in cylindrical drum made of glass (18 cm of diameter × 27 cm of length). The drum with glass rotated with 6 rev·min^{-1}. In the dryer system there was a BL 30P vacuum pump (Tepro, Koszalin, Poland), an MP 211 vacuum gauge (Elvac, Bobolice, Poland) and a compensation reservoir of 0.15 m^3 capacity and a cylindrical tank. In this study, three power levels (240, 360 and 480 W) and pressures ranging from of 4 to 6 kPa were used. The maximum temperature of dried lavender leaves was measured right after the removal from the dryer using an i50 infrared camera (Flir Systems AB, Stockholm, Sweden).

3.2.3. Combined Drying—Pre-Drying by Convective Drying with Vacuum-Microwave Finishing-Drying (CPD-VMFD)

CPD-VMFD performed on samples consisted of CPD at a temperature of 60 °C until a moisture content of leaves was around 0.45 kg·kg^{-1} db, was reached, followed by VMFD at 480 W.

3.3. Modeling of Drying Kinetics

The drying kinetics of CD, VMD and CPD-VMFD were fitted based on the mass losses of the true lavender samples. For CD, weight losses were monitored every 2 min for the initial 20 min and then every 5 min thereafter until the end of the drying process.

VMD samples were monitored every 2, 3 and 4 min for 480, 360 and 240 W. Different drying time intervals were applied in order to ensure a similar energy input between subsequent measurements regardless of the microwave power level.

The moisture ratio (MR) of lavender leaves during drying experiments was calculated using the following equation:

$$MR = \frac{M_{(t)} - M_e}{M_0 - M_e} \tag{1}$$

where $M_{(t)}$ is the moisture content at time τ. M_0 is the initial moisture content, and M_e is the equilibrium moisture content (kg water/kg dry matter). The values of M_e are relatively small comparing to those of $M_{(t)}$ or M_0. The error due to the simplification is negligible [46–48], thus the moisture ratio was calculated as follows:

$$MR = \frac{M_{(t)}}{M_0} \tag{2}$$

Table Curve 2D Windows v2.03 was used to fit the basic drying models to the measured MR determined accordingly to Equation (2). There are several drying models in the literature that can be used to describe the kinetics of drying plant materials. For drying model selection, drying curves were fitted to five well known thin drying models, including the modified Page model. Henderson–Pabis, logarithmic, Midilli-Kucuk, and Weibull ones. The best fit was determined using two parameters: the value of the coefficient of determination (R^2) and root-mean squared error (RMSE). A model fits better if the value of R^2 is closer to 1, and the RMSE value is closer to 0, using the following equations:

$$R^2 = \frac{\sum_{i=1}^{N}(MR_{pre_i} - \overline{MR}_{exp})}{\sum_{i=1}^{N}(MR_{exp_i} - \overline{MR}_{exp})} \tag{3}$$

$$RMSE = \sqrt{\frac{1}{N} \cdot \sum_{i=1}^{N}(MR_{exp_i} - MR_{pre_i})^2} \tag{4}$$

where *MR* is moisture ratio, $\overline{MR}$ is the mean value of moisture ratio, "*pre*" and "*exp*" indicate predicted and experimental values, respectively, while "*i*" indicates subsequent experimental data and *N* is the number of observations.

Tests conducted in this study proved that the best fitting was obtained for the modified Page model:

$$MR = A \exp(-k\tau^n) \tag{5}$$

where *A*, *n*, and *k* are constants.

3.4. Solid-Phase Micro Extraction (SPME) Analysis

HS-SPME analysis (30 min exposure to a 2 cm DVB/CAR/PDMS fiber, Supelco, Bellefonte, PA, USA, followed by analyte desorption at 220 °C for 3 min) was performed on Varian CP-3800/Saturn 2000 apparatus (Varian, Walnut Creek, CA, USA) equipped with a Zebron ZB-5 MSI (30 m × 0.25 mm × 0.25 µm) column (Phenomenex, Shim-Pol, Poland). About 0.100 g of fresh or 0.150 g of dried sample was put in to headspace vials and kept in laboratory water bath at 70 °C. 0.5 mg of 2-undecanone (Sigma Aldrich, Saint Louis, MO, USA) as an internal standard was added.

3.5. GC-MS Analysis

The GC oven temperature was programmed from 50 °C, to 130 °C at rate 4.0°C. then to 180 °C at rate 10.0 °C, then to 280 °C at rate 20.0 °C. Scanning was performed from 35 to 550 m/z in electronic impact (EI) mode at 70 eV. Samples were injected at a 1:10 split ratio and helium gas was used as the carrier gas at a flow rate of 1.0 mL·min^{-1}. Analyses were run in triplicate.

3.6. Hydrodistillation of Essential Oil (EO)

Hydrodistillation of EOs was carried out by applying a Deryng apparatus. About 200 g of fresh sample or 100 g of dried sample was placed in 2 L round bottom flask with 500 mL of added distilled water. Yield was assessed as a measured volume of essential oil.

3.7. Identification and Quantification of Volatile Compounds

Identification of all volatile constituents obtained by HS-SPME analysis and hydrodistillation were based on comparison of experimentally obtained compound mass spectra with mass spectra available in NIST14 database. Also the experimentally obtained retention indeces (RI) by Kovats were compared with RI available in the NIST WebBook and literature data [32]. The quantification analysis was performed using ACD/Spectrus Processor (Advanced Chemistry Development, Inc., Toronto, ON, Canada) through the integration of the peak area of the chromatograms.

3.8. Statistical Analysis

The data from drying kinetics were subjected to the analysis of variance using Tukey's test ($p < 0.05$) and the data from quantitative essential oil and volatile constituents were subjected to the analysis of variance using Duncan's test ($p < 0.05$), all using the STATISTICA 13.3 software for Windows (StatSoft, KrakowPoland).

4. Conclusions

One hundred constituents were identified in the volatile profile of true lavender leaves, with *p*-cymen-8-ol (4.09% ± 0.67), a mixture of borneol and lavndulol (4.66% ± 0.69), *o*-cymene (4.81% ± 0.52), bornyl acetate (5.57% ± 0.82), (*E*)-caryophyllene (6.11% ± 1.48), eucalyptol (7.28% ± 1.06) and γ-cadinene (10.53±1.51) as a major ones. When various methods are applied during the drying process, this profile is strongly affected. The optimal drying method is dependent on the purpose of the product utilization. A most interesting fact is that the drying process may decrease the share of camphor, while increasing the share of linalool and linalyl acetate which are the most desirable in components in true lavender aroma. This result may be a good starting point for considering the improvement of the value of true lavender leaves in comparison to its flowers for flavoring applications.

Supplementary Materials: The following files have been submitted as supplementary materials: mass spectra for unidentified compounds, mentioned in Table 2 Unknown compound mass spectra.pdf.

Author Contributions: Conceptualization, J.Ł. and A.S.; methodology, J.Ł., A.S., K.J. and M.S.; validation. J.Ł., A.S., K.J., M.S., K.M.; formal analysis. J.Ł., K.J. and M.S.; investigation. J.Ł., K.J. and M.S.; resources. A.S.; data curation. J.Ł.; writing—original draft preparation. J.Ł. and K.J.; writing—review and editing. J.Ł. and A.S., K.M.; visualization. J.Ł., K.J., M.S. and K.M.; supervision. A.S.; project administration. J.Ł.

Funding: This research was funded by the Faculty of Biotechnology and Food Science, Wrocław University of Environmental and Life Sciences, grant number B030/0003/1 and under the program of the Minister of Science and Higher Education "Strategy of Excellence University of Research" in 2018–2019 project number 0019/SDU/2018/18 in the amount of PLN 700 000.

Conflicts of Interest: The authors declare no conflict of interest.

References

1. Preedy, V.R. *Essential Oils in Food Preservation, Flavor and Safety*; Elsevier Inc.: London, UK, 2016; ISBN 9780124166417.
2. Prusinowska, R.; Śmigielski, K.B. Composition, biological properties and therapeutic effects of lavender (*Lavandula angustifolia* L). A review. *Herba Pol.* **2014**, *60*, 56–66. [CrossRef]
3. Lesage-Meessen, L.; Bou, M.; Sigoillot, J.C.; Faulds, C.B.; Lomascolo, A. Essential oils and distilled straws of lavender and lavandin: a review of current use and potential application in white biotechnology. *Appl. Microbiol. Biotechnol.* **2015**, *99*, 3375–3385. [CrossRef] [PubMed]
4. Ayaz, M.; Sadiq, A.; Junaid, M.; Ullah, F.; Subhan, F.; Ahmed, J. Neuroprotective and anti-aging potentials of essential oils from aromatic and medicinal plants. *Front. Aging Neurosci.* **2017**, *9*, 1–16. [CrossRef] [PubMed]
5. Ali, B.; Al-Wabel, N.A.; Shams, S.; Ahamad, A.; Khan, S.A.; Anwar, F. Essential oils used in aromatherapy: A systemic review. *Asian Pac. J. Trop. Biomed.* **2015**, *5*, 601–611. [CrossRef]
6. López, V.; Nielsen, B.; Solas, M.; Ramírez, M.J.; Jäger, A.K. Exploring pharmacological mechanisms of lavender (*Lavandula angustifolia*) essential oil on central nervous system targets. *Front. Pharmacol.* **2017**, *8*, 1–8. [CrossRef] [PubMed]
7. Yu, S.H.; Seol, G.H. *Lavandula angustifolia* Mill. Oil and Its Active Constituent Linalyl Acetate Alleviate Pain and Urinary Residual Sense after Colorectal Cancer Surgery: A Randomised Controlled Trial. *Evidence-based Complement. Altern. Med.* **2017**, *2017*, 1–7. [CrossRef] [PubMed]
8. Tomi, K.; Kitao, M.; Murakami, H.; Matsumura, Y.; Hayashi, T. Classification of lavender essential oils: sedative effects of *Lavandula* oils. *J. Essent. Oil Res.* **2018**, *30*, 56–68. [CrossRef]
9. Uritu, C.M.; Mihai, C.T.; Stanciu, G.D.; Dodi, G.; Alexa-Stratulat, T.; Luca, A.; Leon-Constantin, M.M.; Stefanescu, R.; Bild, V.; Melnic, S.; et al. Medicinal plants of the family *Lamiaceae* in pain therapy: A review. *Pain Res. Manag.* **2018**, *2018*, 1–44. [CrossRef]
10. Śmigielski, K.; Prusinowska, R.; Raj, A.; Sikora, M.; Wolińska, K.; Gruska, R. Effect of drying on the composition of essential oil from *Lavandula angustifolia*. *J. Essent. Oil-Bearing Plants* **2011**, *14*, 532–542. [CrossRef]
11. Husnu Can Baser, K.; Buchbauer, G. *Handbook of Essential Oils. Science, Technology and Applications*; CRC Press: Boca Raton, FL, USA, 2009.
12. Marín, I.; Sayas-Barberá, E.; Viuda-Martos, M.; Navarro, C.; Sendra, E. Chemical Composition, Antioxidant and Antimicrobial Activity of Essential Oils from Organic Fennel, Parsley, and Lavender from Spain. *Foods* **2016**, *5*, 18–27. [CrossRef]
13. Hajhashemi, V.; Ghannadi, A.; Sharif, B. Anti-inflammatory and analgesic properties of the leaf extracts and essential oil of *Lavandula angustifolia* Mill. *J. Ethnopharmacol.* **2003**, *89*, 67–71. [CrossRef]
14. Kirimer, N.; Mokhtarzadeh, S.; Demirci, B.; Goger, F.; Khawar, K.M.; Demirci, F. Phytochemical profiling of volatile components of *Lavandula angustifolia* Miller propagated under in vitro conditions. *Ind. Crops Prod.* **2017**, *96*, 120–125. [CrossRef]
15. Mendoza-Poudereux, I.; Kutzner, E.; Huber, C.; Segura, J.; Arrillaga, I.; Eisenreich, W. Dynamics of monoterpene formation in spike lavender plants. *Metabolites* **2017**, *7*, 65. [CrossRef] [PubMed]

16. Rocha, R.P.; Melo, E.C.; Radünz, L.L. Influence of drying process on the quality of medicinal plants: A review. *J. Med. Plants Res.* **2011**, *5*, 7076–7084. [CrossRef]
17. Prusinowska, R.; Śmigielski, K. Losses of essential oils and antioxidants during the drying of herbs and spices. A review. *Eng. Sci. Technol.* **2015**, *2*, 51–62. [CrossRef]
18. Figiel, A.; Michalska, A. Overall Quality of Fruits and Vegetables Products Affected by the Drying Processes with the Assistance of Vacuum-Microwaves. *Int. J. Mol. Sci.* **2017**, *18*, 71. [CrossRef] [PubMed]
19. Figiel, A.; Szumny, A.; Gutiérrez-Ortíz, A.; Carbonell-Barrachina, Á.A. Composition of oregano essential oil (*Origanum vulgare*) as affected by drying method. *J. Food Eng.* **2010**, *98*, 240–247. [CrossRef]
20. Szumny, A.; Figiel, A.; Carbonell-barrachina, A.A. Composition of rosemary essential oil (*Rosmarinus officinalis*) as affected by drying method. *J. Food Eng.* **2010**, *97*, 253–260. [CrossRef]
21. Calín-Sánchez, Á.; Szumny, A.; Figiel, A.; Jałoszyński, K.; Adamski, M.; Carbonell-barrachina, Á.A. Effects of vacuum level and microwave power on rosemary volatile composition during vacuum—microwave drying. *J. Food Eng.* **2011**, *103*, 219–227.
22. Sellami, I.H.; Wannes, W.A.; Bettaieb, I.; Berrima, S.; Chahed, T.; Marzouk, B.; Limam, F. Qualitative and quantitative changes in the essential oil of *Laurus nobilis* L. leaves as affected by different drying methods. *Food Chem.* **2011**, *126*, 691–697. [CrossRef]
23. Motevali, A.; Younji, S.; Chayjan, R.A.; Aghilinategh, N.; Banakar, A. Drying kinetics of dill leaves in a convective dryer. *Int. Agrophysics* **2013**, *27*, 39–47. [CrossRef]
24. Politowicz, J.; Lech, K.; Sánchez-Rodríguez, L.; Szumny, A.; Carbonell-Barrachina, Á.A. Volatile composition and sensory profile of *Cantharellus cibarius* Fr. as affected by drying method. *J. Sci. Food Agric.* **2017**, *97*, 5223–5232. [CrossRef] [PubMed]
25. Tulek, Y. Drying kinetics of oyster mushroom (*Pleurotus ostreatus*) in a convective hot air dryer. *J. Agric. Sci. Technol.* **2011**, *13*, 655–664.
26. Lech, K.; Figiel, A.; Wojdyło, A.; Korzeniowska, M.; Serowik, M.; Szarycz, M. Drying Kinetics and Bioactivity of Beetroot Slices Pretreated in Concentrated Chokeberry Juice and Dried with Vacuum Microwaves. *Dry. Technol.* **2015**, *33*, 1644–1653. [CrossRef]
27. Nöfer, J.; Lech, K.; Figiel, A.; Szumny, A.; Carbonell-Barrachina, Á.A. The Influence of Drying Method on Volatile Composition and Sensory Profile of *Boletus edulis*. *J. Food Qual.* **2018**, *2018*, 1–11. [CrossRef]
28. Figiel, A. Drying kinetics and quality of beetroots dehydrated by combination of convective and vacuum-microwave methods. *J. Food Eng.* **2010**, *98*, 461–470. [CrossRef]
29. Calín-Sánchez, Á.; Figiel, A.; Wojdyło, A.; Szarycz, M.; Carbonell-Barrachina, Á.A. Drying of Garlic Slices Using Convective Pre-drying and Vacuum-Microwave Finishing Drying: Kinetics, Energy Consumption, and Quality Studies. *Food Bioprocess Technol.* **2014**, *7*, 398–408. [CrossRef]
30. Calín-Sanchez, Á.; Figiel, A.; Szarycz, M.; Lech, K.; Nuncio-Jáuregui, N.; Carbonell-Barrachina, Á.A. Drying Kinetics and Energy Consumption in the Dehydration of Pomegranate (*Punica granatum* L.) Arils and Rind. *Food Bioprocess Technol.* **2014**, *7*, 2071–2083. [CrossRef]
31. Wojdyło, A.; Figiel, A.; Lech, K.; Nowicka, P.; Oszmiański, J. Effect of Convective and Vacuum-Microwave Drying on the Bioactive Compounds, Color, and Antioxidant Capacity of Sour Cherries. *Food Bioprocess Technol.* **2014**, *7*, 829–841. [CrossRef]
32. Adams, R.P. *Identification of essential oils by ion trap mass spectroscopy*; Academic Press: San Diego, CA, USA, 2012.
33. Adaszyńska-Skwirzyńska, M.; Śmist, M.; Swarcewicz, M. Comparison of extraction methods for the determination of essential oil content and composition of lavender leaves. Available online: http://ena.lp.edu.ua:8080/handle/ntb/27086 (accessed on 21 January 2019).
34. Hassanpouraghdam, M.B.; Hassani, A.; Vojodi, L.; Asl, B.H.; Rostami, A. Essential oil constituents of *Lavandula ofcinalis* Chaix. from Northwest Iran. *Chemija* **2011**, *22*, 167–171.
35. Nurzyńska-Wierdak, R.; Zawiślak, G. Chemical composition and antioxidant activity of lavender (*Lavandula angustifolia* Mill.) aboveground parts. *Acta Sci. Pol., Hortorum Cultus* **2016**, *15*, 225–241.
36. An, M.; Haig, T.; Hatfield, P. On-site field sampling and analysis of fragrance from living Lavender (*Lavandula angustifolia* L.) flowers by solid-phase microextraction coupled to gas chromatography and ion-trap mass spectrometry. *J. Chromatogr. A* **2001**, *917*, 245–250. [CrossRef]

37. Afifi, F.U.; Abu-Dahab, R.; Beltrán, S.; Alcalde, B.B.; Abaza, I.F. GC-MS composition and antiproliferative activity of *Lavandula angustifolia* Mill. essential oils determined by hydro-distillation, SFE and SPME. *Arab. J. Med. Aromat. Plants* **2016**, *2*, 71–85.
38. Fu, J.; Zhao, J.; Zhu, Y.; Tang, J. Rapid Analysis of the Essential Oil Components in Dried Lavender by Magnetic Microsphere-Assisted Microwave Distillation Coupled with HS-SPME Followed by GC-MS. *Food Anal. Methods* **2017**, *10*, 2373–2382. [CrossRef]
39. Torabbeigi, M.; Aberoomand Azar, P. Analysis of essential oil compositions of *Lavandula angustifolia* by HS-SPME and MAHS-SPME followed by GC and GC-MS. *Acta Chromatogr.* **2013**, *25*, 571–579. [CrossRef]
40. Mirahmadi, S.F.; Norouzi, R. Influence of Thin Layer Drying on the Essential Oil Content and Composition of *Lavandula officinalis*. *J. Essent. Oil-Bearing Plants* **2016**, *19*, 1537–1546. [CrossRef]
41. Milojevic, S.; Radosavljevic, D.; Pavicevic, V.; Pejanovic, S.; Veljkovic, V. Modeling the kinetics of essential oil hydrodistillation from plant materials. *Hem. Ind.* **2013**, *67*, 843–859. [CrossRef]
42. Baydar, H.; Erbaş, S. Effects of harvest time and drying on essential oil properties in lavandin (*Lavandula × intermedia* Emeric ex Loisel.). *Acta Hortic.* **2009**, *826*, 377–382. [CrossRef]
43. Ghasemi, A.; Salehi, S.; Craker, L. Effect of drying methods on qualitative and quantitative properties of essential oil from the aerial parts of coriander. *J. Dermatol. Sci.* **2017**, *4*, 35–40.
44. Kim, N.S.; Lee, D.S. Comparison of different extraction methods for the analysis of fragrances from *Lavandula* species by gas chromatography-mass spectrometry. *J. Chromatogr. A* **2002**, *982*, 31–47. [CrossRef]
45. Da Porto, C.; Decorti, D. Analysis of the volatile compounds of flowers and essential oils from *Lavandula angustifolia* cultivated in northeastern Italy by headspace solid-phase microextraction coupled to gas chromatography-mass spectrometry. *Planta Med.* **2008**, *74*, 182–187. [CrossRef] [PubMed]
46. Aghbashlo, M.; kianmehr, M.H.; Samimi-Akhijahani, H. Influence of drying conditions on the effective moisture diffusivity, energy of activation and energy consumption during the thin-layer drying of berberis fruit (*Berberidaceae*). *Energy Convers. Manag.* **2008**, *49*, 2865–2871. [CrossRef]
47. Alibas, I. Characteristics of chard leaves during microwave, convective, and combined microwave-convective drying. *Dry. Technol.* **2006**, *24*, 1425–1435. [CrossRef]
48. Dadali, G.; Apar, D.K.; Özbek, B. Microwave drying kinetics of okra. *Dry. Technol.* **2007**, *25*, 917–924. [CrossRef]

Sample Availability: Samples of the compounds are not available from the authors.

© 2019 by the authors. Licensee MDPI, Basel, Switzerland. This article is an open access article distributed under the terms and conditions of the Creative Commons Attribution (CC BY) license (http://creativecommons.org/licenses/by/4.0/).

Article

GC-MS and HS-SPME-GC×GC-TOFMS Determination of the Volatile Composition of Essential Oils and Hydrosols (By-Products) from Four *Eucalyptus* Species Cultivated in Tuscany

Francesca Ieri [1], Lorenzo Cecchi [2,*], Elena Giannini [3], Clarissa Clemente [4] and Annalisa Romani [1]

[1] PHYTOLAB-DISIA-Department of Informatics, Statistics and Applications "G. Parenti", University of Florence, Viale Morgagni, 59-50134 Florence, Italy and QuMAP-PIN-Piazza Giovanni Ciardi, 25, 59100 Prato (PO), Italy; francesca.ieri@unifi.it (F.I.); annalisa.romani@unifi.it (A.R.)
[2] Department of NEUROFARBA, University of Florence, Viale Pieraccini 6, 50139 Florence, Italy
[3] Versil Green Società Agricola s.s., via dei Cavalli 96, 55054 Massarosa (LU), Italy; info@elenagiannini.com
[4] Department of Pharmacy, University of Pisa, Via Bonanno Pisano 6, 56126 Pisa, Italy; claramente93@hotmail.it
* Correspondence: lo.cecchi@unifi.it; Tel.: +39-055-457-3707

Academic Editors: Constantinos K. Zacharis and Paraskevas D. Tzanavaras
Received: 3 December 2018; Accepted: 4 January 2019; Published: 9 January 2019

Abstract: Essential oils are widely used as functional ingredients for potential multi-purpose functional uses. Hydrosols, co-products of the distillation of plant material, are used in food and cosmetic industries and in biological agriculture, but their volatile composition is poorly investigated. The volatile fractions of essential oils and hydrosols from four less-studied 1,8-cineol-rich *Eucalyptus* species (*E. parvula* L.A.S. Johnson & K.D. Hill, *E. cinerea* F. Muell, *E. pulverulenta* Sims and *E. pulverulenta* baby blue Sims), cultivated in Tuscany in a system of organic farming, were characterized by solvent dilution (essential oils) or extraction (hydrosols) followed by GC-MS and by HS-SPME-GC×GC-TOFMS analysis. GC-MS analysis showed that essential oils were mainly constituted by oxygenated monoterpenes, particularly 1,8-cineole, with monoterpenes hydrocarbons up to 10.8%. Relative differences in the abundance of minor terpenes as limonene, α-pinene, γ-terpinene, *p*-cymene, terpinen-4-ol, α-terpineol, and alloaromandrene were pointed out and seem to be suitable for differentiation among EOs of the four different *Eucalyptus* species. Hydrosols of these species were characterized for the first time: they were mainly constituted by oxygenated monoterpenes (97.6–98.9%), with 1,8-cineole up to 1.6 g/L, while monoterpene and sesquiterpene hydrocarbons were detected only in traces. HS-SPME-GC×GC-TOFMS analysis also allowed providing metabolic profiling of hydrosols for the direct comparison and visualization of volatile components, pointing out the potentially different uses of these products as functional ingredients in food, beverage, and cosmetic industries.

Keywords: aromatic water; hydrolat; volatile compounds; metabolic fingerprint; eucalyptol

1. Introduction

The discovery of the genus *Eucalyptus* (Myrtaceae) came about when James Cook, an explorer, and Sir Joseph Banks, an expert botanist, travelled in Australia in 1770. This genus comprises more than 800 species of native trees and shrubs from Australia belonging to the Myrtaceae family, which are widely grown in many parts of the world [1].

The aromatic volatile oil (essential oil, EO), which is steam-distilled from the foliage, is among the world's most traded essential oils in terms of volume. The study of EO has attracted much attention for its anti-microbial, antibacterial, antiseptic, fungicidal, and nematicidal activities [2–5]. EO has a long history of use against the effect of cold, flu, sinusitis, rhinitis and other respiratory infections [6]. Tests in vitro showed that EO from *E. globulus* leaves might be exploited as natural antibiotic for the treatment of several infectious diseases caused by *Escherichia coli* and *Staphylococcus aureus* [7]. Treatment of refrigerated pork with EO led to a significant decrease in *Pseudomonas* spp. count and to an increase of customer acceptance [8]. EO from *E. globulus* and its major compound 1,8-cineole were tested against *A. flavus* and *A. parasiticus*: it was found that the antifungal activity was due not only to the 1,8-cineol, but to the whole phytocomplex [9]. A common need is the availability of natural extracts with pleasant taste and/or smell, combined with a preservative action aimed at avoiding lipid deterioration, fungal growth, oxidation and spoilage by microorganisms. The use of essential oils as functional ingredients in food, beverages and cosmetics is gaining increasing interest because of their relatively safe status, their wide acceptance by consumers, and their exploitation for potential multi-purpose functional use [10,11].

To date, the commercial EOs are mainly obtained from the leaves of the most common species of the genus *Eucalyptus* (i.e., *E. globulus*), which, according to the Standards ISO, must contain 1,8-cineole in percentages higher than 80–85%.

Hydrosols (EW), also known as hydrolats, floral waters, distillate waters or aromatic waters, are the co-products or the by-products of hydro- and steam distillation of plant material. Hydrosols are used in food and cosmetic industries for their organoleptic and biological properties. They are also used in biological agriculture against mushrooms, mildew and insects and for fertilization of soils [12]. Commercial EWs from *Eucalyptus* are currently available in the market, even though their volatile compositions have been poorly investigated to date [13,14]. The major components are generally the same present in oxygenated fraction of the corresponding essential oils [15].

In this study, we took into account four less studied 1,8-cineol-rich *Eucalyptus* species cultivated in Tuscany (central Italy) in a system of organic farming, namely *Eucalyptus parvula* L.A.S. Johnson & K.D. Hill, *Eucalyptus cinerea* F. Muell, *Eucalyptus pulverulenta* Sims and *Eucalyptus pulverulenta* baby blue Sims. The characterization of EOs from *Eucalyptus cinerea* [16–18] and *Eucalyptus pulverulenta* [16] has been reported in the literature while, to the authors' knowledge, no reports on EO from *Eucalyptus parvula* have been published, to date. Essential oil obtained from the leaves of these *Eucalyptus* species could potentially be employed for therapeutic ends and as natural additives for use in the food, cosmetics and perfume industries, extending the use of the plant beyond the predominantly ornamental one.

We aimed to evaluate the content and chemical composition of essential oils (EOs) and, for the first time, leaf hydrosols (EWs) obtained by steam distillation of these species. The evaluation of the content and chemical composition of both EOs and EWs was carried out using optimized Gas-Chromatography coupled with Mass Spectrometry (GC-MS). Head-Space Solid Phase Micro Extraction followed by comprehensive two-dimensional Gas-Chromatography (HS-SPME-GC×GC-TOFMS) analyses allowed providing a fast and direct comparison and visualization of the volatile components (fingerprint) of the EWs, also pointing out the presence of some VOCs not detectable only using the GC-MS. To the author's knowledge, this work is the first report regarding the characterization of the aroma components of EW from these four *Eucalyptus* species. Due to the fact that these *Eucalyptus* species are 1,8-cineol-rich, we hypothesized that 1,8-cineol was the main volatile of EOs and EWs, and that the main differences among both EOs and EWs from different species were due to relative amounts of minor volatile compounds.

2. Results and Discussion

Essential oils and leaf hydrosols were analyzed using integrated sampling and chromatographic techniques. In particular, the volatile organic compounds (VOCs) were extracted (from EWs) or diluted (from OEs) with organic solvent and analyzed by GC-MS and the volatile profile of EWs was also analyzed by GC×GC-TOFMS after extraction of VOCs by HS-SPME. GC-MS is the well-recognized technique of choice for analysis of VOCs from plant material and plant extracts [19,20] and, in this work, we applied this technique for the evaluation of the content and chemical composition of both EWs and EOs. HS-SPME is well recognized as a widespread and convenient sampling tool for VOCs and it is increasingly used coupled with GC-MS in analysis of food and more [21]. However, for quantitation purposes, several issues [22,23] (e.g., differences that arise from different absorption capacity of different fibers, changes in sorption temperature, competition between different molecules at different affinities for the absorbing material, fiber wearing) led to the need of using several devices (e.g., the use of a pool of suitable internal standards [23]) to ensure unbiased quantification. For these reasons, we decided to apply HS-SPME-GC×GC-TOFMS analysis on EWs to better elucidate the volatile profile, thus providing a tool for the direct comparison and visualization of plant volatile components and pointing out the presence of molecules not detectable only with GC-MS. Further quantitative evaluation via HS-SPME-GC-MS analysis can be further investigated in future researches. To the authors' knowledge, the characterization of EWs of these *E.* species was not been reported in the literature, to date.

2.1. Chemical Composition of Essential Oils and Hydrolats by GC-MS

Table 1 shows the chemical composition by GC-MS of the EOs and EWs summarized in Table 3 (see experimental section). Overall, 10 monoterpene hydrocarbons, 19 oxygenated monoterpenes, 2 sesquiterpene hydrocarbons, 2 aromatic monoterpenes (one of which oxygenated), 1 ester, 4 ketones, 1 aldehyde and 5 alcohols were identified. Some of the main molecules detected in the EWs and EOs are reported in Figure 1. Relative abundance of each of the molecules identified by GC-MS was calculated as a percentage of the peak area on the total area of the identified peaks. Peak areas from the total ion current were normalized by the use of the area of internal standard (tridecane).

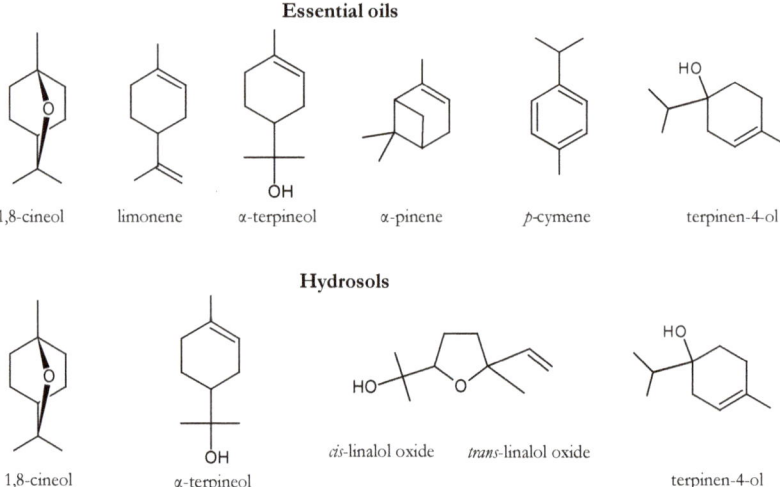

Figure 1. Chemical structure of some of the most abundant molecules in the EWs and EOs.

Table 1. Volatile organic compounds in hydrosols (EW) and essential oils (EO) of four *Eucalyptus* species identified by liquid injection GC-MS analysis as described in paragraph 3.4.1. For each compound, concentration is expressed as area % on the total area after normalization with ISTD. Data are the mean of three determinations. Retention Indices (RI$_{cal}$): Non-isothermal Kovats retention indices from temperature-programming, using the definition of Van den Dool and Kratz, 1963. Retention Indices (RI$_{ref}$): Non-isothermal Kovats retention indices from temperature-programming from Chemistry WebBook. For each compound, different letters indicate significant differences by Fisher's LSD test (z, y, x, w for aromatic waters; a, b, c, d, e for essential oils).

n°	Compound	RI$_{cal}$	RI$_{ref}$	Hydrosols (Area %)								Essential Oils (Area %)					
				1-16-EW	2-16-EW	3-16-EW	4-16-EW	1-17-EW	1-1617-EW	1-16-EO	2-16-EO	3-16-EO	4-16-EO	1-17-EO	1-1617-EO
	Monoterpene Hydrocarbons			0.22	0.02	0.00	0.02	0.01	0.00	8.05	10.76	7.10	8.34	6.74	7.54
1	α-pinene	1030	1026	0.09	-	-	-	-	-	1.20 b	4.45 e	2.38 d	2.15 c	0.95 c	1.12 ab
2	camphene	1067	1065	-	-	-	-	-	-	0.03 bc	0.04 c	0.02 ab	0.02 ab	0.01 a	0.02 ab
3	β-pinene	1122	1118	-	-	-	-	-	-	0.11 ab	0.17 cd	0.14 bc	0.19 d	0.09 a	0.10 a
7	β-myrcene	1168	1167	-	-	-	tr	tr	-	0.12 a	0.22 c	0.17 b	0.25 c	0.09 a	0.11 a
8	α-phellandrene	1177	1177	-	-	-	-	-	-	0.15 b	0.13 b	0.04 a	0.01 a	0.15 b	0.14 b
12	Limonene	1214	1210	0.08 y	0.02 z	tr	0.02 z	0.01 z	-	3.60 b	4.65 d	3.66 b	4.29 c	3.00 a	3.41 b
14	Z-ocimene	1241	1242	-	-	-	-	-	-	0.36 c	0.17 b	0.07 a	0.07 a	0.45 d	0.39 c
15	γ-terpinene	1259	1254	tr	-	-	-	-	Tr	0.54 b	0.13 a	0.10 a	0.07 a	0.48 b	0.50 b
16	*p*-cymene	1286	1281	0.05	-	-	tr	tr	-	1.78 e	0.73 b	0.50 b	1.23 c	1.42 cd	1.63 de
20	alloocimene	1383	1377	-	-	-	-	-	-	0.15 d	0.07 b	0.03 a	0.04 a	0.12 c	0.13 c
	Oxygenated Monoterpenes			98.62	98.50	98.29	97.64	98.91	98.80	91.30	88.74	91.99	90.60	92.71	91.89
11	2,3-dehydro-1,8-cineole	1202	1197	-	-	-	-	-	-	-	0.09 c	0.08 bc	0.07 b	0.05 a	0.05 a
13	1,8-cineol	1225	1221	89.53 y	88.40 z	90.78 x	89.17 zy	90.22 yx	89.35 zy	87.06 b	83.80 a	87.72 bc	85.14 a	88.66 c	87.75 bc
23	*cis*-linalool oxide (furanoid)	1453	1453	0.5 zy	0.77 z	0.47 zy	0.44 z	0.52 zy	0.55 y	-	-	-	-	-	-
25	*trans*-linalool oxide (furanoid)	1481	1482	0.49 y	0.68 x	0.36 z	0.43 zy	0.46 zy	0.52 y	-	-	-	-	-	-
26	linalool	1545	1544	-	-	-	-	-	-	0.05 b	0.09 c	0.09 c	0.11 d	0.02 a	0.03 a
27	fenchyl alcohol	1595	1571	tr	tr	-	tr	tr	-	0.06 b	0.08 c	0.05 b	0.06 b	0.06 b	0.03 a
28	pinocarvone	1600	1575	-	-	-	-	-	-	0.06 b	0.03 a	0.05 b	0.06 b	0.06 b	0.05 b
29	terpinen-4-ol	1617	1612	0.92 y	0.94 yx	1.01 x	1.37 w	0.75 z	0.87 y	0.51 cd	0.47 bc	0.56 d	0.72 e	0.39 a	0.44 b
31	*cis-p*-mentha-2,8-dienol	1637	1642	0.07 zy	0.06 z	0.12 x	0.08 y	0.12 x	0.08 y	0.02 a	0.03 ab	0.06 c	0.08 d	0.04 b	0.04 b
33	*trans*-pinocarveol	1683	1659	0.17 z	0.24 yx	0.23 yx	0.24 yx	0.25 x	0.21 y	0.09 ab	0.13 c	0.11 bc	0.08 a	0.10 ab	0.09 ab
34	*trans-p*-mentha-2,8-dienol	1679	1670	0.11 z	0.10 z	0.11 z	0.12 z	0.13 z	0.10 z	-	-	-	-	-	-
35	terpineol isomer	1682	-	0.25 z	0.40 y	0.37 y	0.42 y	0.25 z	0.24 z	0.14 ab	0.19 c	0.17 bc	0.23 d	0.12 a	0.14 ab
36	citral	1698	1695	0.17 x	0.27 w	0.19 x	0.10 z	0.14 y	0.19 x	0.05 ab	0.07 bc	0.08 c	0.05 ab	0.04 a	0.05 ab
37	α-terpineol	1707	1704	6.02 x	6.24 x	4.19 z	4.50 z	5.53 y	6.21 x	3.11 c	2.77 b	2.00 a	2.20 a	3.02 bc	3.08 bc
38	borneol	1718	1715	tr	0.05 x	-	0.04 y	0.02 z	-	-	-	-	-	-	-
39	α-terpinyl acetate	1722	1721	-	-	-	-	-	-	0.02 a	0.90 b	0.92 b	1.71 c	0.02 a	-
40	*cis*-carveol	1850	1847	0.14 yx	0.11 z	0.12 zy	0.10 z	0.18 w	0.16 xw	-	-	-	-	-	-
42	exo-2-hydroxycineole	1870	1870	0.08 z	0.13 y	0.21 x	0.52 w	0.10 zy	0.11 zy	-	-	-	0.05 ab	0.04 a	-
43	*cis-p*-mentha-1(7),8-dien-2-ol	1905	1888	0.17 y	0.11 z	0.13 z	0.11 z	0.24 w	0.21 x	0.12 bc	0.09 ab	0.09 ab	0.07 a	0.13 c	0.14 c
	Sesquiterpene Hydrocarbons									0.09	0.10	0.47	0.63	0.02	0.03
30	β-caryophyllene	1631	1625	-	-	-	-	-	-	0.07 b	0.07 b	0.03 a	0.03 a	0.02 a	0.03 a
32	alloaromandrene	1640	1645	-	-	-	-	-	-	0.02 a	0.02 a	0.44 b	0.60 c	-	-

Table 1. Cont.

n°	Compound	RI_cal	RI_ref	Hydrosols (Area %)						Essential Oils (Area %)					
				1-16-EW	2-16-EW	3-16-EW	4-16-EW	1-17-EW	1-1617-EW	1-16-EO	2-16-EO	3-16-EO	4-16-EO	1-17-EO	1-1617-EO
	Aromatic Monoterpenes														
24	p-cymenene	1455	1455	-	-	-	-	-	-	0.03 0.03 b	0.04 0.04 c	0.02 0.02 a	0.03 0.03 b	0.04 0.04 c	0.03 0.03 b
	Oxygenated Aromatic Monoterpenes														
41	p-cymen-8-ol	1857	1869	0.06 0.06 z	0.11 0.11 x	0.08 0.08 zy	0.10 0.10 yx	0.10 0.10 yx	0.08 0.08 zy	-	-	-	-	-	-
	Ester														
4	isoamyl acetate	1125	1126	-	-	-	-	-	-	0.04 0.04 a	-	-	-	0.03 0.03 a	0.05 0.05 a
	Ketones														
5	heptan-4-one	1132	-	0.21	0.22	0.20	0.23	0.20	0.27	0.24 0.06 c	0.20 0.03 a	0.25 0.04 ab	0.24 0.04 ab	0.22 0.04 ab	0.22 0.05 bc
6	heptan-3-one	1161	1163	0.05 0.05 z	0.06 0.06 zy	0.06 0.06 zy	0.07 0.07 yx	0.05 0.05 z	0.08 0.08 x	0.06 0.07 a	0.06 0.06 a	0.07 0.07 a	0.06 0.06 a	0.06 0.06 a	0.07 0.07 a
9	heptan-2-one	1190	1185	0.12 0.12 z	0.16 0.16 z	0.14 0.14 z	0.16 0.16 z	0.13 0.13 z	0.16 0.16 z	0.12 0.12 a	0.11 0.11 a	0.14 0.14 a	0.14 0.14 a	0.12 0.12 a	0.11 0.11 a
19	6-methylhept-5-en-2-one	1346	1338	0.04 0.04 x	-	-	-	0.02 0.02 z	0.03 0.03 y	tr	-	-	-	-	-
	aldehyde														
22	nonanal	1405	1401	-	-	tr	tr	-	-	-	-	-	-	-	-
	Alcohols														
10	3-methylbutanol	1198	1210	0.89 0.46 z	1.15 0.76 y	1.43 0.68 y	2.01 1.07 x	0.78 0.45 z	0.85 0.49 z	0.14	0.14	0.15	0.14	0.14	0.16
17	heptan-3-ol	1290	-	0.07 0.07 zy	0.08 0.08 yx	0.06 0.06 z	0.08 0.08 yx	0.06 0.06 z	0.09 0.09 x	-	-	-	-	-	-
18	heptan-2-ol	1314	1318	0.06 0.06 z	0.09 0.09 x	0.08 0.08 yx	0.09 0.09 x	0.07 0.07 zy	0.08 0.08 yx	0.07 0.07 a	0.06 0.06 a	0.08 0.08 a	0.07 0.07 a	0.06 0.06 a	0.08 0.08 a
21	Z-hex-3-en-1-ol	1384	1384	0.14 0.14 x	0.07 0.07 z	0.09 0.09 zy	0.24 0.24 w	0.12 0.12 yx	0.13 0.13 yx	0.07 0.07 a	0.08 0.08 a	0.08 0.08 a	0.07 0.07 a	0.07 0.07 a	0.08 0.08 a
44	2-phenylethanol	1928	1924	0.16 0.16 y	0.15 0.15 y	0.52 0.52 x	0.53 0.53 x	0.08 0.08 z	0.06 0.06 z	-	-	-	-	-	-

Monoterpene hydrocarbons (MH): these terpenes were present in the EO samples with relative abundances between 6.74% and 10.76%. Limonene was the most abundant MH, with similar percentages in all the EOs (3.00–4.65%). Different abundances of the other MHs were pointed out for different *Eucalyptus* species: in *Eucalyptus parvula*, similar amounts of *p*-cymene and α-pinene were detected, followed by lower amounts of γ-terpinene and Z-ocimene. In *Eucalyptus cinerea*, similar amounts of α-pinene and Limonene were detected, followed by lower amounts of *p*-cymene, β-myrcene, β-pinene and Z-ocimene. Regarding the other two species (*Eucalyptus pulverulenta* Sims and *Eucalyptus pulverulenta* baby blue Sims), the α-pinene amount was approximately half the Limonene, with lower amounts of *p*-cymene, β-myrcene, and β-pinene. Other MH (camphene, α-phellandrene, alloocimene) were in low amounts. Noteworthy, the α-phellandrene content in EOs of all these species was lower than 1%, according to the *European Pharmacopoeia* specification for 1,8-cineol-rich E. oils [16].

1,8-cineole (eucalyptol) was by far the main component of the analyzed samples (see the next paragraphs); however, the differences in relative abundance of metabolites present in low amount, or even in trace, play a critical role in mediating different activities for EOs from different *Eucalyptus* species with 1,8-cineole as the main component; indeed, these different activities (i.e., alleophatic [24], protection against *Parthenium hysterophorus* L. [25]) were reported as only due to differences in the relative abundance of minor components [24,25], likely due to the synergistic effect of these latter compounds with other components [26,27].

Regarding the hydrosols, no significant amounts of MH were detected, due to the hydrophobic nature of these molecules.

Oxygenated monoterpenes (OM): 1,8-cineole is the main component of EOs obtained from the leaves of *Eucalyptus globulus*, the most common *Eucalyptus* species [20]. In EOs from these four species, relative abundance of 1,8-cineole ranged between 83.80% and 88.66%, higher than the 80–85% indicated by the standard ISO as the minimum amount of 1,8-cineole for EO from *E. globulus*. Other papers in the literature reported the characterization of EOs from *Eucalyptus cinereal* [16–18] and *Eucalyptus pulverulenta* [16], while, to the authors' knowledge, no reports on EO from *Eucalyptus parvula* have been published, to date. In such papers, the relative abundance of 1,8-cineole showed great variability ranging usually from 58.0 to 69.0% and sometimes reaching 87.8% in *E. cinerea* EOs and being approx. 75% in *E. pulverulenta* EOs. Consequently, also the relative abundances of the other minor terpenes showed a great variability. This variability might be due to the effect of climatic and geographical factors and harvesting season.

In our study, *Eucalyptus parvula* and *Eucalyptus pulverulenta* Sims were the species with the highest amount of 1,8-cineole. Since EOs are totally composed by the volatile fraction, the relative abundance of each compound can be assumed as the amount of this molecule in the oil expressed as g/100g.

Regarding the analyzed hydrosols (EWs), 1,8-cineole was in the range 88.40–90.78%. In order to better characterize these hydrosols, 1,8-cineole was quantified using an external calibration curve, as reported in the experimental section. Table 2 shows that the absolute concentration of 1,8-cineole in the EWs extracts varied in the range 0.74–1.58 g/L, highlighting that this molecule was also recovered in water samples.

Table 2. Content of 1,8-cineole in the hydrosols by GC-MS analysis. Data are expressed in g/L as mean of three independent determinations (SD < 3%). Different letters indicate significant differences at $p < 0.05$.

Sample	Kind of Sample	1,8-cineole (g/L)
1-16-EW	hydrosol	1.58 a
2-16-EW	hydrosol	1.45 b
3-16-EW	hydrosol	1.52 a
4-16-EW	hydrosol	0.74 e
1-17-EW	hydrosol	0.86 d
1-1617-EW	hydrosol	1.20 c

In EWs, OMs constituted 98–99% of the total VOCs, according to their water solubility, higher than MHs. Regarding OMs other than 1,8-cineole, in our samples α-terpineol was the most abundant one (4.19–6.24%), followed by lower amounts of terpinen-4-ol, linalool oxides (furanoid, cis and trans), terpineol isomer, and other minor OMs (<1.5%).

OMs constituted about 90% of the EOs. In these samples, α-terpineol was in the range 2.00–3.11%. Noteworthy, in the EO from *Eucalyptus parvula*, the highest amount of α-terpineol and no presence of its ester, namely α-terpinyl acetate, were detected. In the other three species, α-terpinyl acetate was detected and the sum of the percentages of α-terpineol and α-terpinyl acetate was similar to that of α-terpineol of EO of the *Eucalyptus parvula*. The other OMs didn't exceed 0.72%.

Other terpenes: no sesquiterpene hydrocarbons were identified in EWs, in agreement with their insolubility in water. In EOs, very low percentages of β-caryophyllene ($\leq$0.07%) in all samples, and slightly higher amounts of alloaromandrene in *Eucalyptus pulverulenta* Sims (0.44%) and *Eucalyptus pulverulenta* baby blue Sims (0.60%) were detected.

One aromatic monoterpene, namely p-cymenene, was identified in very low amounts ($\leq$0.04%) only in EOs samples, while one oxygenated aromatic monoterpene (p-cymen-8-ol) was identified in low amounts ($\leq$0.11%) only in EWs, according to their different water solubility.

Other compounds: no significant amounts of esters and aldehydes were identified in our samples, the only exceptions being traces of nonanal in one EW sample and very low amounts of isoamyl acetate in EO from *Eucalyptus parvula*. Ketones (the three linear isomers of heptanone and lower percentages of 6-methylhept-5-en-2-one) were identified in low amounts (0.20–0.27% in both EWs and EOs). Finally, alcohols were identified in very low amounts in EOs (heptan-2-ol and heptan-3-ol for a total amount up to 0.16%), while in EWs they were present in percentages up to 2.01%, with 3-methylbutanol as the main molecule, followed by 2-phenylethanol and Z-hex-3-en-1-ol.

2.2. Fingerprint Analysis by HS-SPME-GC×GC-TOFMS

Solid-phase microextraction (SPME) is a rapid and simple procedure for extraction of volatile fraction from aromatic and medicinal plants [28]. As reported, the divinilbenzene/carboxen/ polydimethylsiloxane (DVB/CAR/PDMS) fiber is the most effective SPME fiber able to isolate the volatile fraction from commercial hydrosols of several plants [12]. HS-SPME and GC×GC-TOFMS fingerprint analysis are ideal tools to analyze complex volatile fraction from several matrices, and to provide a sensitive method for the direct comparison and visualization of plant volatile components. As previously reported, 1,8-cineole was the major component in the EOs and EWs, but differences in the other metabolites present in low amounts are very important. Utilization of comprehensive two-dimensional GC (GC×GC) increases separation power with respect to that of the one-dimensional GC in complex matrices where the presence of low abundant components is critical, such as *Eucalyptus* [29]. To our knowledge, there has been no study reporting the volatile profile of EWs from these *Eucalyptus* species; therefore, hydrosols from the four *Eucalyptus* species were analyzed by HS-SPME-GC×GC-TOFMS to better elucidate the volatile profile of these by-products, also pointing out the presence of molecules not detectable with only GC-MS. HS-SPME-GC×GC-TOFMS analyses of the complex volatile fraction of EWs were submitted to advanced fingerprinting analysis of 2D chromatographic data.

In Figure 2, "contour plots" from HS-SPME-GC×GC-TOFMS analyses of the four *Eucalyptus* species are reported: each 2D-peak corresponds to a single volatile compound. In this case, SPME and comprehensive comparative analysis of 2D chromatographic data showed visual differences among EW samples. 1-16-EW and 2-16-EW showed a larger number and a higher intensity of peaks, with respect to 3-16-EW and 4-16-EW. The most intense peak corresponded to 1,8-cineole.

For example, a total of about 400 compounds was detected by GC×GC analysis in 1-16-EW (estimated from the number of peak contours in 2D plots) and, after subtracting baseline peaks, corresponding to fiber blending or background interferences, 137 peaks/compounds were identified. These results were in agreement with Wong et al. [20], where the 2D rational separation pattern aids the identification of ca. 400 metabolites in *Eucalyptus* spp. leaf oils, 183 of which were identified or tentatively identified and represented percentages between 50.8–90.0% of the total ion count, comprising various chemical families.

HS-SPME-GC×GC-TOFMS provided a high metabolic coverage of VOCs: monoterpenes, oxygenated monoterpenes (the main class), oxygenated monoterpenes acetate, and others (ketones, aldehyde, alcohols).

GC×GC is currently adopted as separation technique not only because of its high separation power and sensitivity, but also for its ability to produce more widely distributed and rationalized peak patterns [30] for chemically correlated group of analytes. Terpenic compounds of *Eucalyptus* hydrosols were organized mainly in three clusters in 2D separation space: monoterpenic hydrocarbons, oxygenated monoterpenes and monoterpenes acetate, except for the 1,8-cineol that wrapped around, resulting in monoterpenes zone (Figure 2A). 1,8-cineol showed high secondary retention and fall outside the range of secondary retention time (wrapped around). As previously reported for volatile oil from leaves of *Eucalyptus dunnii* [31], one molecule that wraps around, does not affect the separation and identification of the compounds, since the more strongly retained components (those that wrap-around) did not overlap peaks that were weakly retained in the subsequent modulation.

Up to 31 peaks/compounds belonging to the class of oxygenated monoterpenes were distributed in a defined part of the contour plot for 1-16-EW (Figure 2A, braced region "b"). The number of oxygenated monoterpenes were 33 for 2-16-EW, 23 for 3-16EW and 24 for 4-16-EW.

An advanced approach known as comprehensive template matching fingerprinting [32] was adopted (Figure 2B). This method considers, as a comparative feature, each individual 2D peak together with its time coordinates, detector response and MS fragmentation pattern, and includes them in a sample template that is created by the analyst and can be used to compare plots from different samples directly and comprehensively. A template could be used to correctly interpret visual differences in further analyses. To create the template, the peak identification was performed by matching the experimental mass spectra against spectra databases combined with GC-MS data.

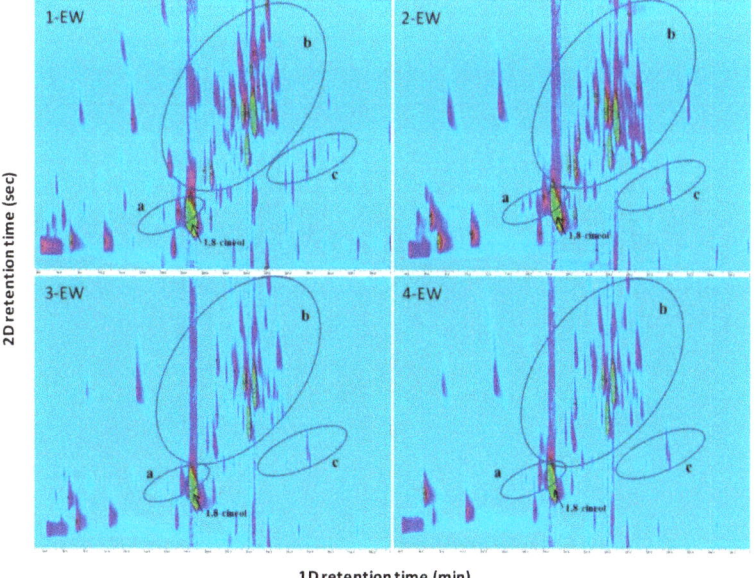

(A)

Figure 2. *Cont.*

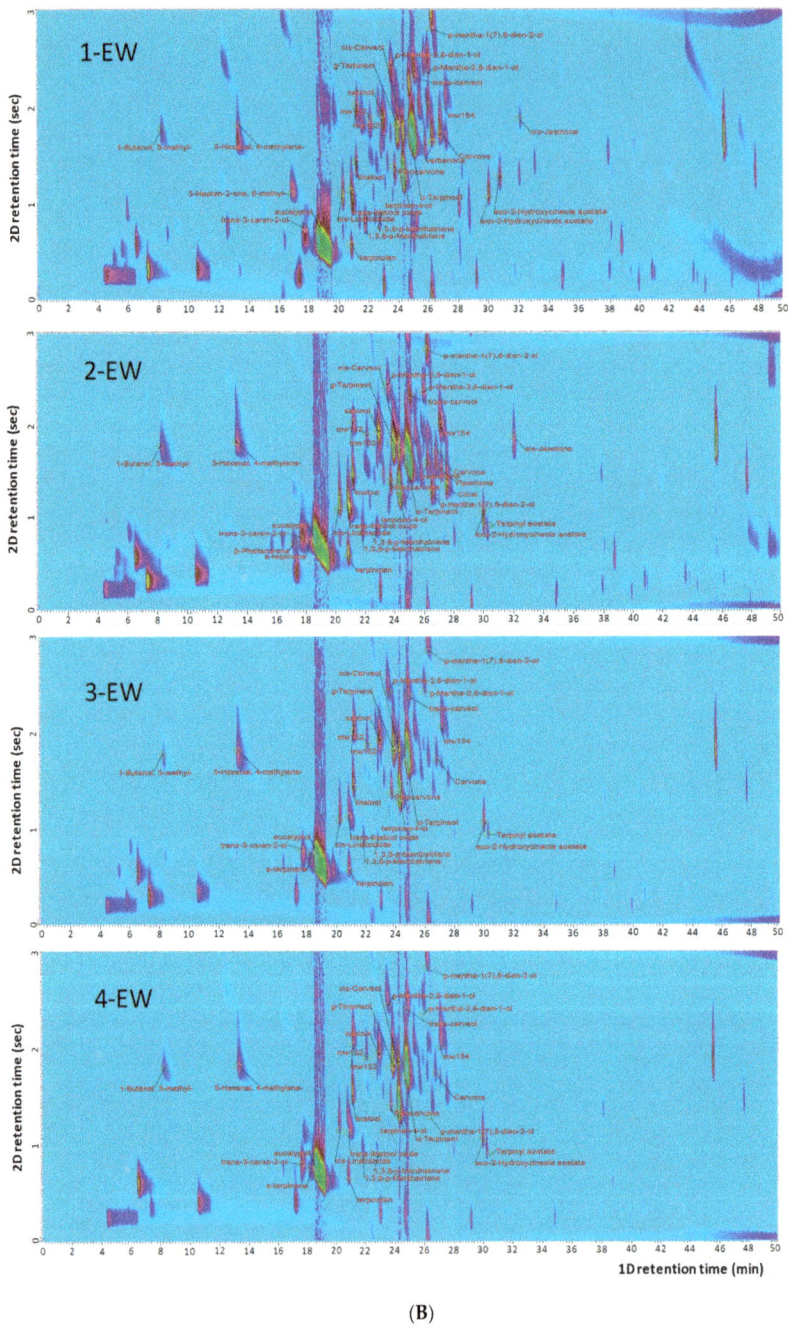

(B)

Figure 2. (**A**) 2D contour plots of the analyzed EWs. Braced region a: monoterpenic hydrocarbons; b: oxygenated monoterpenes; c: oxygenated monoterpenes acetate; (**B**) comprehensive template matching fingerprinting with the main identified volatile compounds of 1-16-EW: *E. parvula* L.A.S. Johnson & K.D. Hill; 2-16-EW; *E. cinerea* F. Muell; 3-16-EW: *E. pulverulenta* Sims; 4-16-EW: *E. pulverulenta* baby blue Sims.

The main differences that emerged between the four varieties could be summarized as follows: 1-16-EW showed the presence of 6-methylhept-5-en-2-one that was not present in the other species (see also Table 1) and the presence of *exo*-2-hydroxycineole acetate isomers. 1-16-EW did not show terpinyl acetate and α-terpinene, which instead were found in the other three species. 2-16-EW was the only species that showed the presence of β-phellandrene, piperitone and citral. 1-16-EW and 2-16-EW showed the presence of *cis*-jasmone and carvone, while 3-16-EW and 4-16-EW didn't show the presence of these molecules. Volatile profiles presented in 2D contour plots allow visual discrimination of the metabolic composition among interspecies of *Eucalyptus* aromatic waters, as reported for leaf oils of other different *Eucalyptus* spp. [20].

3. Materials and Methods

3.1. Chemicals

All chemicals and standards of analytical reagent grade were from Sigma Aldrich (Steinheim, Germany). Tridecane, 1,8-cineole, heptane and a mixture of linear alkanes (C_{10}–C_{26}) in hexane were used. Inert gasses (He and N_2 99.999% purity) were supplied by SOL gas company (Monza, Italy).

3.2. Plant Material

In the littoral area of Versilia and Pisa (North-Tuscany-Italy-Latitude: 43.873651; Longitude: 10.328756), Versil Green Società Agricola s.s., a commercial farm, cultivates several species of *Eucalyptus* for the production of ornamental green fronds and essential oils. The cultivated species are: *Eucalyptus parvula* L.A.S. Johnson & K.D. Hill, *Eucalyptus cinerea* F. Muell, *Eucalyptus pulverulenta* Sims and *Eucalyptus pulverulenta* baby blue Sims (Table 3). Figure 3a–d show a picture of each plant. All samples are grown by organic practices, accredited according to the UNI EN 45011 standard. During the production of ornamental fronds, leaves and little stems were separated from the young branches of healthy plants of two years and distilled as reported in Section 3.3.

Table 3. List of the analyzed samples. EW: hydrosol or aromatic water; EO: essential oil. 1, *E. parvula* L.A.S. Johnson & K.D. Hill; 2, *E. cinerea* F. Muell; 3, *E. pulverulenta* Sims; 4, *E. pulverulenta* baby blue Sims. 16 and 17 indicate the year in which the sample was obtained. 1617 indicates samples obtained as a mixture in equal parts of samples from 2016 and 2017.

Sample Name	Kind of Sample	Eucalyptus Species	Year	Yields %
1-16-EW	aromatic water	*E. parvula* L.A.S. Johnson & K.D. Hill	2016	
2-16-EW	aromatic water	*E. cinerea* F. Muell	2016	
3-16-EW	aromatic water	*E. pulverulenta* Sims	2016	
4-16-EW	aromatic water	*E. pulverulenta* baby blue Sims	2016	
1-17-EW	aromatic water	*E. parvula* L.A.S. Johnson & K.D. Hill	2017	
1-1617-EW	aromatic water	*E. parvula* L.A.S. Johnson & K.D. Hill	2016–2017	
1-16-EO	essential oil	*E. parvula* L.A.S. Johnson & K.D. Hill	2016	1.2
2-16-EO	essential oil	*E. cinerea* F. Muell	2016	1.1
3-16-EO	essential oil	*E. pulverulenta* Sims	2016	1.1
4-16-EO	essential oil	*E. pulverulenta* baby blue Sims	2016	1.1
1-17-EO	essential oil	*E. parvula* L.A.S. Johnson & K.D. Hill	2017	1.3
1-1617-EO	essential oil	*E. parvula* L.A.S. Johnson & K.D. Hill	2016–2017	

3.3. Obtaining of the Essential Oils and Hydrosols

Essential oils and hydrosols were obtained by steam distillation of the *eucalyptus* fresh leaves and little stems, within 24 h after harvesting, using the Essenziale 20 extractor (Tred Technology srl, Italy). The system used low working temperatures (always below 80 °C), thus decreasing the consumption energy and minimizing the degradation of the volatile fraction. The process parameters were continuously checked and adjusted during the distillation, e.g., internal pressure inside the boiler, internal boiler temperature, water temperature, oil and hydrosol flow. The starting material vs hydrosol ratio was 3:1 and the recovery of the corresponding essential oil was in variable percentage depending on the collection period and atmospheric conditions. The mean of yields of EO are from 1.1 to 1.3%, as reported in Table 3.

Figure 3. (**a**): *Eucalyptus parvula* L.A.S. Johnson & K.D. Hill; (**b**): *Eucalyptus cinerea* F. Muell; (**c**): *Eucalyptus pulverulenta* Sims; (**d**): *Eucalyptus pulverulenta* baby blue Sims.

3.4. Analysis of Essential Oils and Aromatic Waters

3.4.1. GC-MS Analysis

EO samples were diluted 10,000 times with heptane, in presence of tridecane (20 ppm), for avoiding saturated signals during the following chromatographic analysis.

Regarding EWs, 0.5 mL of sample were extracted with 0.5 mL of heptane (with 20 ppm, tridecane) for 1 h with an automatic stirrer; water residues were removed using anhydrous sodium sulfate and the obtained organic extracts were diluted 20 times with heptane.

GC-MS analysis of EO and EW solutions, obtained as described above, were carried out by liquid injection on an Agilent 7890a Gas Chromatograph equipped with a Gerstel MPS automatic sampler system and a quadrupole Mass Spectrometer 5975c MSD (Agilent Technologies, Palo Alto, CA, USA) working in split-less mode. The analytes separation was carried out by an Agilent DB InnoWAX column (length, 50 m; i.d., 200 μm, film thickness, 0.4 μm). Initial oven temperature was 40 °C, held for 1 min. Then, it raised to 200 °C at 5 °C min^{-1}, then raised to 260 °C at 10 °C min^{-1} and finally held at 260 °C for 6 min. Injector temperature was 260 °C, while the carrier gas helium, was at a flow rate of 1.2 mL/min. 1 μL of each sample was injected.

Mass spectrometer worked in the mass range 40–350 m/z and with an electron ionization of 70 eV and the Total Ion Current chromatograms were recorded. Compounds were tentatively identified by comparing the mass spectra of each peak with those reported in mass spectral database as the standard NIST08/Wiley98 libraries; when available, standards were used for confirming the nature of the identified molecules: α-pinene, β-pinene, camphene, β-myrcene, α-phellandrene, limonene, Z-ocimene, γ-terpinene, p-cymene, alloocimene, 1,8-cineol, linalool, terpinen-4-ol, citral, α-terpineol, borneol, β-caryophillene, heptan-2-one, 6-methylhept-5-en-2-one, nonanale, 3-methylbytanol, heptan-2-ol and 2-phenylethanol. Peaks identification was then confirmed by comparing their retention index; to this aim, a mixture of linear alkanes (C_{10}–C_{26}) in hexane (Sigma Aldrich, Saint Louis, MI, USA) was injected in the same condition already described for sample analysis and the retention indexes were calculated by the generalized equation [33] and compared with the literature [34] The relative concentration of each identified compound was calculated as peak area on total area of all the identified peaks (peaks areas were normalized using tridecane as internal standard). 1,8 cineole in EWs was quantified by a six point calibration curve, which were built using 1,8 cineole as external standard (range 10–160 ppm, 0.9936 R^2).

3.4.2. HS-SPME-GC×GC-TOFMS Analysis

The EWs from the four *Eucalyptus* species (Table 3) were extracted by solid-phase microextraction (SPME) and analyzed by GC×GC-TOFMS. GC×GC was performed by a flow modulation apparatus consisting on an Agilent 7890B GC (Agilent Technologies, Palo Alto, CA, USA), with flow modulator device for 2D separation, coupled with a time-of-flight mass spectrometer (TOF-DS Markes International Ltd., Llantrisant, UK). After some trials aimed at optimizing amounts of sample, NaCl and water and exposure time and temperature, SPME conditions were set as follow: 1 mL of the EW sample, together with 2 g of NaCl and 4 mL of deionized water were placed into a 20-mL screw cap vial fitted with PTFE/silicone septa. VOCs were absorbed exposing a divinilbenzene/carboxen/polydimethylsiloxane (DVB/CAR/PDMS) 2 cm fiber (Supelco) for 10 min into the vial at 60 °C and then immediately desorbed at 280 °C in a gas chromatograph injection port.

Chromatographic separation was performed using a first dimension (^{1}D) HP-5 column (20 m × 0.18 mm I.D. × 0.18 μm film thickness (df); Agilent Technologies, Palo Alto, CA, USA), and a WAX second dimension (^{2}D) column (5 m × 0.32 mm I.D. × 0.15 μm df; Agilent Technologies, Palo Alto, CA, USA).

Flow modulation was performed with a modulation period of 3 s. Helium was used as carrier gas (99.999% purity) at flow rates of 0.4 and 10 mL/min in first and second dimensions, respectively.

The chromatographic conditions were: oven temperature program, 40 °C, increased at 4 °C/min to 220 °C, increased at 10 °C/min to 260 °C (hold 1 min); injector temperature, 260 °C; split ratio 1:5. The inlet of the ^{2}D column was maintained under vacuum by a deactivated fused silica (0.30 m × 0.10 mm I.D.) placed immediately before the column, after the flow modulator. TOFMS parameters: the ion source temperature was 230 °C; the transfer line temperature was 280 °C; ionization, −70 eV. A mass range of 43–500 Da was used, with data rate of 50 Hz. TOF-DS ^{TM} software, version 2.0 (Markes International Ltd.; Llantrisant, UK, 2016) was used for data acquisition. GC IMAGE version R2.5 GC×GC (64 bit) software (GC IMAGE; LCC-Lincon, NE, USA, 2014) was used for data processing.

3.5. Statistical Analysis

The semi-quantitative (Table 1) and quantitative data (Table 2) are expressed as the mean of three determinations. Statistical significance was evaluate applying one-way ANOVA and F-test ($p < 0.05$) using Microsoft Excel statistical software; means were then compared by Fisher's LSD test using the DSAASTAT excel® VBA macro, version 1.1 (Onofri, A.; Pisa, Italy, 2007).

4. Conclusions

This study reports a first preliminary characterization of the volatile profile of EO and EW of four less studied 1,8-cineol-rich *Eucalyptus* species (*E. parvula* L.A.S. Johnson & K.D. Hill, *E. cinerea* F. Muell, *E. pulverulenta* Sims and *E. pulverulenta* baby blue Sims) cultivated in Tuscany (Italy) and intended to be employed as natural additives in the food, cosmetics and perfume industries, as well as for therapeutic ends, beyond the predominantly ornamental one. Chemical differences in VOCs from EWs and EOs were evidenced, providing products for potentially different uses. Further studies will be necessary for the standardization of commercial EOs and EWs optimizing the technological harvesting period.

Regarding the EOs from these *Eucalyptus* species, GC-MS analysis of diluted samples showed that oxygenated monoterpenes accounted for up to 92.7% and monoterpenes hydrocarbons contributed to volatile fraction for up to 10.8%, with limonene as the most representative MH. The relative abundance of minor terpenes as limonene, α-pinene, γ-terpinene, p-cymene, terpinen-4-ol, α-terpineol and alloaromandrene seems to be suitable for differentiation among EOs of the four different *Eucalyptus* species.

GC-MS analysis allowed pointing out that the volatile fraction of EWs extracts was mainly constituted by oxygenated monoterpenes (97.6–98.9%), with monoterpene and sesquiterpene hydrocarbons detected only in traces. HS-SPME-GC×GC-TOFMS analysis of EWs extracts also allowed for metabolic profiling of EWs for the direct comparison and visualization of volatile components of these not yet investigated co-products. The GC-MS quantitative evaluation of 1,8-cineole in EWs showed amounts of up to 1.6 g/L; consequently, the studied EWs, co-products of steam distillation of fresh leaves and little stems of the four *Eucalyptus* species, can be proposed as functional ingredients for the food, beverage, and cosmetic industries.

Author Contributions: Conceptualization, A.R.; Data curation, F.I., L.C. and C.C.; Formal analysis, F.I. and L.C.; Project administration, E.G. and A.R.; Resources, E.G.; Supervision, A.R.; Writing-original draft, F.I. and L.C.

Acknowledgments: The authors are grateful to ITALCOL Spa, Castelfiorentino (FI), to QuMAP Laboratory of PIN (Prato) and to Fabio Villanelli for technical support.

Conflicts of Interest: The authors declare no conflict of interest

References

1. Coppen, J.J.W. *Eucalyptus: The Genus Eucalyptus*; Taylor & Francis: London, UK, 2002.
2. Ramezani, H.; Singh, H.P.; Batish, D.R.; Kohli, R.K. Antifungal activity of the volatile oil of Eucalyptus citriodora. *Fitoterapia* **2002**, *73*, 261–262. [CrossRef]
3. Cermelli, C.; Fabio, A.; Fabio, G.; Quaglio, P. Effect of eucalyptus essential oil on respiratory bacteria and viruses. *Curr. Microbiol.* **2008**, *56*, 89–92. [CrossRef] [PubMed]
4. Mulyaningsih, S.; Sporer, F.; Zimmermann, S.; Reichling, J.; Wink, M. Synergistic properties of the terpenoids aromadendrene and 1,8-cineole from the essential oil of Eucalyptus globulus against antibiotic-susceptible and antibiotic-resistant pathogens. *Phytomedicine* **2010**, *17*, 1061–1066. [CrossRef] [PubMed]
5. Tyagi, A.K.; Malik, A. Antimicrobial potential and chemical composition of Eucalyptus globulus oil in liquid and vapour phase against food spoilage microorganisms. *Food Chem.* **2011**, *26*, 228–235. [CrossRef]
6. Sadlon, A.E.; Lamson, D.W. Immune-Modifying and Antimicrobical Effects of Eucalyptus Oil and simple Inhalation Devices. *Altern. Med. Rev.* **2010**, *15*, 33–47.
7. Bachir Raho, G.; Benali, M. Antibacterial activity of the essential oils from the leaves of *Eucalyptus globulus* against *Escherichia coli* and *Staphylococcus aureus*. *Asian Pac. J. Trop. Biomed.* **2012**, *2*, 739–742. [CrossRef]
8. Lu, H.; Shao, X.; Cao, J.; Ou, C.; Pan, D. Antimicrobial activity of eucalyptus essential oil against *Pseudomonas* in vitro and potential application in refrigerated storage of pork meat. *Int. J. Food Sci. Technol.* **2016**, *51*, 994–1001. [CrossRef]
9. Vilela, G.R.; de Almeida, G.S.; D'Arce, M.A.B.R.; Moraes, M.H.D.; Brito, J.O.; da Silva, M.F.G.F.; Silva, S.C.; de Stefano Piedade, S.M.; Calori-Domingues, M.A.; da Gloria, E.M. Activity of essential oil and its major compound, 1,8-cineole, from *Eucalyptus globulus* Labill., against the storage fungi *Aspergillus flavus* Link and *Aspergillus parasiticus* Speare. *J. Stored Prod. Res.* **2009**, *45*, 108–111. [CrossRef]
10. Ormancey, X.; Sisalli, S.; Coutiere, P. Formulation of essential oils in functional perfumery. *Parfum. Cosmet. Actual.* **2001**, *157*, 30–40.

11. Sacchetti, G.; Maietti, S.; Muzzoli, M.; Scaglianti, M.; Manfredini, S.; Radice, M.; Bruni, R. Comparative evaluation of 11 essential oils of different origin as functional antioxidants, antiradicals and antimicrobials in foods. *Food Chem.* **2005**, *91*, 621–632. [CrossRef]
12. Paolini, J.; Leandri, C.; Desjobert, J.M.; Barboni, T.; Costa, J. Comparison of liquid–liquid extraction with headspace methods for the characterization of volatile fractions of commercial hydrolats from typically Mediterranean species. *J. Chromatogr. A* **2008**, *1193*, 37–49. [CrossRef] [PubMed]
13. Edris, A.E. Identification and absolute quantification of the major water-soluble aroma components isolated from the hydrosols of some aromatic plants. *J. Essent. Oil-Bear. Plants* **2009**, *12*, 155–161. [CrossRef]
14. Ndiaye, H.B.; Diop, M.B.; Gueye, M.T.; Ndiaye, I.; Diop, S.M.; Fauconnier, M.L.; Lognay, G. Characterization of essential oils and hydrosols from senegalese *Eucalyptus camaldulensis* Dehnh. *J. Essent. Oil-Bear. Plants* **2018**, *30*, 1–11. [CrossRef]
15. Price, L.; Price, S. *Understanding Hydrolats: The Specific Hydrosols for Aromatherapy*; Churchill Livingstone/Elsevier: Amsterdam, The Nederlands, 2004; p. 96.
16. Zrira, S.; Bessiere, J.M.; Menut, C.; Elamrani, A.; Benjilali, B. Chemical composition of the essential oil of nine *Eucalyptus* species growing in Morocco. *Flavour Fragr. J.* **2004**, *19*, 172–175. [CrossRef]
17. Elaissi, A.; Hadj Salah, K.; Mabrouk, S.; Larb, K.K.; Chemli, R.; Harzallah-Skhiri, F. Antibacterial activity and chemical composition of 20 *Eucalyptus* species essential oils. *Food Chem.* **2011**, *129*, 1427–1434. [CrossRef]
18. Kahla, Y.; Zouari-Bouassida, K.; Rezgui, F.; Trigui, M.; Tounsi, S. Efficacy of *Eucalyptus cinerea* as a Source of Bioactive Compounds for Curative Biocontrol of Crown Gall Caused by *Agrobacterium tumefaciens* Strain B6. *BioMed Res. Int.* **2017**. [CrossRef] [PubMed]
19. Ieri, F.; Cecchi, L.; Vignolini, P.; Belcaro, M.F.; Romani, A. HPLC/DAD, GC/MS and GC/GC/TOF analysis of Lemon balm (*Melissa officinalis* L.) sample as standardized raw material for food and nutraceuticals uses. *Adv. Hortic. Sci.* **2017**, *3*, 141–147. [CrossRef]
20. Wong, Y.F.; Perlmutter, P.; Marriott, P.J. Untargeted metabolic profiling of *Eucalyptus* spp. leaf oils using comprehensive two-dimensional gas chromatography with high resolution mass spectrometry: Expanding the metabolic profile. *Metabolomics* **2017**, *13*, 46. [CrossRef]
21. Calamai, L.; Villanelli, F.; Bartolucci, G.; Pieraccini, G.; Moneti, G. Sample preparation for direct ms analysis of food. In *Comprehensive sampling and sample preparation*; Elsevier: Amsterdam, The Nederlands, 2012.
22. Oliver-Pozo, C.; Aparicio-Ruiz, R.; Romero, I.; Garcia-Gonzalez, D.L. Analysis of volatile markers for virgin olive oil aroma defects by SPME-GC/FID: Possible sources of incorrect data. *J. Agric. Food Chem.* **2015**, *63*, 10477–10483. [CrossRef]
23. Fortini, M.; Migliorini, M.; Cherubini, C.; Cecchi, L.; Calamai, L. Multiple internal standard normalization for improving HS-SPME-GC-MS quantitation in virgin olive oil volatile compounds (VOO-VOCs) profile. *Talanta* **2017**, *165*, 641–652. [CrossRef]
24. May, F.; Ash, J. An assessment of the allelopathic potential of *Eucalyptus*. *Aust. J. Botany* **1990**, *38*, 245–254. [CrossRef]
25. Kohli, R.K.; Batish, D.R.; Singh, H.P. Eucalypt oils for the control of *Parthenium* (*Perthenium hysterophorus* L.). *Crop Prot.* **1998**, *17*, 119–122. [CrossRef]
26. Hummelbrunner, L.A.; Isman, M.B. Acute, sublethal, antifeedant, and synergistic effects of monoterpenoid essential oil compounds on the tobacco cutworm, *Spodoptera litura* (Lep., Noctuidae). *J. Agric. Food Chem.* **2001**, *49*, 715–720. [CrossRef] [PubMed]
27. Nerio, L.S.; Olivero-Verbel, J.; Stashenko, E. Repellent activity of essential oils: A review. *Bioresour. Technol.* **2010**, *101*, 372–378. [CrossRef] [PubMed]
28. Spietelun, A.; Marcinkowski, L.; De la Guardia, M.; Namieśnik, J. Review recent developments and future trends in solid phase microextraction techniques towards green analytical chemistry. *J. Chromatogr. A* **2013**, *1321*, 1–13. [CrossRef] [PubMed]
29. Hantao, W.L.; Toledo, B.R.; de Lima Ribeiro, F.A.; Pizetta, M.; Geraldi Pierozzi, C.; Luiz Furtado, E.; Augusto, F. Comprehensive two-dimensional gas chromatography combined to multivariate data analysis for detection of disease-resistant clones of Eucalyptus. *Talanta* **2013**, *116*, 1079–1084. [CrossRef]
30. Cordero, C.; Bicchi, C.; Rubiol, P. Group-type and fingerprint analysis of roasted food matrices (coffee and hazelnut samples) by comprehensive two-dimensional gas chromatography. *J. Agric. Food Chem.* **2008**, *56*, 7655–7666. [CrossRef]

31. Von Mühlen, C.; Alcaraz Zini, C.; Bastos Caramaob, E.; Marriott, P.J. Comparative study of Eucalyptus dunnii volatile oil composition using retention indices and comprehensive two-dimensional gas chromatography coupled to time-of-flight and quadrupole mass spectrometry. *J. Chromatogr. A* **2008**, *1200*, 34–42. [CrossRef]
32. Cordero, C.; Atsbaha Zebelo, S.; Gnavi, G.; Griglione, A.; Bicchi, C.; Maffei, M.E.; Biolo, P. HS-SPME-GC× GC-qMS volatile metabolite profiling of *Chrysolina herbacea* frass and *Mentha* spp. Leaves. *Anal. Bioanal. Chem.* **2012**, *402*, 1941–1952. [CrossRef]
33. Van den Dool, H.; Kratz, P.D. A generalization of the retention index system including linear temperature programmed gas-liquid partition chromatography. *J. Chromatogr.* **1963**, *11*, 463–471. [CrossRef]
34. Chemistry WebBook. Available online: http://www.nist.gov/index.html (accessed on 1 December 2018).

Sample Availability: Samples of the compounds are not available from the authors.

© 2019 by the authors. Licensee MDPI, Basel, Switzerland. This article is an open access article distributed under the terms and conditions of the Creative Commons Attribution (CC BY) license (http://creativecommons.org/licenses/by/4.0/).

Article

Profiling Volatile Constituents of Homemade Preserved Foods Prepared in Early 1950s South Dakota (USA) Using Solid-Phase Microextraction (SPME) with Gas Chromatography–Mass Spectrometry (GC-MS) Determination

Lucas J. Leinen [1,†], Vaille A. Swenson [2,†], Hope L. Juntunen [1,2], Scott E. McKay [1], Samantha M. O'Hanlon [3], Patrick Videau [4,*] and Michael O. Gaylor [1,*]

1. Department of Chemistry, Dakota State University, Madison, SD 57042, USA; ljleinen@pluto.dsu.edu (L.J.L.); hope.juntunen@trojans.dsu.edu (H.L.J.); Scott.McKay@dsu.edu (S.E.M.)
2. Department of Biology, Dakota State University, Madison, SD 57042, USA; Vaille.Swenson@trojans.dsu.edu
3. School of Psychological Science, Oregon State University, Corvallis, OR 97331, USA; ohanlons@oregonstate.edu
4. Department of Biology, Southern Oregon University, Ashland, OR 97520, USA
*   Correspondence: videaup@sou.edu (P.V.) michael.gaylor@dsu.edu (M.O.G.); Tel.: +1-541-552-6788 (P.V.); +1-605-256-5822 (M.O.G.); Fax: +1-541-552-7127 (P.V.); +1-605-256-5643 (M.O.G.)
†   These authors contributed equally to the work.

Academic Editors: Constantinos K. Zacharis and Paraskevas D. Tzanavaras
Received: 30 December 2018; Accepted: 8 February 2019; Published: 13 February 2019

Abstract: An essential dimension of food tasting (i.e., flavor) is olfactory stimulation by volatile organic compounds (VOCs) emitted therefrom. Here, we developed a novel analytical method based on solid-phase microextraction (SPME) sampling in argon-filled gas sampling bags with direct gas chromatography–mass spectrometry (GC-MS) determination to profile the volatile constituents of 31 homemade preserves prepared in South Dakota (USA) during the period 1950–1953. Volatile profiles varied considerably, but generally decreased in detected compounds, complexity, and intensity over three successive 2-h SPME sampling periods. Volatile profiles were generally predominated by aldehydes, alcohols, esters, ketones, and organic acids, with terpenoids constituting much of the pickled cucumber volatiles. Bisphenol-A (BPA) was also serendipitously detected and then quantified in 29 samples, at levels ranging from 3.4 to 19.2 µg/kg, within the range of levels known to induce endocrine disruption effects. Absence of BPA in two samples was attributed to their lids lacking plastic liners. As the timing of their preparation coincides with the beginning of BPA incorporation into consumer products, these jars may be some of the first BPA-containing products in the USA. To the best of our knowledge, this is the first effort to characterize BPA in and volatile profiles of rare historical foods with SPME.

Keywords: historical foods; preserves; volatile organic compounds (VOCs); bisphenol-A (BPA)

1. Introduction

Human olfactory sensing of volatile organic compounds (VOCs) released from foodstuffs (i.e., their volatile profile) is an essential component of the perception of flavor [1]. When sealed foods are opened and prepared, VOCs spanning a multitude of functional classes interact synergistically to generate the aromatic and gustatory properties of the food. Given the variety of food types and associated volatile inventories (and the potential complexity thereof), it is difficult to directly relate human perceptions of, for example, "sweet" and "savory" tastes to specific volatile functionalities.

However, based on general trends in the literature, the sweet aromas of fruits and vegetables appear to be defined primarily by volatile profiles enriched in, for example, aldehydes, alcohols, esters, ketones, quinones, and terpenes [2,3], while savory aromas, e.g., of meats and cheeses, tend to be defined by volatile profiles more enriched in sulfur- and nitrogen-containing compounds (e.g., thiols, sulfides, and amines) [4,5].

Evidence of food preservation by humans using primitive techniques (e.g., burial storage and sun drying, etc.) dates to 20,000 BC [6], with the first evidence of liquid preservation via pickling dating to ancient Greek and Egyptian cultures [7]. Many contemporary food preservation techniques are now liquid-based, and necessitate boiling and subsequent sealing of metal or glass vessels, and appear to be derived from those developed in more recent times (circa 300 years ago) [8]. These methods—generically referred to as "canning"—involve the immersion of foods in acidic brine and/or sugar solutions with subsequent heating and sealing to prevent microbial degradation. Such techniques have since been practiced and evolved by cultures around the globe, including the early pioneer settlers of the circa mid-19th-century North American frontier, who required simple, effective methods for longer term food storage prior to the advent of in-home refrigeration [9].

It is unlikely that food preservation was also intended as a method of preserving records of past chemical environments, but it occurred to us during conception and design of this study that it may be a fortuitous outcome and worthy of exploration. During the canning process, some chemical constituents originating from that point in time (e.g., via atmospheric deposition or plant-associated soil components, etc.) may be preserved along with the foods, creating a sort of chemical "time capsule" that might yield interesting insights into the nature of those environments [10,11]. To the best of our knowledge, there are no published reports of solid-phase microextraction (SPME) applied to archaeological studies of the volatile constituents of canned or other prepared foodstuffs. However, SPME applications in archaeology appear to be increasing [12–14], and reports of other analytical/chemical methods applied to archaeological studies of historical foods and food remains dates back to the late 20th century [10,11,15–20].

Inspired by the potential to obtain new insights into the food preservation techniques and possibly the environmental conditions of the North American Great Plains region during this time, and to address the lack of SPME studies of historical foods, we sought to profile the volatile inventories of a variety of homemade preserved foods prepared in Moody County, South Dakota (SD; USA) during the period 1950–1953. A secondary aim of the study was to quantify burdens of the toxic plasticizer bisphenol-A (BPA), discovered serendipitously in pursuit of the above primary study aims. To the best of our knowledge, this is the first reported effort to use SPME to profile the volatile constituents of rare preserved historical foods sealed for more than three generations, as well as the first report of a SPME-detectable toxic compound therein.

2. Results and Discussion

2.1. SPME Method Development

Once the initial batch of 1950s preserves was collected, it quickly became evident that there were few studies of, and no published methods to assess, the VOCs from these types of rare historical food samples. To design and validate a method to profile the volatile constituents from these irreplaceable historical preserves, SPME sampling was used to define the optimal sampling time for preserved foods, and to verify that the compounds identified indeed originated from the samples being assessed. Store-bought preserved foods were used to assess the time to equilibrium uptake of VOCs in the AtmosBag because these foods were locally available, are presumed to have highly consistent volatile profiles across individual jars based on the batch preparation of such foods, and could be purchased in sufficient quantities to permit replicate single samplings. Lacking a priori knowledge of the composition of their volatile profiles, savory and sweet preserved foods (dill pickles and maraschino cherries) were selected to encompass a wide range of representative compound functionalities during method development. Store-bought dill pickle and cherry preserves were sealed in the sampling bag along with the SPME device, and sampled in triplicate for 30, 60, 120, 240, and 360 min as described in the Experimental Section (Figure S1). Equilibrium uptake (as measured by maximum total integrated peak area) was achieved by 120 min for both pickles and cherries (Figure 1). This sampling time was used for all subsequent SPME samplings. Sampling efficacy was further validated by assessing uptake of authentic bisphenol-A (BPA) standards spiked into both store-bought matrices and sampled with SPME for 120 min. Reproducibility between samples and within triplicate resampling replicates differed by less than 5% (Figure 1), confirming method efficacy.

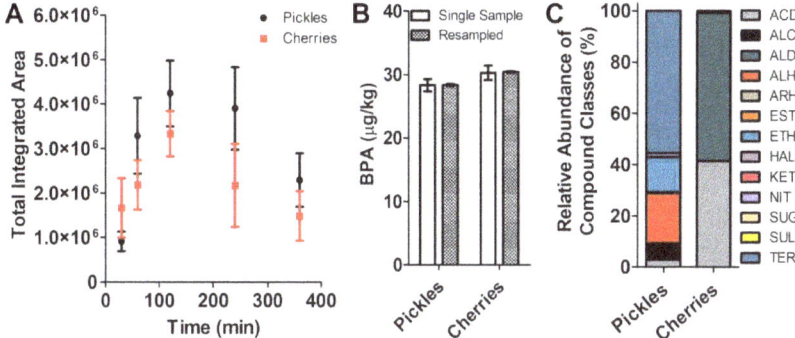

Figure 1. Validation of a solid-phase microextraction (SPME) method to assess the volatile inventories of preserved foods using commercially available dill pickles and maraschino cherries. A time course from 30–360 min of SPME sampling was conducted for both pickles (black circles) and cherries (red squares) in triplicate. Total integrated peak area is presented as the average ± standard deviation (**A**). As the 120 min (2 h) timepoint produced the greatest total integrated peak area for both preserved foods, the reproducibility of the sampling protocol was assessed based on the level of BPA measured following spiking 1000 µg into jars of pickles and cherries (**B**). Five jars each of pickles and cherries were sampled a single time, and three different jars of each food were sampled in three successive replicates. The concentration of BPA was determined by comparison to a standard curve and is presented as the average ± standard deviation for single samplings (clear bars) and average ± standard error of the mean for replicate samplings (dotted bars). The relative abundance of the assessed compound classes (as percent of total) is presented for the pickles and cherries sampled for 2 h with SPME (**C**). Compound classes are defined as follows: ACD = acids; ALC = alcohols; ALD = aldehydes; ALH = aliphatic hydrocarbons; ARH = aromatic hydrocarbons; EST = esters; ETH = ethers; HAL = halogen-containing; KET = ketones; NIT = nitrogen-containing; SUG = sugar alcohols; SUL = sulfur-containing; TER = terpenoids.

The SPME method permitted detection of 20 unique compounds, constituting the Vlasic dill pickle volatile inventory, that conformed to our analytical detection and identification parameters (see Experimental Section). The volatile profile of the Vlasic dill pickles was dominated by acetic acid, ammonium acetate, alcohols, esters, and ketones, and the terpenoids tentatively identified as β-myrcene, α-phellandrene, 4-carene, p-cymene, limonene, α-terpineol, and D-carvone (Table S1). The absence of unsaturated aldehydes, e.g., (E,Z)-2,6-nonadienal and (E)-2-nonenal, in the Vlasic volatile profile was surprising, as these compounds have been reported to be essential aroma compounds in pickled cucumber preparations [21]. In contrast to the 20 compounds identified from Vlasic dill pickles, the maraschino cherry volatile inventory was comprised of only 4 compounds, tentatively identified as benzaldehyde, (Z)-3-hexenol acetate, ascorbic acid, and benzoic acid (Supplementary Data File 1), which is consistent with known cherry volatile profiles [22,23].

To validate the integrity of the AtmosBag, further assess the efficacy of the SPME method, and assess the extent to which the individual components of a pickled cucumber preparation contributed compounds to the measured volatile inventory, we characterized the VOCs of store-bought cucumbers (sliced into spears and stored in clean Vlasic dill pickle jars to keep system dimensions constant), a lab-prepared artificial pickling brine (used to approximate the Vlasic dill pickle brine), and a lab preparation of the cucumber spears pickled in the artificial brine for 7 days (see Experimental Section). Each of the components and the mixture thereof was analyzed in triplicate with the SPME method. As published volatile profiles for commercial pickle preparations are lacking, this was done to show that the major compounds detected in the Vlasic pickles were consistent with those detected in the individual components, and not sampling artifacts or environmentally-derived compounds, providing additional verification of method efficacy. The cucumber spears alone produced only 2 compounds, tentatively identified as limonene and 3,6-dimethyl-2,3,3a,4,5,7a-hexahydrobenzofuran, while the artificial brine and the cucumber spears pickled therein combined to produce a volatile inventory enriched with most of the major compounds detected in the Vlasic dill pickles with the established method (Table S2). The 20 compounds detected here are consistent with the 21 compounds reported to comprise the volatile inventory of a competing brand of dill pickles [24]. However, only a few compounds detected here were reported in that study (acetic acid, ethyl acetate, α-Terpineol). This is likely due to differences in brine composition and SPME fiber phase used (75 μm Carboxen-polydimethylsiloxane (PDMS) phase).

Important aroma/flavor volatiles detected in the Vlasic pickles included acetic acid, ammonium acetate, β-myrcene, α-phellandrene, 4-carene, p-cymene, limonene, and D-carvone, further confirming the efficacy of our SPME method to capture the major volatile constituents of this commercial dill pickle preparation. Though these compounds appear typical of preserved cucumber preparations, the relatively simpler volatile profile was also a bit surprising, in that previous studies have reported abundances of, for example, longer chain alcohols and unsaturated aldehydes in cucumber volatile emissions [25]. This is presumably due to the different SPME fiber phase and sampling approaches, as well as the different cultivars, used in that study. The presence of limonene may be an artifact of the chitosan coatings commonly applied to fresh cucumbers to extend shelf life [26], or perhaps a metabolic response to stress induced by the physical process of preparing the cucumbers for study [27]. This experiment could not be performed with the maraschino cherries, as it was not possible to obtain store-bought Marasca cherries (or other types used for this recipe), and the complex ingredients listed on the store-bought cherries were not reproducible in the lab. However, the capacity of the SPME method to reproducibly detect major compounds reported to comprise the volatile inventories of Marasca cherry [23] and commercial dill pickle preparations [24] over all sampling times and replicate sampling periods provided compelling support for the validity and efficacy of the method developed here to assess the major compounds of the historical preserves. Additional confirmation of method validity was attained via the BPA spiking experiments (Figure 1).

2.2. SPME Analysis of 1950s Preserves

After validating the SPME sampling method for jars of commercially preserved foods from the present day, the technique was applied to the historical samples for which it was developed (31 jars of historical preserves; Figure S2). The volatile profiles of the 1950s preserves varied considerably in number of compounds detected, class composition, and intensity over the sample set and over three successive SPME samplings. Total ion chromatograms (TICs) showing the volatile profiles of representative low, medium, and high complexity determined during the first SPME sampling are presented in Figure 2B–D. Numbers of compounds varied from as few as 2 (rhubarb, sample 24) to as many as 67 (sweet pickle, sample 31) detected in the first sampling period, with compound numbers, total integrated peak area, and volatile profile complexity generally decreasing over successive SPME samplings (Figure 3, Figure S3). The first SPME sampling period was of particular interest, because it is presumed to represent the total VOCs from the production time period before exogenous compounds (i.e., gases) were introduced upon repeated samplings. Of the samples, 74% displayed a lower total integrated peak area in the third sampling than in the first, and 26% of these had a higher total integrated peak area in the second sampling than in the first sampling. Interestingly, 22% of the preserves showing decreased integrated peak area in successive samplings produced a higher number of compounds in the third sampling than in the first sampling, which indicates that new compounds were evolved from these samples over successive SPME sampling periods. Compound class composition and diversity was also highly variable, as one might expect from homemade food preserves sealed for three generations (Figure 3, Figure S3). Compound class diversity was lowest in tomato preserves (sample 16) and highest in apricot preserves (sample 27).

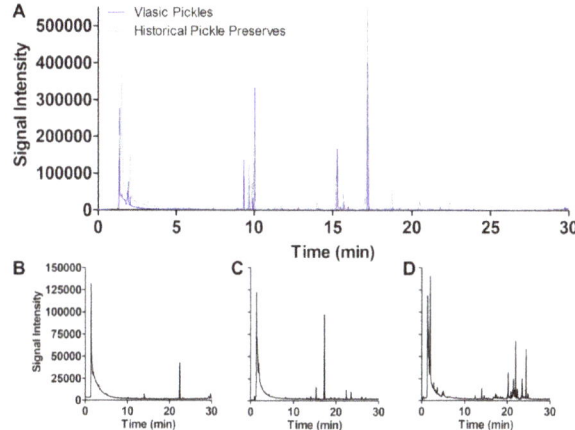

Figure 2. Total ion chromatograms (TICs) derived from 2 h SPME sampling of representative samples. Vlasic dill pickle (blue line) and historical preserve dill pickle type 2 (sample 29; gray line) TICs are overlaid to show similarities and differences in compounds detected and their intensities (**A**). Representative TICs of low (tomato sample 2; **B**), medium (dill pickle sample 30; **C**), and high (sweet pickle sample 31; **D**) complexity samples. The *y*-axis labels and intensity presented in the panel B TIC are the same for panels C and D.

TICs of the first SPME sampling of the Vlasic dill pickle volatiles were compared to a 1950s dill pickle volatile profile to assess compound relatedness (an overlay of their TICs is shown in Figure 2A). Similar to the Vlasic pickles, the 1950s dill pickle volatile profile was predominated by acids, alcohols, esters, and ketones, but the latter produced nearly twice the number of compounds, including a larger inventory of aliphatic and aromatic hydrocarbons. However, the 1950s sample produced a comparable inventory of terpenoid compounds (Table S2). This was also one of only two historical preserve samples to produce detectable levels of the compound tentatively identified as apiol (a phenylpropanoid isolated from parsley with known abortive effects), the other being a sweet pickle preserve.

Consistent with their bulk acidic nature (Table 1), the volatile profiles of nearly all the preserves contained organic acids, but acid species predominated in apple, apricots, dill pickle, rhubarb, and tomato preserves. Acetic acid was the dominant acid in the "less sweet" dill pickle, rhubarb, and tomato preserves. The volatile profiles of sweet preserves were generally predominated by esters, followed by aldehydes, consistent with studies of modern sweet foodstuffs [22]. Three of the four rhubarb samples were exceptions to this trend, consistent with our taste test classifications of "slightly sweet" (Table 1). Much of the ester inventory is presumed to arise from esterification reactions involving abundant free fatty acids likely produced and solubilized to higher concentrations during heating of the foods prior to canning [28]. The prevalence of esters followed by aldehydes in dill pickles is generally consistent with those reported for the volatile inventories of modern cucumbers and pickled cucumbers [21,24].

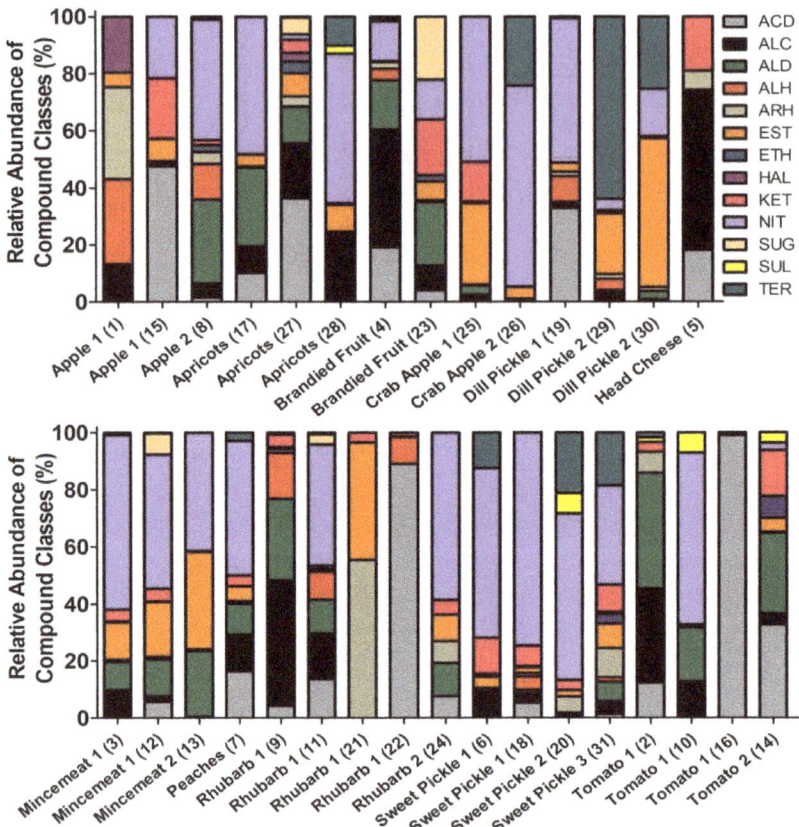

Figure 3. The relative abundance of the assessed compound classes (as percent of total) in the historical preserves. Samples are ordered alphabetically by type, and the sample number is in parenthesis after the sample type. Compound classes presented as colors are defined as follows: ACD = acids; ALC = alcohols; ALD = aldehydes; ALH = aliphatic hydrocarbons; ARH = aromatic hydrocarbons; EST = esters; ETH = ethers; HAL = halogen-containing; KET = ketones; NIT = nitrogen-containing; SUG = sugar alcohols; SUL = sulfur-containing; TER = terpenoids.

Aldehydes are recognized as impactful aroma components in foods, which may derive from Maillard reactions occurring during the heating of foods prior to sealing. Aldehydes may also arise from oxidation of unsaturated fatty acids via peroxidation, which can drive production of, for example, the increased alcohol and hydrocarbon inventories [28] also detected in these samples. In support of this explanation, peroxides were also detected in the aldehyde enriched preserves, but did not meet the established criteria for positive compound identification (described in Experimental Section). Aliphatic hydrocarbons were also detected in nearly all samples, but were prevalent in apples, brandied fruit, dill pickles, and rhubarb, several of which also contained appreciable aldehydes. Alkanes constituted the majority of the aliphatics, which may also be explained in part by degradative reaction cascades originating with Maillard reactions [28].

Table 1. Quantitative and qualitative data for the preserved foods assessed in this study.

Sample [a]	Sample Number	N [b]	pH	Sample Description [c]	Taste [d]
Apple 1	1, 15	2	3.1–3.4	Sliced and shredded apple in opaque liquid with apple sauce consistency	Sweet
Apple 2	8	1	3.0	Sliced apples stacked in clear liquor	Sweet
Apricots	17, 27, 28	3	2.9–3.5	Spherical, fleshy fruit bodies in dark liquid	Sweet
Brandied Fruit	4, 23	2	2.9–3.3	Dark gelatinous fruit bodies resembling plums in dark, viscous liquor	Sweet/alcohol
Crab Apple 1	25	1	3.4	Round fleshy fruit bodies with a single stem and many small seeds in the center in dark liquid	Sweet
Crab Apple 2	26	1	3.8	Round fleshy fruit bodies with no stem and many small seeds in the center in dark liquid	Sweet
Dill Pickle 1	19	1	3.8	Whole cucumbers with dill plants and seeds in clear liquid—smaller jar	Salty
Dill Pickle 2	29, 30	2	3.2–3.3	Cucumbers quartered longitudinally with dill plants and seeds in clear liquid—larger jar	Salty
Head Cheese	5	1	3.7	Cylindrical mass of spongy material resembling cheese in amber liquor	Salty/savory
Mincemeat 1	3, 12	2	3.9–4.0	Fleshy fruit bodies resembling currants, raisins, leafage, and fleshy chunks resembling meat in small volume of clear liquid—larger jar	Sweet/savory
Mincemeat 2	13	1	4.0	Fleshy fruit bodies resembling currants, raisins, leafage, and fleshy chunks resembling meat in small volume of clear liquid—smaller jar	Sweet/savory
Peaches	7	1	3.6	Dark, spherical, fleshy fruit bodies with a single solid pit in the center in dark liquid	Sweet
Rhubarb 1	9, 11, 21, 22	4	2.9–3.0	Shredded vegetative material in dark liquid	Slightly sweet
Rhubarb 2	24	1	3.1	Cylindrical mass of vegetative material in dark liquid	Slightly sweet
Sweet Pickle 1	6, 18	2	2.5–3.2	Halved cucumbers in dark liquid	Sweet
Sweet Pickle 2	20	1	3.2	Cucumber slices in dark liquid—smaller jar	Sweet
Sweet Pickle 3	31	1	3.2	Cucumber slices in dark liquid—larger jar	Sweet
Tomato 1	2, 10, 16	3	4.1–4.2	Whole and shredded tomato in opaque, red liquor with tomato soup consistency	Sweet/salty
Tomato 2	14	1	4.1	Whole and shredded tomato in red liquor with spaghetti sauce consistency	Sweet/salty

[a] Sample identification and determination of unique sample types and replicates of the same sample type was achieved via combination of anecdotal accounts of method/time of preparation by the donating family and our own examinations of appearance, smell, and taste. [b] Represents the number of replicates of the same sample type. [c] Sample descriptions are derived from visual inspections of the samples by the authors to arrive at consensus characterizations. [d] Taste descriptions are derived from tastings performed by author MOG (Supplemental Video V1).

Alcohols are also recognized as important aroma compounds in many food types, but are generally less abundant in food volatiles relative to, for example, aldehydes, esters, and terpenes. Alcohols are also important precursors for aldehyde and ester production, which may account for their lesser abundance in the volatile inventories of foods [28]. In this study, alcohols were prevalent in apricots, brandied fruit, head cheese, rhubarb, and tomato samples. The preponderance of alcohols in head cheese is consistent with the volatile inventories reported for some types of meats [29]. Like alcohols, ketones may be essential aroma compounds in foods, but are generally less abundant relative to, for example, aldehydes, esters, and terpenes [22,30,31]. In this study, ketones were prevalent in apple, brandied fruit, crab apple, head cheese, mincemeat, peaches, rhubarb, sweet pickle, and tomato preserves, consistent with ketone inventories reported for other fruit- and meat-based foodstuffs [29,32].

Terpenoids were detected sporadically in multiple preserves, but were prominent constituents in some dill pickle, sweet pickle, and crab apple preserves (Figure 3, Figure S3, Table S2), consistent with previous reports of volatile profiles of analogous foods [21,24,33]. Terpenoids produce aromas in foods previously described as, for example, "sweet," "rose-like," "green," "fruity," "citrus," "piney," "floral," "resinous," "lemon," and "lemon-like" [34]. These descriptions are generally consistent with our own smell and taste characterization of the preserves (see Supplemental Video V1). Surprisingly, these terpenoids were absent in the other crab apple (sample 25), dill pickle (sample 19), and sweet pickle (sample 18) preserves. It is not clear why such substantial compositional differences would be detected in such seemingly similar preserves, but this may be indicative of quite different preparation methods, or variations in the individual fruits/vegetables harvested from the rural home garden during this time period.

Nitrogen-containing compounds were prevalent in most of the preserves, but were conspicuously absent from apple (sample 1), head cheese (sample 5), rhubarb (samples 21, 22), and tomato (samples 2, 16, 14) (Figure 3). The most commonly occurring compounds were tentatively identified as hydrazides of acetic acid (detected in apple, apricots, dill pickle, mincemeat, rhubarb, sweet pickle, tomato), butyric acid (detected in apple, mincemeat, rhubarb), and formic acid (detected in mincemeat, rhubarb, sweet pickle), ammonium acetate (detected in dill pickle, mincemeat, sweet pickle), N,N-dibutylformamide (detected in apricots, brandied fruit, peaches, rhubarb, tomato), (Z)-9-octadecenamide (detected in rhubarb), and hydrazine derivatives (detected in apricots, brandied fruit, crab apples, mincemeat, sweet pickle, tomato). Reports of nitrogen-containing volatiles in preserved foods are limited, but are presumed to be driven by Maillard reactions between amino acids and sugar alcohols during preparation, and other degradative reactions involving amines, amino acids, peptides, and alkaloids thereafter. Sulfur compounds were minor constituents of only a few preserves (detected in apricots, sweet pickle, tomato), and were tentatively identified as sulfurous acid esters and alkyl sulfides, consistent with sulfur compounds reported in sweet, tangy, and savory foods [35].

2.3. BPA in 1950s Preserves

Detection of BPA in all the preserves was surprising, as this compound was just being introduced into consumer products in the early 1950s as an additive in the protective epoxy resin coatings of food containers [36]. These jars may thus represent some of the earliest BPA-containing consumer products in the USA. BPA was detected in the volatile profiles of all preserves (but below the quantitation limit of 1 µg/kg in samples 20 and 28—sweet pickle and apricots), and was detected in all sampling replicates at levels ranging from 3.4 to 19.2 µg/kg (Figure 4). Such low BPA levels in samples 20 and 28 was interesting, and is presumed to be the result of different lids (lacking plastic liners) being used to seal these preserves. These preserves were not sealed using the Ball canning lids used on all other preserves (Figure S2), supporting the idea that the incurred BPA was derived from the plastic liners of the Ball lids.

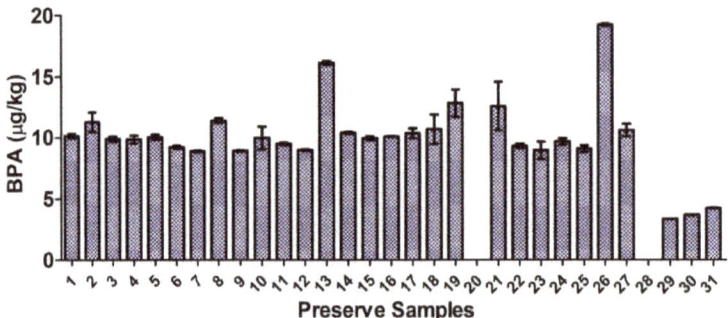

Figure 4. The concentration of bisphenol-A (BPA) in historical preserves is presented as the average ± standard deviation determined from three successive samplings of each historical preserve.

These samples notwithstanding, detected BPA levels were in the range of those reported to induce endocrine disruption (e.g., antiandrogenic) effects in biochemical assays [37]. BPA levels were fairly constant over the entire sample set and over the three successive samplings of each preserve (Figure S4), consistent with reported persistence of release pathways from consumer polymers [38] and our data from resampling experiments using BPA spiked into store-bought preserves (Figure 1). Though significant differences in SPME-detected BPA burdens were detected ($p < 0.05$; paired t-test), if one considers the absolute mass of BPA sampled by the SPME fiber, detected masses were considerably more uniform over the data set (Figure 4; Figure S4; Table S3). Such uniform volatilization from the preserves over lengthy successive SPME sampling periods provides potentially useful insight into human exposure risks prior to regulatory controls on BPA. Indeed, these may be the first such insights available in the literature.

Using spiked BPA uptake data, we estimated naturally incurred BPA burdens (presumed to have leached from the protective plastic lid liners during storage) to be of the order of 1000 µg sample^{-1}. This estimate is within the range of measured concentrations of BPA leached from consumer bottles lined with epoxy resin-based protective coatings [39]. The absolute mass of spiked BPA sampled by the SPME fiber over the entire sample set was calculated to be $2.5 \pm 0.2\%$ of the total spiking mass (1000 µg), which is consistent with BPA's low vapor pressure and the Henry's Law constant. It is also in general agreement with uptake efficiencies predicted for the 100 µm PDMS phase possessing an absorptive sampling volume of the order of 10^{-9} L [40] (sampling volume of the AtmosBag is a billion times larger), and those previously reported for structurally analogous aromatic pollutants [41].

The data presented indicate that the compound class compositions profiled in these rare historical preserves are in general agreement with those determined in many studies of modern preserved food types. This is particularly interesting given that the historical preserves assessed here were prepared before the widespread use of ultra-pasteurization, UV sterilization, or even the freeze-drying methods used to preserve many commercial foods. The general similarities in class compositions suggests that the method(s) used to prepare these historical preserves and those described elsewhere were not sufficiently different to produce radically distinct SPME-detectable volatile profiles. Alternatively, if modern food preservation methods are vastly different from those used to prepare these preserves during the period 1950–1953, those differences were not manifested in wholly different SPME-detectable volatile profiles. It is also possible that, because the overall class compositional profiles were comparable between modern and historical preserves, the cultivars preserved were not different enough to evolve entirely disparate volatile profiles. Though differences at the level of individual compounds were certainly apparent, the trends in compound classes detected with the SPME method suggests overarching similarities with modern preserved foods.

Detection of BPA in all of these historical foodstuffs provides an unprecedented glimpse into what may have been some of the earliest incidences of human exposure to this toxic compound, originally believed to be tightly sequestered in consumer polymers and thus biologically inaccessible. Finally, to the best of our knowledge, this is the first account of a simple, reproducible, and robust SPME method for profiling volatile inventories and industrial pollutant burdens of rare canned historical foodstuffs. Taken together, these study findings indicate that SPME may be a powerful, but so far underexplored, analytical tool for elucidating the chemical constituents of archaeological samples more generally.

3. Experimental Section

3.1. Chemicals, Equipment, and Ingredients

All solvents and reagents were of analytical grade (purity > 99%) and were purchased from Sigma-Aldrich (Saint Louis, MO, USA). The SPME sampler, SPME fibers, and sampling bags (Aldrich® AtmosBag) were also purchased from Sigma-Aldrich (product number Z564427). A 24 gauge, 100 µm polydimethylsiloxane (PDMS) SPME fiber was used to sample all preserves. This phase was selected because it has been shown to be one of the most versatile and rugged fiber phases for sampling a wide range of compound functionalities in demanding environmental sampling applications (e.g., field analyses of challenging environmental matrices in Tedlar bags) [42]. This phase has also proven effective for characterizations of the volatile and semi-volatile profiles of food and beverage matrices [5,43]. Additionally, PDMS is similar to the GC column phase (DB-5) used here, has greater film thickness (permitting increased uptake capacity), is generally more reproducible over larger numbers of samplings, and exhibits greater thermal resilience over numerous/longer high temperature injection times (personal communication, Sigma-Aldrich applications chemists; [44]). PDMS has also been described as a good general phase for profiling volatiles/semi-volatiles when one is uncertain as to which fiber phase is most appropriate for a given analysis and suite of compounds (personal communication, Sigma-Aldrich chemists; [40,42]). Finally, prior to profiling the preserves with this fiber, we conducted preliminary sampling trials of store-bought dill pickle and maraschino cherry preserves with the 100 and 7 µm PDMS and 85 µm polyacrylate fibers (purchased as an assortment kit from Sigma-Aldrich), and found that the 100 µm PDMS fiber sampled the greatest number of compounds and produced the largest total chromatographic area for these sample types.

Analytical grade (99.999% purity) helium and argon were purchased from A-OX Welding Supply (Sioux Falls, SD, USA). Dill pickles (Vlasic Kosher Dill Spears), maraschino cherries (Best Choice), cucumbers, and brine ingredients (white vinegar (ShurFine), dill seeds (McCormick), and table salt (ShurFine)) used to develop and validate the SPME method were purchased from a local supermarket (Sunshine Foods, Madison, SD, USA). Sample pH was determined using an Accumet Excel XL15 pH meter (Fisher Scientific, Waltham, MA, USA).

3.2. Sample Collection, Curation, and Preparation

Following discussions with the Moody County Historical Society (Flandreau, SD, USA) Director and its Board of Trustees, access to a circa 1895 constructed bare-earth basement (near Flandreau, SD) housing the preserved foods was arranged by a local family in August 2016 and again in August 2018. Therein, an estimated 125 glass jars of varying types and sizes containing preserved foods were stacked on hand-made wooden shelves assembled along the north-facing wall (Supplemental Video V2). The homeowner indicated that her mother had prepared the preserves during the period 1950–1953, but, because she was a young child at the time, she was unable to recall the precise method(s) by which the preserves were prepared. Other than intermittent storage, the musty-smelling basement was largely unused, and a heavy layer of dust had accumulated on the preserves (Supplemental Video V2). Preserves were chosen for analysis if they showed no evidence of desiccation or microbial growth inside the jar and, secondarily, had a label with a date. Based on these criteria, a total of 31 sealed glass jars of preserved foods were identified and returned to the lab for SPME analysis. Following photo documentation of the preserves (Figure S2), all preserves were stored at room temperature (~19–21 °C) in plastic totes to shield them from light. After SPME sampling, preserves were stored in a refrigerator at 4 °C until the qualitative analysis described below was conducted.

Preserves were tentatively identified initially by visual examination in consultation with the family that donated them and staff members of the Moody County Museum, who possess considerable expertise in the foods traditionally canned in the region during this period. The identities of some preserves (e.g., pickled dill and sweet cucumbers, tomatoes, and apples) were confirmed via visual inspection relatively easily; however, roughly half of the preserves required further identification by smell and taste after SPME analysis was completed. A compiled tasting video of representative tests is presented in Supplemental Materials (Supplemental Video V1; Table S4). Unique and replicate sample types were established via smell and taste assessment in consultation with family members and Moody County Museum staff. Nearly all of the preserves were confirmed to a reasonable degree of certainty after smell and taste tests, but those we have identified as "head cheese," "apricots," and "crab apples" remain tentative (Table 1, Figure S2).

3.3. SPME Analysis

Prior to SPME sampling, the SPME fiber was conditioned according to the manufacturer's instructions, and all glass jars were wiped with a large Kimwipe soaked with 95% ethanol to remove dust and external contaminants. The ethanol was allowed to evaporate in a fume hood for 15 min prior to SPME sampling to avoid it being absorbed onto the SPME fiber. Cleaned jars of preserve were placed in the middle of a 1.75 ft^3 (~50 L) AtmosBag directly adjacent to a 2 L beaker with an adjustable lab clamp affixed to it for holding the SPME during sampling (Figure S1). A beaker was used in this instance because ring stands proved too tall and tended to damage the bag during manipulation of the jars for SPME sampling within the evacuated bag. The sampling bag was sealed and clamped shut after sample introduction according to manufacturer recommendations, and the entrained air was pumped out through a gastight vacuum hose connected to the bag. To ensure that the SPME did not absorb atmospheric contaminants, and to create a small volume (~1 L) inside the bag to permit easier manipulation of the sample and the SPME apparatus, the bag was inflated with ~1 L of argon gas. Once the bag was inflated, the lid was removed from the sample and the SPME fiber was deployed into the mouth of the jar about 2 centimeters above the sample surface (Figure S1). After 120 min, the SPME fiber was retracted, the SPME removed from the bag, and the SPME was immediately inserted into the GC-MS injector port for analysis. During the GC-MS run that followed (30 min), the sampling bag was purged with high velocity lab air to remove any residual moisture and/or volatile residues before the next SPME sampling period. Once the GC-MS run was complete, the same sample was again sealed in the bag and resampled with SPME as described above. This was repeated three times for each sample, with the aim of assessing volatile intensity and functional class composition with time. After three successive SPME samplings (6 h total sampling time), and prior to tasting, the pH of the preserves was determined (Table 1).

The SPME sampling method was first developed and validated using store-bought savory and sweet preserves (Vlasic dill pickle spears and ShurFine maraschino cherries), to encompass a suite of representative compounds and compound classes believed relevant to those likely to comprise the volatile inventories of the historical preserves. To assess method sensitivity and reproducibility, authentic BPA standards were prepared and spiked into store-bought pickles and cherries and sampled with SPME. Informed by preliminary spiking trials and determinations of incurred BPA concentrations in selected preserves obtained from the basement and sacrificed for this purpose, and reviews of the relevant literature, 1000 μg of BPA (solubilized in ethanol) was deemed optimal and was spiked directly into the pickles/cherries. BPA-spiked preserves were then sealed inside the sampling bag as described above and sampled with SPME at 30, 60, 120, 240, and 360 min in triplicate to assess time to equilibrium uptake. Each of the triplicate samples was derived from SPME sampling of a freshly opened jar of each preserve to more accurately represent the sampling conditions under which the historical preserves were opened for the first time at the time of sampling. Equilibrium uptake, as measured by maximum total integrated chromatographic peak area, was reached by 120 min for both pickles and cherries (Figure 1A). This sampling time was used for SPME analysis of all the historical preserves as described above. The between- and within-sample precision of the method was assessed by SPME sampling of BPA-spiked pickles (N = 5) and cherries (N = 5), and then by three successive SPME samplings of three jars of each these same preserves (Figure 1B).

As there have been no reports of the volatile inventory of Vlasic dill pickles against which we could compare our SPME-detected volatile inventory, we sought to further assess the efficacy of the method for this product by using it to profile volatiles emitted from the separate components of a pickled cucumber mixture. To achieve this, store-bought cucumbers were cut into spears of the same dimensions as the Vlasic dill pickle spears (sold in the 16 ounce jar), placed into washed and solvent-rinsed Vlasic jars, and then sampled via SPME as described above. To ensure there was no compound carryover from the original Vlasic pickle mixture, the recycled Vlasic jars were first sampled with SPME while empty. The remainder of the cucumber spears were placed into cleaned Vlasic jars (N = 5), which were filled to the neck with a lab-prepared approximation of the Vlasic dill brine.

To ensure the lab-prepared dill brine was as close in composition to the Vlasic brine as possible, the concentration of vinegar (acetic acid) in the latter was determined via titration with sodium hydroxide (first standardized via potassium hydrogen phthalate (KHP)) and found to be 0.915 ± 0.0353 % v/v. The lab-prepared brine was then prepared by adding 0.184 L white vinegar to 0.816 L of distilled water into which 4.220 g and 0.465 g of dill seeds and table salt, respectively, were added. The lab brine was stirred aggressively with a magnetic stir bar set close to its maximum setting for 90 min. The lab-prepared brine solution was then used to pickle the fresh cucumber spears at room temperature (~19–21 °C) for 7 days prior to SPME analysis. The lab-prepared brine and the pickled cucumber spears were then each individually sampled with the SPME method in triplicate, as described above, to define their volatile profiles (Table S2).

3.4. SPME QA/QC

To ensure no cross contamination between samples and sampling replicates, the sampling bags were purged with high velocity lab air for 30 min before and after each jar was sampled with SPME (and between replicate samplings of the same jar). Each sampling bag containing only argon gas, and the SPME apparatus was sampled with SPME for 120 min after every five jars analyzed to serve as a procedural blank. To ensure bag integrity, sample bags were discarded after each set of seven preserves analyzed. Butylhydroxytoluene (*m/z* 220) was consistently detected in the procedural blanks, determined as direct off gassing from the bag and from the adhesive used to affix the lab clamp (holding the SPME sampler) to the beaker. This compound was excluded from the analysis. To ensure the SPME fiber was free of contaminants prior to each SPME sampling period, the fiber was heated at 250 °C in the GC injector for 30 min and a TIC was generated to track contaminant desorption from the fiber. SPME fibers never required more than a single heating period in the GC injector to remove all contaminants accumulated while stored between sampling periods.

3.5. GC-MS Analysis

VOCs were desorbed from the SPME fiber via direct insertion into the heated injector port of a QP2010 SE GC-MS (Shimadzu, Kyoto, Japan), equipped with a SPME injector liner (0.75 mm; Restek Corporation, Bellefonte, PA, USA) and Rxi-5ms capillary column (30 m x 0.25 mm with a 0.25 µm 5% diphenyl 95% dimethyl polysiloxane phase; Restek). Helium was used as the carrier gas. The SPME fiber was desorbed for 30 min. GC injector temperature was maintained at 250 °C and operated in the splitless mode with a helium flow rate of 1.15 mL min^{-1} through the column (8.4 mL min^{-1} total flow). The initial GC column temperature was set to 40 °C, ramped to 100 °C at 5 °C min^{-1} with a 3 min hold, and then ramped to 250 °C at 10 °C min^{-1}. The MS was operated in the electron impact ionization mode at 70 eV and 0.1 kV detector voltage. The ion source and MS interface temperatures were maintained at 260 °C and 270 °C, respectively. Mass spectra were obtained in the full scan mode, with the mass range 30–500 amu, scanned at a rate of 2000 scans/s.

3.6. Data Analysis

TICs of preserve samples and procedural blanks were initially overlaid for preliminary comparison and identification of compounds occurring in both. Compounds present in both the samples and blanks were excluded from analysis. TIC and mass spectral analyses were performed using the Shimadzu GCMSsolution software (Version 4.11; Shimadzu, Kyoto, Japan). Individual compounds were identified using authentic standards and Kovats retention indices in conjunction with comparisons of measured mass spectra with reference spectra contained in the NIST/EPA/NIH Mass Spectral Library (NIST 14, Version 2.2, Gaithersburg, MD, USA). Only TIC peaks with S/N > 3 and mass spectral library similarity index > 80% were considered positively identified compounds. All other TIC peaks were excluded from analysis, which necessitated an estimated 40% of detected compounds being excluded from analysis. A unique compound was defined as the singular detection of that compound in each sample conforming to these detection criteria. Percentage contributions of sampled compound classes and unique compounds to total integrated TIC peak areas were computed using only the sum of all peak areas conforming to these detection criteria. Spiked and naturally incurred BPA concentrations were quantified using as 6 point calibration curve prepared from serial dilutions of authentic standards (Figure S5). All data reduction and statistical analysis was performed with Excel (Microsoft, Redmond, WA, USA), StatPlus (AnalystSoft, Walnut, CA, USA), and Prism (GraphPad, San Diego, CA, USA) software.

4. Conclusions

The results presented here demonstrate that SPME sampling is a viable and reproducible method for volatile profiling of rare historical preserved foods. Additionally, the use of an AtmosBag as a chamber that can be sealed, vacuumed, and purged presents a relatively simple and inexpensive means of assessing volatile emissions from any large and/or oddly-shaped container. As the method was developed with modern store-bought preserved foods and was applied to historical preserves, it likely represents a generally accessible means of sampling the volatile inventory of any food type stored in a container. The jars assessed in this work opened with screw-top lids, so additional modifications may be necessary if foods sealed by other means were to be assessed using this method. This work represents one of the first efforts to profile the volatile inventories of rare historical preserved foods, and presents one of the oldest instances of BPA detection in foodstuffs.

Supplementary Materials: The following are available online: Schematic of SPME sampling in the AtmosBag (Figure S1); photodocumentation of the 31 historical preserves sampled in this work (Figure S2); analysis of the total integrated area and the breakdown of compound classes for each of the historical preserves assessed (Figure S3); amount of BPA per sample (Figure S4); standard curve to quantify BPA (Figure S5); raw data of compound classes used to create Figure 3 (Table S1); compounds listed are those detected during three independent 2 h SPME samplings and aggregated to display the complexity of the volatile profile (Table S2); the masses of the historical preserves (Table S3); sample numbers and video start times for each of the historical preserves taste tested in Supplementary Video V1 (Table S4); all data collected in this study (Supplementary Data File 1); visual documentation of taste tests conducted by author M.O.G. on representative historical preserve samples (Supplementary Video V1); video documentation of the basement where the historical preserves were stored since the period 1950–1953 (Supplementary Video V2).

Author Contributions: L.J.L. contributed to study design, funding, sample analysis, data analysis, and manuscript editing. V.A.S. contributed to study design, funding, data analysis, and manuscript editing. H.L.J. contributed to study design, funding, data analysis, and manuscript editing. S.E.M. contributed to funding and manuscript editing and provided invaluable analytical insights in support of the project. S.M.O. contributed to sample collection, funding, data analysis, and manuscript editing. P.V. and M.O.G. wrote the manuscript and contributed to sample collection, study design, funding, sample analysis, data analysis, and manuscript editing.

Funding: This work was supported by a crowd funding campaign conducted on Experiment.com. L.J.L. was supported by an undergraduate research fellowship award from the Dakota State University (DSU) Provost's Office. V.A.S. was supported by a Barry M. Goldwater Scholarship award. H.L.J. was supported by a Barry M. Goldwater Scholarship award and an undergraduate research grant from the DSU Provost's Office. M.O.G. was supported by DSU College of Arts and Sciences faculty research grants and a DSU Faculty Research Initiative (FRI) grant. P.V. was supported by faculty startup funds from Southern Oregon University. This work was further supported by the DSU College of Arts and Sciences as part of an undergraduate research-in-teaching initiative in M.O.G.'s Organic Chemistry I & II (CHEM 326 & CHEM 328) courses.

Acknowledgments: The authors thank Berdyene Bowen and her family for providing the preserves and for insightful anecdotal accounts of their history, preparation, and storage. We thank the Moody County Historical Society (Flandreau, SD, USA) and its Board of Trustees for their interest in and enthusiastic support and promotion of this project within the local historical community. We thank then Moody County Museum Director, Virginia Hazlewood-Gaylor, for her tireless efforts to liaise with our group, the Bowen family, the Moody County Museum and its Board of Trustees, and community residents; and to educate our group about regional preserve types and preservation history and techniques. We thank anonymous reviewers for providing detailed feedback that significantly improved the quality of the manuscript. We are also indebted to Dr. Dale Droge (DSU) for many insightful conversations and to Nancy Presuhn (DSU) for exceptional administrative assistance.

Conflicts of Interest: The authors declare no conflict of interest.

References

1. Lawless, H.T. *Blackwell Handbook of Sensation and Perception*; Goldstein, E.B., Ed.; Blackwell Publishing: Malden, MA, USA, 2005; pp. 601–635.
2. Ma, X.; Su, M.; Wu, H.; Zhou, Y.; Wang, S. Analysis of the Volatile Profile of Core Chinese Mango Germplasm by Headspace Solid-Phase Microextraction Coupled with Gas Chromatography-Mass Spectrometry. *Molecules* **2018**, *23*, 1480. [CrossRef]
3. Costa, F.; Cappellin, L.; Zini, E.; Patocchi, A.; Kellerhals, M.; Komjanc, M.; Gessler, C.; Biasioli, F. QTL Validation and Stability for Volatile Organic Compounds (VOCs) in Apple. *Plant Sci.* **2013**, *211*, 1–7. [CrossRef] [PubMed]
4. Rowe, D.J. Aroma Chemicals for Savory Flavors. *Perfume. Flavor.* **1998**, *23*, 9–18.
5. Merkle, S.; Kleeberg, K.K.; Fritsche, J. Recent Developments and Applications of Solid Phase Microextraction (SPME) in Food and Environmental Analysis—A Review. *Chromatography* **2015**, *2*, 293–381. [CrossRef]
6. Hayashi, H. Drying Technologies of Foods—Their History and Future. *Dry. Technol.* **1989**, *7*, 315–369. [CrossRef]
7. Prajapati, J.B.; Nair, B.M. The History of Fermented Foods. In *Handbook of Fermented Functional Foods*; CRC Press: Boca Raton, FL, USA, 2003; pp. 1–22.
8. Graham, J.C. The French Connection in the Early History of Canning. *J. Royal Soc. Med.* **1981**, *74*, 374–381.
9. Kreidberg, M. *Food on the Frontier*; Minnesota Historical Society: Saint Paul, MN, USA, 1975; p. 324.
10. Evershed, R.P.; Heron, C.; Charters, S.; Goad, L.G. The Survival of Food Residues: New Methods of Analysis, Interpretation and Application. *Proc. British Acad.* **1991**, *77*, 187–208.
11. Evershed, R.P.; Bland, H.A.; van Bergen, P.F.; Carter, J.F.; Horton, M.C.; Rowley-Conwy, P.A. Volatile Compounds in Archaeological Plant Remains and the Maillard Reaction During Decay of Organic Matter. *Science* **1997**, *278*, 432–433. [CrossRef]
12. Perrault, K.A.; Stefanuto, P.; Dubois, L.; Cnuts, D.; Rots, V.; Focant, J. A New Approach for the Characterization of Organic Residues from Stone Tools Using GCGC-TOFMS. *Separations* **2016**, *3*, 16. [CrossRef]
13. Regert, M.; Alexandre, V.; Thomas, N.; Lattuati-Derieux, A. Molecular Characterisation of Birch Bark Tar by Headspace Solid-Phase Microextraction Gas Chromatography–Mass Spectrometry: A New Way for Identifying Archaeological Glues. *J. Chromatogr.* **2006**, *1101*, 245–253. [CrossRef]
14. Cnuts, D.; Perrault, K.A.; Stefanuto, P.-H.; Dubois, L.M.; Focant, J.-F.; Rots, V. Fingerprinting Glues Using HS-SPME GCxGC-HRTOFMS: A New Powerful Method Allows Tracking Glues Back in Time. *Archaeometry* **2018**, *60*, 1361–1376. [CrossRef]
15. Evershed, R.P. Organic Residue Analysis in Archaeology: The Archaeology Biomarker Revolution. *Archaeometry* **2008**, *50*, 895–924. [CrossRef]
16. Evershed, R.P.; Heron, C.; Goad, L.G. Epicuticular Wax Components Preserved in Potsherds as Chemical Indicators of Leafy Vegetables in Ancient Diets. *Antiquity* **1991**, *65*, 540–544. [CrossRef]
17. Evershed, R.P.; Heron, C.; Goad, L.J. Analysis of Organic Residues of Archaeological Origin by High Temperature Gas Chromatography/Mass Spectrometry. *Analyst* **1990**, *115*, 1339–1342. [CrossRef]
18. Craig, O.E.; Chapman, J.; Heron, C.; Willis, L.H.; Bartosiewicz, L.; Taylor, G.; Whittle, A.; Collins, M. Did the First Farmers of Central and Eastern Europe Produce Dairy Foods? *Antiquity* **2005**, *79*, 882–894. [CrossRef]
19. Oudemans, T.F.M.; Boon, J.J. Molecular Archaeology: Analysis of Charred Food Remains from Prehistoric Pottery by Pyrolysis-Gas Chromatography/Mass Spectrometry. *J. Anal. Appl. Pyrol.* **1991**, *20*, 197–227. [CrossRef]
20. Barnard, H.; Ambrose, S.H.; Beehr, D.E.; Forster, M.D.; Lanehart, R.E.; Malainey, M.E.; Parr, R.E.; Rider, M.; Solazzo, C.; Yohe, R.M. Mixed Results of Seven Methods for Organic Residue Analysis Applied to One Vessel with the Residue of a Known Foodstuff. *J. Archaeol. Sci.* **2007**, *34*, 28–37. [CrossRef]
21. Palma-Harris, C.; McFeeters, R.F.; Fleming, H.P. Solid-Phase Microextraction (SPME) Technique for Measurement of Generation of Fresh Cucumber Flavor Compounds. *J. Agricult. Food Chem.* **2001**, *49*, 4203–4207. [CrossRef]
22. Legua, P.; Domenech, A.; Martinez, J.J.; Sanchez-Rodriguez, L.; Hernandez, F.; Carbonell-Barrachina, A.A.; Melgarejo, P. Bioactive and Volatile Compounds in Sweet Cherry Cultivars. *J. Food Nutr. Res.* **2017**, *5*, 844–851. [CrossRef]

23. Levaj, B.; Dragovic-Uzelac, V.; Delonga, K.; Ganic, K.K.; Banovic, M.; Kovacevic, D.B. Polyphenols and Volatiles in Fruits of Two Sour Cherry Cultivars, Some Berry Fruits and Their Jams. *Food Technol. Biotechnol.* **2010**, *48*, 538–547.
24. Marsili, R.T.; Miller, N. Determination of Major Aroma Impact Compounds in Fermented Cucumbers by Solid-Phase Microextraction–Gas Chromatography–Mass Spectrometry–Olfactometry Detection. *J. Chromatogr. Sci.* **2000**, *38*, 307–314. [CrossRef] [PubMed]
25. Guler, Z.; Karaca, F.; Yetisir, H. Volatile Compounds in the Peel and Flesh of Cucumber (*Cucumis sativus* L.) Grafted onto Bottle Gourd (*Lagenaria siceraria*) Rootstock. *J. Horticult. Sci. Biotechnol.* **2013**, *88*, 123–128. [CrossRef]
26. Maleki, G.; Sedaghat, N.; Woltering, E.J.; Farhoodi, M.; Mohebbi, M. Chitosan-Limonene Coating in Combination with Modified Atmosphere Packaging Preserve Postharvest Quality of Cucumber During Storage. *J. Food Meas. Character.* **2018**, *12*, 1610–1621. [CrossRef]
27. Bruni, R.; Bianchi, A.; Bellardi, M.G. Essential Oil Composition of *Agastache anethiodora* Britton (Lamiaceae) Infected by Cucumber Mosaic Virus (CMV). *Flav. Fragr. J.* **2007**, *22*, 66–70. [CrossRef]
28. Iranmanesh, M.; Ezzatpanah, H.; Akbari-Adergani, B.; Torshizi, M.A.K. SPME/GC-MS Characterization of Volatile Compounds of Iranian Traditional Dried Kashk. *Int. J. Food Prop.* **2018**, *21*, 1067–1079. [CrossRef]
29. Shahidi, F.; Rubin, L.J.; D'Souza, L.A.; Teranishi, R.; Buttery, R.G. Meat Flavor Volatiles: A Review of the Composition, Techniques of Analysis, and Sensory Evaluation. *CRC Crit. Rev. Food Sci. Nutr.* **1986**, *24*, 141–243. [CrossRef] [PubMed]
30. Manyi-Loh, C.E.; Ndip, R.N.; Clarke, A.M. Volatile Compounds in Honey: A Review on Their Involvement in Aroma, Botanical Origin Determination and Potential Biomedical Activities. *Int. J. Molec. Sci.* **2011**, *12*, 9514–9532. [CrossRef] [PubMed]
31. Beaulieu, J.C.; Lea, J.M. Characterization and Semiquantitative Analysis of Volatiles in Seedless Watermelon Varieties Using Solid-Phase Microextraction. *J. Agric. Food Chem.* **2006**, *54*, 7789–7793. [CrossRef]
32. Malorni, L.; Martignetti, A.; Cozzolino, R. Volatile Compound Profiles by HS GC-MS for the Evaluation of Postharvest Conditions of a Peach Cultivar. *Ann. Chromatogr. Sep. Tech.* **2015**, *1*, 1007.
33. Ferreira, L.; Perestrelo, R.; Caldeira, M.; Camara, J.S. Characterization of Volatile Substances in Apples from *Rosaceae* Family by Headspace Solid-Phase Microextraction Followed by GC-qMS. *J. Sep. Sci.* **2009**, *32*, 1875–1888. [CrossRef]
34. Wu, Y.; Duan, S.; Zhao, L.; Gao, Z.; Luo, M.; Song, S.; Xu, W.; Zhang, C.; Ma, C.; Wang, S. Aroma Characterization Based on Aromatic Series Analysis in Table Grapes. *Sci. Rep.* **2016**, *6*, 31116. [CrossRef] [PubMed]
35. Cannon, R.J.; Ho, C. Volatile Sulfur Compounds in Tropical Fruits. *J. Food Drug Anal.* **2018**, *26*, 445–468. [CrossRef] [PubMed]
36. Vogel, S.A. The Politics of Plastics: The Making and Unmaking of Bisphenol A "Safety". *Am. J. Pub. Health* **2009**, *99*, S559–S566. [CrossRef] [PubMed]
37. Chen, D.; Kannan, K.; Tan, H.; Zheng, Z.; Feng, Y.; Wu, Y.; Widelka, M. Bisphenol Analogues Other Than BPA: Environmental Occurrence, Human Exposure, and Toxicity-A Review. *Environ. Sci. Technol.* **2016**, *50*, 5438–5453. [CrossRef] [PubMed]
38. Halden, R.U. Plastics and Health Risks. *Ann. Rev. Pub. Health* **2010**, *31*, 179–194. [CrossRef] [PubMed]
39. Cooper, J.E.; Kendig, E.L.; Belcher, S.M. Assessment of Bisphenol A Released from Reusable Plastic, Aluminium and Stainless Steel Water Bottles. *Chemosphere* **2011**, *85*, 943–947. [CrossRef] [PubMed]
40. Koziel, J.A.; Pawliszyn, J. Air Sampling and Analysis of Volatile Organic Compounds with Solid Phase Microextraction. *J. Air Waste Manag. Assoc.* **2001**, *51*, 173–184. [CrossRef]
41. Tuduri, L.; Desauziers, V.; Fanlo, J.L. Potential of Solid-Phase Microextraction Fibers for the Analysis of Volatile Organic Compounds in Air. *J. Chromatogr. Sci.* **2001**, *39*, 521–529. [CrossRef]
42. Augusto, F.; Valente, A.L.P. Applications of Solid-Phase Microextraction to Chemical Analysis of Live Biological Samples. *Trends Anal. Chem.* **2002**, *21*, 428–438. [CrossRef]

43. Wardencki, W.; Michulec, M.; Curylo, J. A Review of Theoretical and Practical Aspects of Sold-Phase Microextraction in Food Analysis. *Int. J. Food Sci. Technol.* **2004**, *39*, 703–717. [CrossRef]
44. Balasubramanian, S.; Panigrahi, S. Solid-Phase Microextraction (SPME) Techniques for Quality Characterization of Food Products: A Review. *Food Bioproc. Technol.* **2011**, *4*, 1–26. [CrossRef]

Sample Availability: Samples of the historical preserves assessed in this study are not available from the authors.

© 2019 by the authors. Licensee MDPI, Basel, Switzerland. This article is an open access article distributed under the terms and conditions of the Creative Commons Attribution (CC BY) license (http://creativecommons.org/licenses/by/4.0/).

MDPI
St. Alban-Anlage 66
4052 Basel
Switzerland
Tel. +41 61 683 77 34
Fax +41 61 302 89 18
www.mdpi.com

Molecules Editorial Office
E-mail: molecules@mdpi.com
www.mdpi.com/journal/molecules

www.ingramcontent.com/pod-product-compliance
Lightning Source LLC
LaVergne TN
LVHW071940080526
838202LV00064B/6647